Cambridge Geographical Series.

GENERAL EDITOR: F. H. H. GUILLEMARD, M.D.,

FORMERLY LECTURER IN GEOGRAPHY IN THE UNIVERSITY OF CAMBRIDGE.

A GEOGRAPHICAL

HISTORY OF MAMMALS.

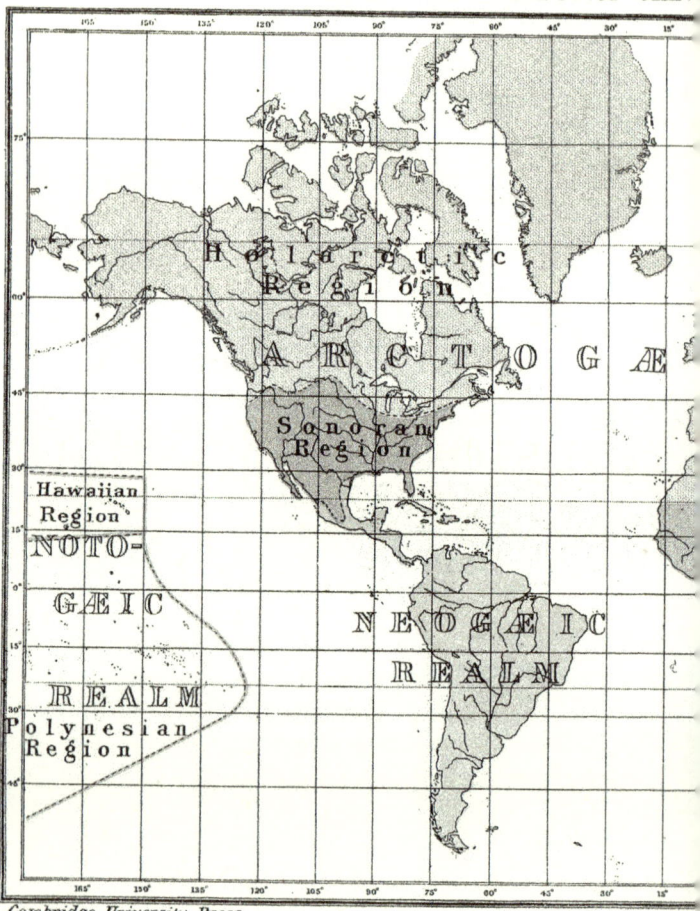

MAMMALIAN GEOGRAPHICAL REALMS & REGIONS.

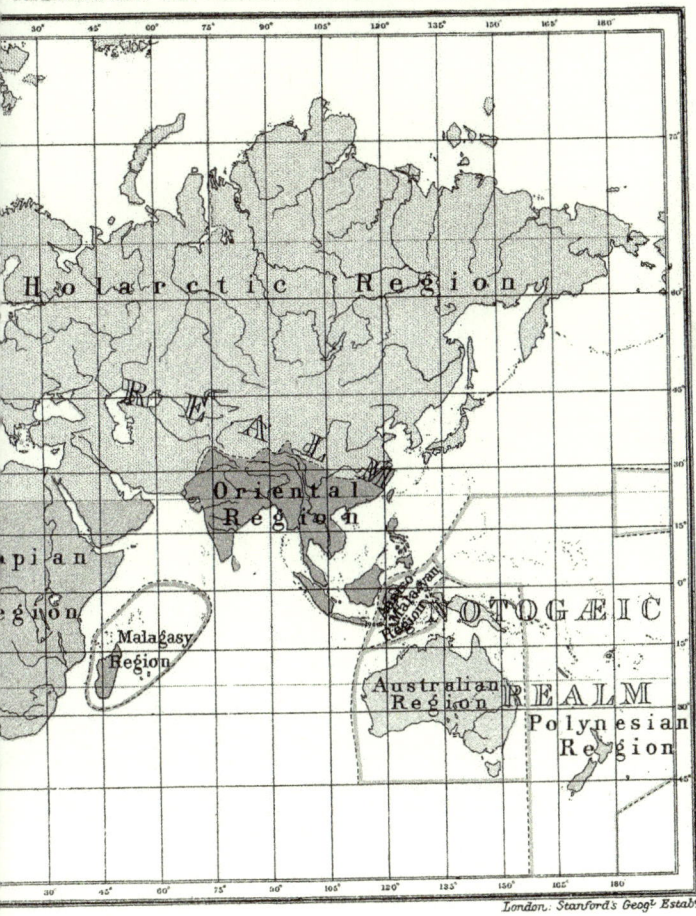

A GEOGRAPHICAL

HISTORY OF MAMMALS.

BY

R. LYDEKKER, B.A., F.R.S., V.P.G.S. ETC.

CAMBRIDGE:

AT THE UNIVERSITY PRESS.

1896

CAMBRIDGE UNIVERSITY PRESS

Cambridge, New York, Melbourne, Madrid, Cape Town,
Singapore, São Paulo, Delhi, Tokyo, Mexico City

Cambridge University Press
The Edinburgh Building, Cambridge CB2 8RU, UK

Published in the United States of America by
Cambridge University Press, New York

www.cambridge.org
Information on this title: www.cambridge.org/9781107600164

First published 1896
First paperback edition 2011

A catalogue record for this publication is available from the British Library

ISBN 978-1-107-60016-4 Paperback

PREFACE.

SINCE the publication of Dr Wallace's book on the geographical distribution of animals in general, the only works which have appeared relating to Mammals from the same point of view are the small volume by Mr F. E. Beddard and the series of papers by Mr W. L. Sclater, referred to in the Appendix.

Both the latter admittedly take but little account of fossil forms; and there is accordingly ample room for a work which should collect and arrange the information on this subject, and indicate the deductions which may be drawn therefrom. This task has been attempted in the present volume. The subject is, however, so vast, and the information relating to it scattered through so many publications, that it is probable many points of interest or importance have not been noticed. From the mode of arrangement of the work, a considerable amount of repetition was inevitable.

The long time that the volume was in the press will account for its containing no allusion to certain papers which have appeared recently. It may therefore be mentioned here that recent discoveries have shown a slight intermixture of northern types in the Palæozoic flora of Argentina, so that the isolation of the northern and southern floras is not so nearly complete as was supposed. A paper just published by M. Boule[1] indi-

[1] *Comptes Rendus*, vol. cxxii (1896).

cates that the relationship between *Cadurcotherium* and the Astrapotheria is closer than suggested on page 82. Hence there is further evidence that South America received its Tertiary ungulates by way of Africa. The extinct Patagonian birds *Phororhachis* and *Brontornis* appear, according to Mr C. W. Andrews, to be nearly allied to the existing South American seriema (*Cariama*), and since *Filholornis*, of the French Phosphorites, has been shown by Prof. A. Milne-Edwards to be related to the hoatzin (*Opisthocomus*), it would seem that all these peculiar South American types of birds have a history somewhat similar to that of the extinct ungulates of the same region.

R. LYDEKKER.

Harpenden.
  June 1st, 1896.

CONTENTS.

CHAPTER I.

INTRODUCTORY.

CHAPTER II.

THE NOTOGÆIC REALM.

CHAPTER III.

THE NEOGÆIC REALM.

CHAPTER IX.

THE HOLARCTIC REGION.

CHAPTER X.

THE SONORAN REGION.

APPENDIX.

LIST OF ILLUSTRATIONS.

LIST OF ILLUSTRATIONS.

Figs. 1, 4, 22, 28, 32, 48, 49, 50, 52, 53, 54, 55, 57, 59, 60, 63, 65, 66, 68, 70, 71 are from *The Study of Mammals, Living and Extinct*, by Flower and Lydekker. Figs. 13, 14, 15, 16, 17, 18, 19, 20, 21, 23, 24 are taken from the author's work on Argentine Fossil Mammals, published in the *An. Mus. La Plata*. Figs. 9, 25, 26, 27 are from Señor Ameghino. Fig. 67 is from a photograph by Mr Abel Chapman. Fig. 82 is from Prof. Cope, fig. 81 from Prof. Earle, fig. 40 from Prof. Boyd Dawkins, fig. 30 from Prof. Fraas, and figs. 7, 8 from Prof. Goodyear. Fig. 46 is taken from a plate in Hutchinson's *Extinct Monsters*; 58 from a photograph by Mr Eccles, of Reading, and 75 from one in the possession of Mr J. E. Harting. Figs. 6, 35, 41, 43 are from Prof. Marsh; and 11, 12, 31, 33, 37, 78, 79, 80 from Prof. H. F. Osborn. For the blocks of figs. 72, 73, 74, the author is indebted to Dr C. H. Merriam. Fig. 45 is from Nicholson and Lydekker's *Manual of Palæontology*; fig. 29 from Owen; fig. 64 from Dr Sclater in the *Proc. Zool. Soc.*; fig. 77 from Dr Wortman, and 39 from Prof. von Zittel.

A GEOGRAPHICAL
HISTORY OF MAMMALS.

CHAPTER I.

INTRODUCTORY.

THAT there are differences in the animals and plants of different
districts and different countries is a fact apparent to every person
who has travelled at all; while to those who have travelled extensively it will further be evident that the amount of this difference
is by no means correlated with the distance of one country from
another, the fauna of Japan, for instance, being much more like
that of England and France than is the fauna of Eastern Africa to
that of the adjacent island of Madagascar. Unfortunately, among
persons who are not conversant with the principles of zoological
science, the distinction between the faunas of different countries
has been much obscured by the practice common to almost all
the old voyagers and colonists of bestowing upon the animals of
new countries the names of such Old World creatures as they
appeared most nearly to resemble. The puma of America was,
for instance, called the lion, and the jaguar of the same country,
the tiger; while the koala of Australia was christened the native
bear, and its marsupial allies the dasyures are still commonly
spoken of as native cats. To students of the science of

Geographical Distribution it is, therefore, essential to discard such misleading popular titles, and to speak of animals by their correct names.

Apart from the specific or generic distinctions between the animals of one country and another, the observer will not fail to notice more or less well-marked differences between those inhabiting different districts of a single country; such differences being most intensified when a country presents great variation in its physical features. An excellent instance of this is afforded by South America, where there are the open grassy plains of the Argentine, the dense tropical forests of Paraguay and Brazil, and the snow-clad heights of the Andes. In the former tract the traveller will meet with the peculiar rodents known as viscachas, the Patagonian cavy, a species of deer, numerous armadillos, and the rhea (miscalled the American ostrich). In the Brazilian forests, on the other hand, he will find monkeys, marmosets, tapirs, tree-porcupines, sloths, and anteaters, together with certain armadillos which are for the most part specifically or generically distinct from those of the pampas. If, on the other hand, he ascend high on the Andes, he will leave behind the animals of the forest, to be confronted with chinchillas, guanacos and vicuñas. Different, however, as are the animals of these various districts, yet an acquaintance with their zoological affinities will prove that many of them belong to closely allied groups, some of which are met with in no other parts of the world. This will serve to show that they belong to what is known as one zoological province or region, and that the differences between the faunas of different districts of that province are due to the physical variations between its component districts.

Perhaps this point may be still better illustrated by the cases where the same species of animal is restricted to different districts of one country or continent. For instance, the common squirrel is only found in the wooded districts of Europe, and is entirely absent from open plains. The chamois, again, is only met with in the isolated mountain ranges of the Pyrenees, the Alps, and the Caucasus; while the Siberian ibex of the Altai reappears in Tibet and the Himalaya, but is wanting in the intervening tracts. In these instances Europe would be spoken of as the *distributional*

Distributional Area and Station.

area of the squirrel and the chamois, and Central Asia as that of the Siberian ibex; but the particular districts suited to the existence of each would be termed its *station*. And here it may be mentioned that whereas the distributional area of a species is generally continuous, its various stations may be partially or completely isolated, as in the instances of the aforesaid mountain animals, which cannot live on the plains below.

Station is thus seen to be very intimately connected with temperature; and of this a very striking example may be found among the mammals of South America. As already mentioned, the llama-like animals respectively known as vicuñas and guanacos are met with in company on the highlands of the Cordillera in Peru and Ecuador, but as we go further south the latter are found on the plains of southern Argentina and Patagonia, as well as in the island of Tierra-del-Fuego, at the sea-level. Here, then, there is a clear proof of the intimate connection existing between temperature and station; the guanaco, being an animal which can live only in cold or temperate climates, finds suitable conditions for its existence in tropical latitudes solely at a height of many thousands of feet, although further south it is able to thrive at the sea-level.

This being so, it is obvious that temperature must likewise exert a very considerable influence on the whole distributional area of many animals. Of this, the most marked instance is found in the fauna of the Arctic regions, which forms a circumpolar zone of animals more or less markedly distinct from those dwelling further south. And if the whole land-area of the world were connected, and not broken up by mountain-chains, its faunas might probably be divided into zones or belts, whose limits would mainly depend upon temperature.

In this connection it may be mentioned that the instance of the range of the guanaco is of considerable importance in regard to a decided difference between the Old and New Worlds in respect to the influence of mountains on the present distribution of the animals of the two areas. In the Old World the chief mountain-ranges, such as the Pyrenees, Alps, Carpathians, Caucasus, Hindu-Koh, Himalaya, Thian Shan, and Altai, run in a more or less decided east-and-west direction; whereas in America, and more especially in the southern half of that continent, they have a north-and-south trend. Consequently, whereas in the former area the mountain-ranges have acted as barriers to the dispersal of animal life, there has been no such obstacle to diffusion in America, and animals have been able to distribute themselves according to temperature-conditions. It is in consequence of this physical feature that a single species of cold-loving animal like the guanaco can range at the present day from the equator to latitude 55° south, whereas in the Old World the common ibex is now restricted to the isolated mountain-ranges of Europe, and the Siberian ibex is confined to the systems of the Himalaya and the Altai. Such isolated stations could, of course, only have been reached during a period when the general temperature of the northern hemisphere was much colder than it is at present, and the animals were thus enabled to cross the lowlands from one chain to the other ; and that such a cold period formerly existed, there is abundant evidence in the traces of an extensive glaciation found over a large portion of Europe and Asia. Although, therefore, strictly speaking, temperature has really had as much effect in the distribution of animals in the Eastern Hemisphere as in the Western, yet, since the glacial epoch, its influences have been considerably masked by the trend of the chief mountain-ranges, and likewise by the greater isolation of the different countries of the former area as compared with the latter.

These essential differences thus render it impossible to mark out the Old World in the zones of animal distribution which have been attempted for North America ; and they will likewise serve largely to explain the divergence of views on this point which may be noticed between the writings of Drs Wallace and Merriam.

The latter writer[1], whose conclusions are mainly based on the evidence of North American animals and plants, is of opinion that in the northern hemisphere " animals and plants are restricted in southward distribution by the mean temperature of a brief period covering the hottest part of the year "; and it is added that in certain districts the mingling of essentially northern types with those characteristic of a more southerly zone is due to the mean temperature of the hottest part of the year being sufficiently low

[1] Appendix, No. 20.

for the existence of the former, while the total quantity of heat suffices for the latter. In other words, there is a low summer temperature combined with a high total sum of heat.

Among the secondary causes affecting distribution, humidity, according to the same observer, may occupy the first place. "Humidity," he writes, "governs details of distribution of numerous species of plants, reptiles and birds, and of a few species of mammals, within the several temperature-zones. ...Humidity and other secondary causes determine the presence or absence of particular species in particular localities within their appropriate zones, but temperature predetermines the possibilities of distribution; it fixes the limits beyond which species cannot pass; it defines broad trans-continental belts within which certain forms may thrive if other conditions permit, but outside of which they cannot exist, be the other conditions never so favourable."

Important as the influence of temperature and, in a smaller degree, that of humidity, has undoubtedly been in determining the distributional limits of the species or genera of animals and plants now inhabiting particular countries, and large as has been the part played by the glacial epoch in producing the present condition of things, it is evident that temperature has been by no means the only, even if it be the chief factor in distribution. In the first place there are several species, more especially among the carnivorous mammals, which seem quite independent of both station and temperature, the New World puma ranging from Patagonia to Canada, while the tiger inhabits alike the burning jungles of India and Burma, and the Arctic tundras of Siberia. Striking as such cases are, they are, however, to be regarded merely as examples of the individual adaptability of certain species, which, like the carnivores named, are able to obtain suitable food in any part of the world, and they do not throw any discredit on the power of temperature as a controlling factor in animal distribution generally.

Of the utmost importance in this respect are the changes which the surface of the globe itself has undergone in past epochs, whereby continents that are now more or less completely sundered from one another were formerly connected, while what are now islands were once parts of continents, and *vice versâ*. Such connections and disconnections, by allowing migrations at one time, and preventing them from taking place in the reverse direction at a subsequent epoch, have been the chief factors which have resulted in the present very remarkable difference in the faunas of different parts of the globe. And it is solely due to such changes that many of the lower types of mammalian life, like the marsupials of Australia, and the lemurs and insectivores of Madagascar, have been preserved at the present day; their insulation having afforded them protection from the invasion of the larger and more specialised mammals of other parts of the world, by which they would inevitably have been swept away, had the two groups ever come in contact. It is in regard to these migratory movements of animals and changes in the land-surface of the globe that zoology and geography are brought into such close relationship; the former science sometimes helping to explain the alterations that have taken place in the contours of the land, while in other cases the present distribution of the land explains the past history of the animals by which it is inhabited.

To understand rightly the present distribution of animals, it is, however, essential to study their past history as recorded by the preservation of their fossilised remains in the strata of the earth's crust; as without such history it would be quite impossible to grasp the reason of many apparent anomalies in their present distribution. How, for instance, without the aid of palæontology would it be possible to understand how it came about that tapirs are now found only in tropical America and the Malayan countries, or that marsupials occur solely in America and Australasia at the present day? And here it may be well to mention that the science of geographical distribution depends essentially upon a belief on the part of the student that all animals are genetically connected one with another, and that the existing forms have originated from earlier kinds by some mode of evolution. Were this belief not accepted, the whole science of distribution would fall to pieces; as if animals were separately created, there would be nothing calling for special explanation in the fact of tapirs being restricted to the two areas mentioned.

Importance of Palæonto-logy.

To those readers who may not be geologists, the following table of the leading divisions into which the strata of the earth's crust have been divided will probably be advantageous. Commencing with the highest beds, the series will run in descending order as follows, viz. :—

Tertiary.   PLISTOCENE.—Cavern and River Deposits.
PLIOCENE.—The " Crags " of the East Coast.
MIOCENE.—Œningen beds of Baden.
OLIGOCENE.—Gypsum of Paris Basin ; Phosphorites of Central France.
EOCENE.—London Clay.

Secondary.   CRETACEOUS.—Chalk, Upper Greensand, Gault, Lower Greensand, and Wealden.
JURASSIC.—Purbeck beds, Portland series, Kimmeridge Clay, Coral Rag, Oxford Clay, Great Oolite, Stonesfield Slate, Inferior Oolite, Lias.
TRIASSIC.—New Red Sandstone of Cheshire.

Palæozoic.   PERMIAN.—Red Marls.
CARBONIFEROUS.—Coal-Measures and Mountain-Limestone.
SILURIAN.
ORDOVICIAN.   } Older rocks of Wales and the Lake-District.
CAMBRIAN.

Before attempting to draw any conclusions as to the former configuration of the surface of the earth from the distribution of the animals now inhabiting its different countries, it is essential to understand that the different classes into which vertebrate animals are divided (and these only will be taken into consideration in the present volume) have a very different past history ; the lower groups, such as fishes, reptiles, and amphibians being much older types than mammals and birds, and having attained their maximum development at a time when the two latter formed but a small minority of the earth's population.

Inequality in the Ages of different Groups of Animals.

There is a considerable probability that at least a very large proportion of the animals that have populated the globe in the later geological epochs originated high up in the northern hemisphere, if not, indeed, in the neighbourhood of the pole itself (which is known to have enjoyed a genial climate during the Tertiary period), and that they gradually migrated southwards in a series of waves, probably under pressure of the development of new and higher types in high latitudes; and it is to such southerly migrations that the present marked differentiation of the fauna of different parts of the earth's surface is chiefly due. Whether such a northerly origin held good for the terrestrial life of the Secondary epoch, there are no means of determining; but it would appear that the higher animals (which were chiefly reptiles) of that epoch were very similar throughout the world, and that the differentiation of faunas had scarcely, if at all, commenced. Instances of this are afforded, as noticed in the sequel, by the occurrence of an identical genus of mammal (*Tritylodon*) in the lower Secondary rocks of Europe and South Africa; as well as by the close alliance between the dinosaurian reptiles from the Jurassic rocks of Europe, North America, Argentina, India, and Madagascar (the genera being in some cases identical), and likewise between the anomodont reptiles of the Trias of Europe and the early Secondary rocks of South Africa and India.

Reptiles belonging to orders still existing, such as crocodiles and chelonians (tortoises and turtles), had already attained a high degree of development in the Eocene division of the Tertiary period, when many genera now living had already made their appearance, whereas at that time the mammals were quite different from the modern forms. At the same time the side-necked tortoises (Pleurodira) were the dominant forms in the northern hemisphere, whereas they have now all migrated to southern lands, their place in the north being taken by the more specialised S-necked group (Cryptodira). This, however, is not all, for the rhynchocephalians, of which the sole existing representative is the New Zealand tuatera (*Sphenodon*), attained their maximum development in the northern hemisphere during the early part of the Secondary epoch, and their southern migration must have taken place during some portion of the same period. The palæontological history of amphibians is still very imperfectly known, but since the group as a whole is an ancient one, the migrations of the earlier forms must likewise probably have taken place at an early epoch.

With mammals the case is very different. The earliest known forms, which date from the Triassic and Jurassic rocks, are chiefly marsupials and forms apparently allied to the monotremes, and it is probable that most of the descendants of these, as is more fully indicated in the sequel, migrated southwards during the early part of the Tertiary epoch, to find in Australasia a refuge from the competition of higher forms. Of the higher placental mammals, none of the modern types make their appearance before the Oligocene and Miocene periods, while many do not antedate the Pliocene. Their southern migrations accordingly took place later on in the Tertiary period, one of the earliest movements being the wandering of lemuroids, insectivores, and civet-like carnivores into South Africa and Madagascar. On the other hand, many other higher types, such as the hippopotami, giraffes, and antelopes, which were abundant in Europe and southern Asia during the Pliocene, only left their more northern homes to find a permanent abiding place in Africa at a very late epoch in the earth's history.

Although the glacial epoch probably had a large share in the southern movements of the later Tertiary mammals, some cause with which we are unacquainted would appear to have been the impelling power at earlier epochs. But be this as it may, it is quite evident that a continuous series of waves of migrations of animal life has taken place throughout a very long portion of the earth's history. Similar migrations are also evident in the case of birds (which are likewise a modern group), many forms, such as secretary-birds and trogons, now exclusively southern in their distribution, being represented in Europe during the middle part of the Tertiary period.

From this inequality in the ages, and consequently in the date of migration, of different groups of animals, it is manifest that there will be great differences in the present distribution of such groups; and hence it will be evident that zoological provinces indicated by one group will not hold good for others. Notable instances of this are afforded by the very different divisions into which the globe is divided by those who take mammals, reptiles,

Different Groups have different Geographical Distribution.

or amphibians as their standards. Between birds and mammals, as might have been expected from the comparatively recent high development of these groups, there is much greater accord. Birds, however, differ from mammals (with the exception of bats) in their power of flight, which enables many of them to cross wide ocean-tracts, and therefore renders them less valuable as indicative of the changes that have taken place in the distribution of land and water than the latter, which, as a rule, require direct means of land transit for their wanderings.

Excluding man (and for the most part bats) and likewise the aquatic forms, such as seals, whales, and porpoises, mammals are the animals best adapted for parcelling out the globe into zoological provinces for two chief reasons. Firstly, they form a group which only attained its maximum development at a comparatively late epoch of the earth's history; and, secondly, their movements are mainly limited by the extent of the land-surfaces of the globe which were in actual communication at the time of such migrations. Consequently they afford the safest and truest indications of the latest changes which have taken place in the distribution of land and water.

Importance of Mammals in Geographical Distribution.

As reference in the following pages will constantly have to be made to the various groups of mammals, it will be well to give a list in this place of the chief ordinal and subordinal groups into which the class is divided ; such as are now extinct being indicated by an asterisk. The list stands as follows, viz. :—

Classification of Mammals.

i.   Order PRIMATES.—Apes, Monkeys, and Lemurs.
     1 Suborder ANTHROPOIDEA.—Apes and Monkeys.
     2   ,,   LEMUROIDEA.—Lemurs.

ii.   Order CHIROPTERA.—Bats.
     1 Suborder MEGACHIROPTERA.—Fruit-bats.
     2   ,,   MICROCHIROPTERA.—Insectivorous Bats.

iii.   Order INSECTIVORA.—Insectivores.
     1 Suborder DERMOPTERA.—Flying-Lemurs.
     2   ,,   INSECTIVORA VERA.— Shrews, Moles, Hedgehogs, etc.

iv. Order CARNIVORA.—Carnivores.
  1 Suborder CARNIVORA VERA.—Cats, Dogs, Bears, Weasels, etc.
  2   ,,   PINNIPEDIA.—Seals and Walruses.
  3   ,,   *CREODONTA.—*Hyænodon*, etc.

v. Order RODENTIA.—Rodents.
  1 Suborder SCIUROMORPHA.—Squirrels, Marmots, and Beavers.
  2   ,,   MYOMORPHA.—Dormice, Mice, and Jerboas.
  3   ,,   HYSTRICOMORPHA.—Porcupines, Agutis, Cavies, etc.
  4   ,,   LAGOMORPHA.—Picas and Hares.

vi. Order UNGULATA.—Hoofed Mammals.
  1 Suborder ARTIODACTYLA.—Antelopes, Camels, Pigs, etc.
  2   ,,   PERISSODACTYLA.—Horses, Tapirs, and Rhinoceroses.
  3   ,,   *LITOPTERNA.—*Macrauchenia*, etc.
  4   ,,   *ASTRAPOTHERIA.—*Astrapotherium*, etc.
  5   ,,   *PYROTHERIA.—*Pyrotherium*.
  6   ,,   *TOXODONTIA.—*Toxodon*, etc.
  7   ,,   HYRACOIDEA.—Hyraces.
  8   ,,   PROBOSCIDEA.—Elephants and Mastodons.
  9   ,,   *AMBLYPODA.—*Coryphodon*, etc.

vii. Order SIRENIA.—Manatis and Dugongs.
viii.   ,,   CETACEA.—Whales and Porpoises.
ix.   ,,   EDENTATA.—Edentates.
x.   ,,   EFFODIENTIA.—Aard-varks and Pangolins.
xi.   ,,   MARSUPIALIA.—Pouched Mammals.
  1 Suborder DIPROTODONTIA.—Kangaroos, Phalangers, and Wombats.
  2   ,,   POLYPROTODONTIA.—Dasyures, Bandicoots, etc.

xii. Order MONOTREMATA.—Egg-laying Mammals.
xiii.   ,,   *MULTITUBERCULATA.—*Plagiaulax*, *Tritylodon*, etc.

These orders are further subdivided into families and genera. In regard to the number of the latter of these, there is still a considerable difference of opinion among naturalists ; but in the present work those adopted in Flower and Lydekker's *Study of Mammals*[1] will be in the main adhered to, with such corrections and additions as are necessary owing to recent emendations in nomenclature or to the discovery of new forms.

Omitting from consideration the purely aquatic and volant members of the class, the most effectual barriers to the dispersal of mammals are formed by channels of fresh or salt water of such width as to preclude their being crossed by swimming. And it is this inability to traverse any extent of water that renders what are known as oceanic islands practically devoid of all mammalian life, with the exception of a few bats and small rodents ; the latter animals having apparently some means of dispersal not common to other members of this class. Oceanic islands, it may be explained, are such as rise from great depths in the ocean, and are composed, almost invariably, either of volcanic rocks or of coral. They show, for the most part, no decisive evidence of having been connected with any continental land, and thus have never been enabled to receive a mammalian fauna[2]. In marked distinction to these are the so-called continental islands, such as Madagascar and Great Britain, which, both from the evidence of their mammalian fauna and their geological conformation, have indubitably been in direct communication with the adjacent continent at no very distant epoch. As a rule, the channels between such islands and the mainland are comparatively shallow, so that a moderate degree of upheaval would place the two in direct connection.

Barriers to Dispersal of Mammals.

The relative depth of the channel between two islands, or between an island and a continent is indeed of much more importance in regard to the dispersal of mammals than is its width. This is best exemplified by the well-known case of "Wallace's line" in the Malayan archipelago ; that name being applied to

[1] London, 1891.

[2] There is a possibility that some oceanic islands may have been connected with continents, and that their original mammalian fauna has been destroyed by submergence.

the narrow strait separating the islands of Bali and Lombok, and its northward continuation, the Makassar Strait, dividing Borneo from Celebes. Although the two former islands are extremely close together, while Celebes is much less widely separated from Borneo than is the latter from Sumatra, yet the faunas of Lombok and Celebes are markedly distinct from those of the islands lying to the north and west of Wallace's line. Soundings show that the Makassar Strait, and likewise the Bali-Lombok Strait are of greater depth than the channels separating the other islands of the archipelago; and consequently that Wallace's line indicates a very old barrier which has long been impassable to the majority of mammals.

That continental islands have received the great bulk of their mammalian fauna by means of a more or less complete land-connection with the mainland, is perfectly evident. Nevertheless, there are cases where certain mammals have crossed the intervening channel, either by swimming, or by having been carried across on natural rafts of some kind; an instance of this nature being exemplified by the occurrence of an African type of pig in Madagascar. It thus becomes a question of considerable interest to ascertain what stretches of sea large mammals are capable of crossing. It is stated that the jaguar has been known to swim across the Rio de la Plata, which at its mouth is something like eighty miles across; and a polar bear has been observed swimming at a distance of twenty miles from land in Bering Strait. The tiger frequently crosses the narrower channels in the Sandarbans of Lower Bengal; and both deer, pigs, and elephants are good swimmers. The latter animals have, indeed, been known to swim for six hours at a stretch, and, with a rest, for upwards of nine; but their rate of progress is extremely slow, and probably exceeds but little, if at all, a mile an hour. The Palk Strait, which is considerably less than forty miles in width at its narrowest part, has formed an effectual barrier to the passage of the tiger from India into Ceylon; and it may accordingly be assumed that about twenty miles is the utmost limit which mammals are likely to cross by swimming, even when favoured by currents.

Such passages as these must, however, be of very rare occurrence, for a terrestrial mammal is not likely to take it into its head to swim straight out to sea in an unknown direction. Moreover, supposing a mammal new to a particular island to have arrived there by swimming, unless it happen to be a pregnant female, or unless another individual of the same species but of the opposite sex should arrive soon after (a most unlikely event), it would in due course die without being able to propagate its kind. And even if it should happen to be a pregnant female, there would be no certainty that its offspring, if but one in number, should be of the opposite sex to its parent. Accordingly, it would seem that the population of islands by mammals that have arrived by swimming must be a very rare event indeed. Rafts may be of more importance. Mr Aplin, in the *Proc. Zool. Soc.* for 1894, mentions that jaguars and pumas are frequently transported by them from one side of the Rio de la Plata to the other; and in the rainy season many are to be seen off the northern coast of Borneo, some of which may be as much as thirty yards in length. Still such rafts are not likely to cross straits, except when there is a current setting from one bank to the other.

Before leaving this part of the subject, it must be mentioned that the degree of difference between the fauna of an island and that of the adjacent continent, or between the faunas of two islands, affords a most important clue as to the relative date of the land-connection between them. Madagascar, for instance, has a mammalian fauna which although clearly derived from Africa, is yet so different from that now inhabiting the mainland, that the land-connection between the two must have been broken up at an epoch comparatively remote. Ceylon and India present a condition in respect of their faunas intermediate between the last and that of Great Britain as compared to the Continent. In the latter instance, with the exception of a single Irish weasel, all the mammals are identical with species now inhabiting the Continent, thus proving that the connection has been a comparatively recent one; although the peculiar weasel indicates that the separation between Ireland and Britain has been of sufficient duration to admit of the development of a distinct specific type in the former country.

Although large rivers like the Amazon and La Plata un-doubtedly form serious barriers to the migration of mammals, yet these are not so insuperable as might at first sight be supposed, owing to the fact that in districts where vegetation is luxuriant, huge natural rafts are formed by the trunks of trees intermingled with vegetable matter, upon which numbers of animals may be borne down stream, and thus transferred from one bank to the other. Nevertheless, in treeless districts, or near their mouths, large rivers afford absolutely impassable barriers to the movements of mammals. In South America, for instance, even such an aquatic creature as the carpincho, or capibara (*Hydrochœrus*) has been unable to cross the Plata river from Uruguay into the Argentine, while, conversely the viscacha (*Lagostomus*) of Argentina is prevented by the same river from reaching Uruguay.

Deserts are, perhaps, even more impracticable than rivers; the Sahara—which was long supposed to have been the site of an ancient sea, although it has really been a desert since very remote ages—having apparently formed a barrier preventing the fusion of the mammals of North Africa with those to the south of that tract since at least the Pliocene epoch. It must not, however, be supposed that what are desert-tracts at the present day have always been such, the existence of a fossil chimpanzee in North-Western India during the Pliocene period indicating that the open sandy plains of the Punjab were at that time covered with dense tropical forests, and probably that the same was the case with parts of Syria and Arabia.

Before taking leave of seas and deserts, it should be mentioned that in the polar regions ice may act in lieu of a land-connection to enable mammals to pass from one country to another. On this subject Dr Heilprin writes that "the reindeer is stated to cross the Bering Strait by way of the Aleutian Islands and the Frozen Sea, and in a somewhat similar manner the musk-ox finds its way to Melville Island; it is, however, somewhat singular that the last-named animal, despite its long ice-journeys, never manages to reach either the continent of Asia or Greenland."

High mountain ranges form an effectual barrier to the migration of mammals, not only on account of the physical difficulties of crossing them, but likewise by the lowness of the temperature at great altitudes, coupled with the absence of proper food, being fatal to the existence of many. As already stated, however, mountain-ranges are much more far-reaching in their effects on such migrations when, as in the Old World, their trend is from west to east, than when, as in America, they run from north to south, and thus do not interfere with the free movements of plain-dwelling animals on either side. In many instances the mammals inhabiting each of the isolated mountain-chains, as those of Europe, are to a great extent specifically identical one with another, not having had time to become modified into distinct species since they reached their present haunts at the close of the glacial period. Even among these, however, there are indications of the commencement of specific differences, the chamois of the Caucasus forming a variety differing somewhat from the typical Alpine race. Where the isolation has been longer, as in the case of the fauna of the highlands of Tibet, the difference is much more strongly marked; the mammalian fauna in this instance being as peculiar and distinct as that of many ancient continental islands.

Probably ever since man has existed in any numbers on the globe he has been exerting a more or less strongly-marked influence on the distribution of animals, either by destroying them, or by conveying them to countries or districts which are not their natural home. By the involuntary aid of man the common rat and mouse, which belong to a genus unknown in the New World, have been conveyed to every country in the globe; while the rabbit has been carried to the Antipodes, where it has flourished and increased in an unprecedented manner. Cattle and horses have been introduced into South America, Australia, and other countries where they were naturally unknown, and by their rapid increase have shown that the absence of particular animals from particular districts is not necessarily due to their being unsuited to live there, but rather to the fact that they have been unable to find their way thither. The fallow-deer, again, has been imported from its Mediterranean home into England and other countries of northern Europe; while goats and pigs have been carried to a number of oceanic islands, where they have done irreparable harm in exterminating the native fauna and flora.

In all these instances the fact of the introduction has always been more or less clearly known, and therefore no difficulty arises as to what are native and what are introduced forms. Very

different, however, is the case with the islands of the Malay Archipelago, where the natives, who have a wonderful facility for
taming animals, have carried a species peculiar to one district or
island to localities where it is quite unknown as a native; and in
consequence of this transportation and acclimatisation it is probable that several mammals have been given a habitat to which
they have not the most remote right.   To the Malays is due the
introduction of the small civet known as the rasse into Madagascar.   Whether the dingo, or native dog of Australia, was introduced at an exceedingly remote era by the original colonisers of
that island, or whether it is truly indigenous, is a question that
will probably never be decisively answered.   It is likewise quite
impossible to say what part man may have played in the extermination of the large mammals that inhabited Europe about the
close of the glacial period, but it seems quite probable that he
may have had a considerable share in their destruction.   Be this
as it may, the domestication of certain mammals has undoubtedly
had the effect of destroying the wild race, as is remarkably exemplified by the two existing species of camel, of neither of which
do we know the original habitat.   The original European wild ox
—unless, indeed, the half-wild cattle of the British parks be its
direct descendants—has likewise disappeared at some unknown
epoch owing to the hand of man.   Although other mammals, such
as the quagga (*Equus quagga*), Burchell's rhinoceros (*Rhinoceros
simus*), and the blaubok (*Hippotragus leucophæus*) have been
almost or completely exterminated by human agency in South
Africa, while the American bison has been practically swept away
from its native prairies, yet in all these instances there is a more
or less full record of the original range of the creatures.   In
other cases also mammals have been utterly exterminated by
human agency from countries of which they were originally inhabitants, as is exemplified by the disappearance from the British
Islands of the bear, the wolf, the beaver, and the wild boar within
the historic period, although they still survive in other parts of
their habitat.   In these particular instances there is fortunately
full evidence as to the former existence of these animals in
Britain; but it is highly probable that in more remote countries
mammals have been exterminated without any record being left

of their existence, so that the full extent of their range, if they be
surviving forms, can now never be ascertained.

To quote the words of Dr Wallace, it is evident that in the
present day we live in an impoverished epoch, so
far as the larger mammals are concerned, as compared with the Plistocene era; this being true not
only as regards the northern half of the Old World,
but likewise North and South America, as well as Australia.
From the northern half of the Old World have disappeared the
mammoth, the elasmothere, the woolly and other species of
rhinoceros, the sabre-toothed tigers, etc.; North America has lost
the megalonyx and the Ohio mastodon ; from South America the
glyptodonts, mylodons, the megalothere, and the macrauchenia have
been swept away; while Australia no longer possesses the diprotodon and various gigantic species of kangaroos and wombats.
In the northern hemisphere this impoverishment of the fauna has
been very generally attributed to the effects of the glacial period,
but although this may have been a partial cause, it can hardly be
the only one. The mammoth, for instance, certainly lived during
a considerable portion of the glacial epoch, and if it survived thus
far, why should it have disappeared at the close ? Moreover, all
the European mastodons and the southern elephant (*Elephas
meridionalis*) died out before the incoming of glacial conditions;
and the same is true of all the extinct elephants and mastodons of
southern Asia. Further, a large number of English geologists
believe the brick-earths of the Thames valley, which contain
remains of rhinoceroses and elephants in abundance, to be of
post-glacial age. As regards the southern hemisphere, it can
hardly be contended that glacial conditions prevailed there at the
same time as in the northern half of the world.

It is thus evident that although a very great number of large
mammals were exterminated (perhaps partly by the aid of human
agency) at the close of the Plistocene period, when the group
had attained its maximum development as regards the bodily size
of its members, yet other large forms had been steadily dying
out in previous epochs. And it would seem that there must
be some general deep-seated cause affecting the life of a species
with which we are at present unacquainted. Indeed, as there

Marginal note: Extinction of the larger Plistocene Mammals.

is a term to the life of an individual, what is more natural than that there should also be one to the existence of a species? It still remains, indeed, to account for the fact that the larger Plistocene mammals had no successors in the greater part of the world, but perhaps this is in some way connected with the advent of man.

Before coming to the consideration of the zoological divisions into which, from the present geographical distribution of mammals, the world may be mapped out, it is necessary to devote a brief space to the consideration of two other points ; the first relating to the relative size of the distributional area of genera and species, and the second to the permanency of ocean-basins and continents.

Distribu-
tional Areas of
Genera and
Species.

As regards the first point, it appears to be true in the case of mammals (although not of all other groups) that every species has a continuous distributional area, except where this has been broken up by human destructiveness. It is not meant by this that every part of such area is inhabited by the particular species, as "station" renders this impracticable; but merely that the whole area is ranged over by the species in such spots as are suited to its particular mode of life. Great variation obtains, however, in regard to the size of such distributional area; and it will be obvious that the size of the area varies directly as the adaptability of the species to different climatic and other physical conditions. Perhaps the most important condition of all is the possibility of obtaining suitable food ; and in this respect carnivorous mammals are in a far better position than any other members of their class, since the kind of animal on which they prey is immaterial. This will readily account for the extensive geographical ranges enjoyed by the puma and the tiger, which, as stated on page 5, embrace almost every degree of latitude. Animals with such a wide distribution are of but little use to the student of geographical distribution. Moreover, it will generally be found that species with a wide range belong to large genera having a still more extensive distributional area; this being markedly the case with the puma and the tiger ; the genus *Felis* being one of the largest in the class, and ranging over the whole world with the exception of Australasia. Such cosmopolitan genera are likewise almost valueless to the distributionist.

On the other hand, species with a small distributional area usually belong to small genera, of which they may be the only representatives. Instances of this nature are afforded by the panda (*Ælurus*) and binturong (*Arctictis*) of the eastern Himalaya, and the parti-coloured bear (*Æluropus*) and chiru antelope (*Pantholops*) of the Tibetan plateau. Although such single representatives of genera are highly important to the study of distributional zoology, of vastly greater importance are small genera having two or more species living in widely separated areas. Examples of such are to be found among the porcupines of the genus *Atherura*, of which one species is Malayan and the other two West and Central African; in the mice of the genus *Golunda*, with one African and one Indian representative; and likewise by the tapirs (*Tapirus*), of which there is one Malayan, and several tropical American species. These examples of "discontinuous distribution" among genera are of the very highest import to the science; since they clearly indicate that some of the lands lying between its present disconnected distributional areas must have formerly been the habitat of the genus, and thus enable important conclusions to be drawn as to the former land-connections between such areas. Both in the case of the tapirs and of the brush-tailed porcupines, remains of extinct species have been discovered in the intermediate areas.

Equally important are families, either large or small, which contain two or more closely allied small genera respectively confined to distant areas. As an instance of a large family containing such allied genera, may be cited the *Viverridæ*, among which the true linsangs (*Linsanga*) are represented by several species from the Eastern Himalaya and the Malayan countries, while the closely-allied *Poiana* is confined to West Africa. The chevrotains (*Tragulidæ*), on the other hand, form a small family with a discontinuous distribution; one genus (*Tragulus*) being now Oriental, while the other (*Dorcatherium*) is West African. Here it is quite evident, of course, that the distributional area of the family must once have been continuous; and, as a matter of fact, remains of both genera occur in the Pliocene of India, those of the latter being also found in the European Miocene.

In other families with a discontinuous distribution, as in the rodent family *Octodontidæ*, which is now mainly confined to Africa south of the Sahara, and Central and South America, the genera may be less closely allied, although sufficiently so to indicate a continuous distributional area, or rather a common centre of dispersal, at no very remote epoch.

Allied families, with a small number of genera, severally confined to distant localities are likewise of the highest value in building up the former history of the globe. As examples of this nature may be cited the tree-shrews (*Tupaiidæ*) of the Oriental countries and the jumping shrews (*Macroscelididæ*) of Africa on the one hand, and the *Solenodontidæ* of the West India islands, and the *Centetidæ* of Madagascar on the other. Such families must clearly have had a common centre of origin and dispersal; the available evidence suggesting that in the case of the two former such centre was Europe.

Although of far less common occurrence than among families or genera, discontinuous distribution in an order is perhaps of even more importance than either of the other cases, as it implies a greater interval of time since the original dispersal took place, and, therefore, carries back such conclusions as can be drawn in regard to former land-connections to a still earlier epoch. Among mammals the only instance of this nature is to be found in the marsupials[1], of which two families are American (and mainly South and Central American), while all the others are confined to Australasia and some of the adjacent Malayan islands. In this case also there is abundant evidence of the wide distribution of the whole group in former epochs of the earth's history.

This last instance leads on to the consideration of what may be termed "centres of evolution." In a previous paragraph it has been stated that, according to the **Centres of evolution.** available evidence, a very large proportion, if not the whole, of the terrestrial mammalian life of the globe has originated in the northern hemisphere, from which it has spread southwards in a continuous successive series of waves. When, however, certain groups of mammals had once reached the more

---

[1] In this work the Effodientia are separated from the Edentata; but when these are united, there is a second instance.

remote parts of the southern hemisphere, where they were free from the competition of the higher forms, and met with favourable conditions, they seemed to take a new lease of life, and attained a fulness and variety of development which they had never reached before. As a rule, more or less complete isolation has been a dominant feature of this development; of which the best and most striking instance is that of the marsupials in Australasia. That area may accordingly be called the marsupial evolutionary centre. Scarcely less striking is the instance of the edentates (of which the original derivation is unknown) in South America, where, in company with certain peculiar extinct groups of ungulates, they attained an extraordinary development, both as regards the number of specific, generic, and family types, and likewise in respect of the bodily size of some of its members. This second area may be termed the evolutionary centre of the edentates. A third great centre is constituted by Europe, Asia, and North America, which appear to have been the main developmental centre of the higher mammals, and may accordingly be named the placental evolutionary centre. Two other minor centres are respectively indicated by Madagascar and Africa south of the Sahara: the former as being the headquarters of the lemurs, may appropriately be spoken of as the lemuroid centre, while the great development of the antelopes in Africa suggests the name of the antelopine evolutionary centre for that continent.

The circumstance that throughout the greater part of North America and Europe a very large proportion of the continents are built up of sedimentary strata of marine origin, naturally led geologists in the early days of their science to the conclusion that every part of the land had at one time been deep ocean, and every stretch of ocean dry land. More careful study led, however, to the belief that this idea was not founded on fact, and that although it was perfectly true that what are now continents had been many times under the sea, yet that such areas had never formed abyssal ocean-depths; and, conversely, that such ocean-depths had never been dry land. In addition to many other lines of evidence, this view of the permanency of continents and ocean-basins is strongly supported by the circumstance that nearly all

Permanency of Continents and Ocean-Basins.

oceanic islands are either of volcanic or coral origin, and do not contain sedimentary rocks; and also that deposits analogous to those laid down in the deepest ocean beds are generally wanting from among the sedimentary series of rocks of which the continents and islands are composed. A further argument was afforded by the discovery that the greater portion of peninsular India and South Africa has been dry land since the Palæozoic epoch.

As is so commonly the case in similar instances, the promulgators of the doctrine of the permanency of continents and ocean-basins pressed their hypothesis too far; and it is now evident that although the doctrine is true as a whole, and more especially as regards the later stages of the earth's history, yet it requires very considerable modification from the original form in which it was advanced. In the first place, it has been shown that crystalline granitic and gneissic rocks occur in the Seychelles, which were formerly regarded as true oceanic islands; and, secondly, deep-sea deposits have been discovered in the West Indies and the Solomon Islands. Moreover, various lines of evidence indicate that during the Jurassic and Cretaceous epochs there was a continuous land-connection between Africa (by way of Madagascar and the Seychelles) and India; while at some time in the Secondary era, in the opinion of Drs Neumayr and Blanford, South America and South Africa were in communication across the South Atlantic. The latter connection appears, indeed, to have been a survival from an older Palæozoic girdle of land which, from the evidence of fossil floras, seems to have existed in low latitudes round nearly three-quarters the circumference of the globe, and which was cut off from the land to the north. There is, moreover, the possibility of a Tertiary connection of Australia with Patagonia by way of Polynesia, to which allusion is made in the third chapter. Then, again, the recent investigations of Dr J. W. Gregory[1] on the fossil corals of the West Indies have afforded strong support to the view that the Atlantic is of comparatively recent origin. After referring to the remarkable resemblance between the existing fauna of the West Indian seas and that of the Miocene deposits of the Mediterranean basin, Dr Gregory[2] observes that the sea-urchins, or

---

[1] *Quart. Journ. Geol. Soc.* vol. LI. pp. 255—312 (1895).

[2] *Ibid.*, pp. 306, 307.

echinoderms, yield still more conclusive evidence. "As I have previously pointed out," he writes, "the intimate affinity between those of the West Indies and the Mediterranean can only be explained by the assumption of the existence of a shallow-water connection across the Central Atlantic in—at latest—Miocene times. That the fauna did not follow along the shores of the North Atlantic basin, is shown by its absence from the northern Miocenes of Europe and North America. The evidence now adduced from the fossil corals of Barbados lends support to this view, as showing that the West Indian fauna is only a fragment of that of the Mediterranean Miocene, and has received nothing from the Pacific. This is in full agreement with Prof. Suess's theory that the Atlantic is of comparatively recent geological age, and arose by the gradual enlargement of two bays which ran north and south from a sea that once extended across the Mid-Atlantic from Europe to America, including both the Mediterranean and the Caribbean Sea."

The question of the southward extension of America, Africa, and Australia to join the Antarctic continent during Tertiary times is alluded to in the sequel.

Summing up the evidence in regard to the permanency of oceans and continents, Dr Blanford[1] several years ago observed "that whilst the general permanence of ocean-basins and continental areas cannot be said to be established on anything like firm proof, the general evidence in favour of this view is very strong. But there is no evidence whatever in favour of the extreme view accepted by some physicists and geologists that every ocean-bed now more than 1000 fathoms deep has always been ocean, and that no part of the continental area has ever been beneath the deep sea. Not only is there clear proof that some land-areas lying within continental limits have at a comparatively recent date been submerged over 1000 fathoms, whilst sea-bottoms now over 1000 fathoms deep must have been land in part of the Tertiary era, but there are a mass of facts both geological and biological in favour of land-connection having formerly existed in certain cases across what are now broad and deep oceans."

[1] Appendix, No. 8, p. 107.

Although much previous work had been done on the subject, the first real attempt to divide the land-areas of the globe into zoological provinces, or regions, was made by Dr P. L. Sclater[1] in 1858. According to this scheme, which was mainly based on the study of Passerine birds, the world was parcelled out into the following six zoological regions, viz.:—

Zoological Realms and Regions.

1. *Palæarctic;* Europe, Northern Africa, Northern and Central Asia.

2. *Ethiopian ;* Africa south of the Atlas, and Madagascar.

3. *Indian,* renamed *Oriental* by Dr Wallace; India, South-eastern Asia, and part of the Malay Archipelago.

4. *Australian ;* Australia, with New Guinea and the adjacent islands, New Zealand, and Polynesia.

5. *Nearctic ;* America as far south as Mexico.

6. *Neotropical ;* Central and South America, with the West Indies.

This scheme, which has been adopted and developed in the brilliant writings of Dr Wallace, has the important merit that it coincides to a great extent with the leading geographical divisions of the globe. It has, however, the serious drawback that it gives no greater rank to Australasia and South America than to the other divisions ; whilst the remarkable difference between the fauna of Africa and Madagascar is overlooked. Further, the northern parts of America are widely separated from those of Europe and Asia to which they are faunistically extremely close.

It should be added that in Dr Sclater's scheme the first four regions, or those belonging to the Old World, were brigaded together under the title of PALÆOGÆA, while the two last, or New World regions, were bracketed as NEOGÆA.

The next important classification was one propounded in 1868 by Professor Huxley[2], who, basing his conclusions on the distribution of the game-birds, divided the world into a northern and a southern division, taking the name of ARCTOGÆA for the former, and NOTOGÆA for the latter ; Notogæa being further sub-divided into a Novo-Zelanian (New Zealand), Australian, and Austro-

¹ Appendix, No. 26.   ² *Ibid.,* No. 18.

Columbian region, the latter being equivalent to the Neotropical of Sclater.

Six years later, Dr Sclater[1], who had by this time turned his attention to the distribution of mammals, proposed to group the regions he had previously named under three larger divisions, making a fourth division for New Zealand and Polynesia. This scheme is as follows, viz.:—

I.   ARCTOGÆA.—Palæarctic, Nearctic, Oriental, and Ethiopian regions.

II. DENDROGÆA.—Neotropical region.

III. ANTARCTOGÆA.—Australian region (exclusive of New Zealand and Polynesia).

IV. ORNITHOGÆA.—New Zealand and Polynesian region.

So far as mammals are concerned, this scheme was a great advance on the first one, although the distinctness of Madagascar was not recognised, while the Palæarctic and Nearctic regions were still maintained. Most of the names for the major divisions are, however, open to objection.

In 1878 Dr Heilprin[2], who does not employ these larger groups, proposed, after a suggestion of Professor A. Newton, to unite Dr Sclater's Palæarctic and Nearctic regions under the common title of the Holarctic region; separating, however, from the former a "transitional" Mediterranean region, and from the latter a similar Sonoran region.

A further step was made in 1890 by Dr Blanford[3], who proposed the following scheme, viz.:—

I.   *Australian* region.

II. *South American* region.

III. *Arctogæan* region; this being divided into Malagasy, Ethiopian, Oriental, Aquilonian (= Palæarctic and northern part of Nearctic), and Medio-Columbian (= Sonoran).

Several other minor modifications have been suggested from mammalian evidence, Dr Allen[4] in 1892 reviving the view that the Oriental and Ethiopian regions should be united, under the name of the Indo-African; but the next most important memoir is that

[1] Appendix, No. 27.   [2] *Ibid.*, No. 17.
[3] *Ibid.*, No. 8, p. 76.   [4] *Ibid.*, No. 2.

of Dr Hart Merriam[1] in 1892, whose views are fully discussed in the sequel.

It will accordingly suffice for our present purpose to say that in 1893 an anonymous writer[2] proposed to take the terms NOTOGÆA, NEOGÆA, and ARCTOGÆA to indicate the three major divisions of Dr Blanford's classification; the same terms being used by Mr W. L. Sclater[3] in a nearly similar sense.

The following scheme is the one adopted in the present volume, viz. :—

I. NOTOGÆIC REALM.—1. Australian Region.
        2. Polynesian Region.
        3. Hawaiian Region.
        4. Austro-Malayan Region.

II. NEOGÆIC REALM.—Neotropical Region.

III. ARCTOGÆIC REALM.—1. Malagasy Region.
        2. Ethiopian Region.
        3. Oriental Region.
        4. Holarctic Region.
        5. Sonoran Region.

It will be noticed that the three realms correspond to the three great evolutionary centres of mammals alluded to in an earlier page.

It may be added that, in a work expressly devoted to the geographical distribution of mammals, it will be unnecessary to allude to the schemes proposed on the evidence of other groups of animals, and we may accordingly proceed forthwith to the consideration of the distinctive features of the various realms and regions here adopted.

[1] Appendix, No. 19. [2] *Ibid.*, No. 4, p. 289. [3] *Ibid.*, No. 28.

CHAPTER II.

THE NOTOGÆIC REALM.

Definition and Characters of the Realm—Australian Region—Monotremes—Marsupials—Rodents—Carnivores—Ungulates—Bats—List of Australian and Papuan Genera—Polynesian Region—Hawaiian Region—Austro-Malayan Region—Palæontological History of Marsupials—How Australia received its Fauna.

THE term Notogæa was first proposed, as stated in the preceding chapte , by Professor Huxley[1], to include not only the Australian region of Dr Sclater, but likewise the Neotropical region (Austro-Columbia); but an anonymous writer[2] appears to have been the first to restrict it to the former of these areas[3]. This view, as being, on the whole, the most convenient, is adopted here; and the Notogæic realm may accordingly be taken as the first of the three primary zoological divisions of the globe, and as equivalent to the Australian region of Drs Sclater and Wallace. According to the latter writer[4], "its central and most important masses consist of Australia and New Guinea, in which the main features of the region are fully developed. To the north-west it extends to Celebes, in which a large proportion of the Australian characters have disappeared, while Oriental types are mingled with them to such an extent that it is rather difficult to determine where to locate it. To the south-east it includes New Zealand, which is in some respects so peculiar that it has even been proposed to constitute it a distinct region. On the east it embraces the whole of Oceania [Polynesia] to the Marquesas and Sandwich Islands,

[1] Appendix, No. 18.   [2] *Ibid.*, No. 4.

[3] The term Antarctogæa has been proposed by Dr Sclater (Appendix, No. 27, p. 214), for this area, but it is not a happy one.

[4] Appendix, No. 32, vol. i., p. 387.

where a very scanty and often peculiar fauna must be affiliated to the general Australian type." To the north-east the line of demarcation of the realm from the Oriental region of Arctogæa has been finally fixed at the deep channel separating the islands of Celebes and Lombok on the one side from those of Borneo and Bali (at the extremity of Java) on the other; this division being now well known under the name of "Wallace's line."

All writers are, however, by no means agreed as to the right of the whole of the area thus indicated to form a single zoological division. Before the publication of Dr Wallace's great work, Professor Huxley had proposed to separate New Zealand as a region of equal rank with his Australasian region. At a later date Professor Heilprin[1] suggested that the Australian realm should include only Australia, Tasmania, New Guinea, with the smaller Papuan islands, and New Zealand; Polynesia, including all the islands lying to the east of the Coral Sea, being raised to the rank of a distinct realm (the Polynesian), while the Austro-Malayan islands were regarded as forming a transitional tract between the Australian realm and what is here termed the Oriental region. In this connection it may be well to notice that the Austro-Malayan sub-region of Dr Wallace is by no means coterminous with the Austro-Malayan transition-tract of Heilprin, the former including, and the latter excluding New Guinea.

So far as mammals alone are concerned, Notogæa is widely separated from the whole of the rest of the world by being the sole habitat (both now and in the past) of the typical diprotodont marsupials and the monotremes; although it must not be supposed that either of these groups is distributed over the entire area. As a matter of fact, apart from introduced rodents, Polynesia is devoid of mammalian life with the exception of bats and a rat[2], while New Zealand has but two representatives of the former group, and a rat which may or may not be indigenous. But wherever we meet with a fully developed mammalian fauna, as in the transitional Austro-Malayan islands, there a certain number of marsupials are met with, although the monotremes

---

[1] Appendix, No. 17.

[2] *Mus exulans*, see *Proc. Zool. Soc.* 1895, p. 338.

are restricted to Australia, with Tasmania; and New Guinea, with the adjacent islands, such as the Aru group.

The Notogæic realm, as defined above, may be conveniently divided into four distinct regions, as follows. Firstly the Australian region, comprising Australia, Tasmania, New Guinea and the adjacent Papuan islands; characterised by marsupials and monotremes forming by far the predominant element in the mammalian fauna. Secondly the Austro-Malayan region, embracing Lombok, Celebes and the other islands lying between them and the Australian region; this area being characterised by the absence of monotremes and by the marsupials (all of which belong to the diprotodont division of the order) forming only a small minority of the mammalian fauna. Thirdly, there is the Hawaiian region, including only the Sandwich Islands; and, lastly, the Polynesian region, which may be taken to include all the islands, save those last named, lying to the eastward of the Coral Sea, together with New Zealand, and is characterised by the general absence of terrestrial mammals. There is some difficulty in deciding whether the islands of the Solomon group should be included in this region, or classed with the Papuan division of the Australian region, seeing that, in addition to a considerable number of bats, they have four species of mice, and one diprotodont marsupial (*Phalanger*)[1]. When, however, the poverty of this fauna as compared with that of Papua is taken into consideration, and it is also borne in mind that Mr C. Hedley[2] has come to the conclusion that the Solomon Islands, together with the New Hebrides, New Caledonia, Fiji, and Norfolk Island, are closely connected by means of their flora with New Zealand, and have but little in common with Australia and New Guinea, it seems preferable to include the former group in the Polynesian region. On the same grounds, New Zealand is regarded, in accordance with the views of Heilprin, as also forming a portion of the same zoological region, and not as the representative of a separate region by itself. The fauna of the Solomon Islands has doubtless been derived directly from that of the Duke of York group, which clearly belongs to the same region as New Guinea, and shows a much more strongly marked Papuan facies, having three species of mice, and four marsupials.

[1] See Thomas, Appendix, No. 30.   [2] Appendix, No. 16.

Although the northern half of Australia lies within the tropics, yet few portions of this great island present that luxuriance of vegetation which we are accustomed to associate with tropical scenery; and large tracts **Australian region.** of the interior, owing doubtless to the absence of elevated mountain ranges in the central districts, form arid sandy deserts more or less unsuited to the maintenance of animal life. The coast regions and the borders of the larger rivers are accordingly those where vegetation flourishes best; the finest tracts of pasture-country, well supplied with water, lying to the east and south-east, and Victoria possessing a mountain range whose summits are perpetually clothed with snow. Mountains also occur in the dry and hot western districts. Although Tasmania enjoys moister conditions, Australia as a whole is characterised by the lack of water and the general dryness of its climate; and it is probable that to this aridity the number of jumping animals, such as kangaroos, rat-kangaroos, and jerboa-rats,—now characteristic of this part of the region—is due, since such creatures are admirably adapted for traversing long distances in search of food and water. On the other hand, New Guinea, together with the Papuan islands, has a moist tropical climate, essentially different from that of Australia, but similar to the conditions obtaining in a large portion of the Austro-Malayan islands. Hence it is not to be wondered at that the mammals of New Guinea differ very markedly from those of Australia; this being especially noticeable in the paucity of typical jumping kangaroos, and the proportionately large number of arboreal members of this group. Nevertheless the mammalian fauna of Queensland and North Australia exhibits a marked approximation to that of New Guinea, one species of kangaroo, as well as a cuscus, a striped phalanger (*Dactylopsila*), a flying phalanger (*Petaurus*), a pouched-mouse (*Phascologale*), and an echidna, being common to the two areas, and it is in these countries alone that tree-kangaroos are met with. From these resemblances in their faunas—and especially from the restriction of the monotremes to these two areas,—there can be no question as to the propriety of including Australia and New Guinea in the same zoological region, and thus separating the latter country from the Austro-Malayan region.

The egg-laying mammals, or monotremes, constitute not only
a distinct order (Monotremata), but likewise a
separate sub-class (Prototheria); and are broadly
distinguished from all other members of their class by laying eggs,
from which the young are in due course hatched; as they are likewise
by the milk-glands of the female opening on the surface of the
skin by means of a number of minute perforations, without being
furnished with nipples. The group is represented by three genera,

FIG. 1.   THE DUCKBILL.   (*Ornithorhynchus anatinus.*)

one of which is widely different from the other two and forms a
family by itself, while the latter constitute a second family. The
duckbill (*Ornithorhynchus anatinus*), as the single representative
of the first family (*Ornithorhynchidæ*) is commonly termed, is an
aquatic, somewhat mole-like, burrowing animal, easily recognised
by the expansion of the muzzle into a broad duck-like beak
covered during life with a sensitive skin, and also by the broadly
webbed feet, of which the soles are naked and devoid of pads.

Although in the adult the mouth is furnished only with horny plates, in young individuals the sides of the jaws are provided with three pairs of molar teeth, quite unlike those of any other living mammals.   At the present day the duckbill is confined to Queensland south of latitude 18°, New South Wales, Victoria, South Australia, and Tasmania; it is represented by an extinct species from the Plistocene of Queensland, but otherwise the palæontological record of the group is a complete blank.

Just the same is the case with the echidnas, or spiny anteaters (*Echidnidæ*), of which the only fossil remains known have been obtained from the superficial deposits of New South Wales. Terrestrial and fossorial in their habits, the echidnas differ from the duckbill in having the muzzle in the form of an exceedingly slender cylindrical toothless beak, furnished with an extensile worm-like tongue; while the fur is thickly mingled with short spines, the tail being rudimental, and the unwebbed toes provided with extremely powerful claws.   Of the two species the common five-clawed echidna (*Echidna aculeata*) extends from south-eastern New Guinea throughout Australia to Tasmania; whereas the three-clawed echidna (*Proechidna*[1] *bruijni*) is restricted to New Guinea.

With the exception of the Plistocene forms already alluded to, no fossil monotremes whatever are known to science.   It is, how-ever, not improbable that certain extinct mammals from the Secondary and lower Tertiary rocks of Arctogæa, commonly termed Multituberculata, which will be more fully alluded to in the sequel, may indicate a second order of the sub-class Prototheria.   Both the extinct and the living groups are, however, of a highly special-ised type, so that the one cannot apparently be regarded as ancestral to the other; but if the presumed distant relationship between the two be substantiated, it will indicate that we are to look to a northern origin for the existing monotremes.

The marsupials, which likewise represent both a separate order (Marsupialia) and a sub-class (Metatheria) by themselves, differ from the monotremes by **Marsupials.** producing living young, and by the milk-glands of the female

---

[1] It has recently been proposed to substitute the term *Zaglossus*, which is stated to be earlier, for this genus.

discharging their secretion by means of nipples. From the higher mammals (Eutheria) they are distinguished by the imperfectly developed condition of the newly-born young, and the absence of any prenatal connection between the vascular system of the fœtus and the maternal parent by means of the organ known as the placenta. Very generally the young are carried about for some time after birth in a pouch situated on the abdomen of the parent, where they at first remain immovably fixed to the nipples, the milk being injected into their throats by the action of a special muscle. In the carnivorous and insectivorous forms the number of incisor teeth in the upper jaw usually exceeds the three pairs which form the general maximum limit in the higher mammals. A further peculiarity of the order is to be found in the replacement of the teeth. Instead of the whole or nearly the whole of the first, or milk-set of teeth in advance of the true molars or hinder cheek-teeth being replaced by a second set of permanent teeth, only one tooth is thus (and that by no means invariably) replaced. The tooth thus replaced was long regarded as corresponding to the last or fourth milk-molar of the higher mammals, while the apparently replacing tooth was identified with the last or fourth premolar of the same. From recent researches, however, it would seem that in reality this is not a case of true replacement, and that the tooth which makes its appearance late in life is a retarded premolar, representing the fourth in that series, while the replacing tooth is the fifth.

Marsupials may be divided into two main sections or sub-orders, readily distinguished from one another by their dentition, both of which are represented in Notogæa. In the first of these, or Diprotodont sub-order, which is the more specialised of the two, the incisor teeth are separated by a gap from those of the cheek-series, and do not usually exceed three in number on each side of the upper jaw[1], and in the lower jaw are generally reduced to a single pair, while the tusks, or canines (c), are either small or wanting. In their habits the members of this section are more or less exclusively herbivorous. On the other hand, in the Polyprotodont mar-supials, all of which are mainly carnivorous or insectivorous in

---

[1] The only exception to this occurs in the South-American forms.

their diet, the incisor teeth are numerous and pointed, the canines are large and well developed, and the whole of the anterior teeth form a series more or less nearly continuous with those on the sides of the jaws. The typical diprotodonts, or those in which two of the toes of the hind foot are enclosed in a common integument[1], are exclusively confined to the Notogæic realm, where they attain their maximum development in the Australian region; but the polyprotodonts and an aberrant group of diprotodonts are still represented in the Neogæic realm, while during the Secondary and earlier part of the Tertiary period the former were widely

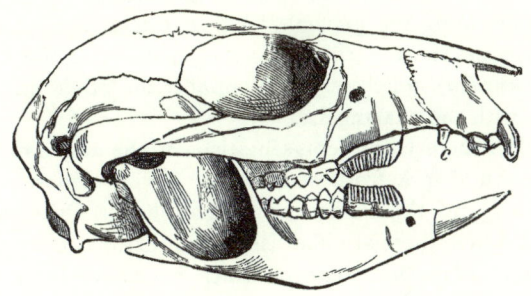

FIG. 2.   SKULL OF RAT-KANGAROO.
(*To exhibit Diprotodont type of dentition.*)

spread over Arctogæa. In the Australian region marsupials play the part of the eutherians of other regions, and show a remarkable diversity of external form and structure, adapting them to all modes of life with the exception of the aquatic. And it is fairly evident that within the limits of this region the diprotodonts were originally evolved from the more generalised polyprotodonts.

Of the three existing family groups into which the Notogæic diprotodonts are divided the first is the *Macropodidæ*, or the kangaroos and their allies; this being in some respects the most specialised group of all, and characterised by certain peculiar features in the skull and dentition. Among these the typical genus *Macropus*, including the true kangaroos, comprises a total of twenty-three species, out of which twenty are confined to Australia and

---

[1] The term syndactylous is applied to this type of foot.

Tasmania, one (*M. agilis*) is common to Australia and Queensland, and two others (*M. bruijni* and *M. browni*) are confined to
New Guinea or the adjacent islands.   Of the six species of rock-
kangaroos (*Petrogale*), none are found out of continental Australia,
and the same is true with regard to the three representatives of
the nail-tailed wallabies (*Onychogale*), and likewise with the three
hare-wallabies (*Lagorchestes*). On the other hand, the three kinds of
dorca kangaroo (*Dorcopsis*) are exclusively Papuan ; while of the
climbing tree-kangaroos (*Dendrolagus*), three are from Papua
and two from Queensland.   The single species of banded wallaby
(*Lagostrophus*) is Australian ; as are also the whole of the rat-
kangaroos, forming the genera *Potorous, Caloprymnus, Bettongia*,
and *Æpyprymnus*, and likewise the peculiar musk-kangaroo
(*Hypsiprymnodon*), which serves to connect the other members of
the family with the phalangers.

Several of the existing representatives of the above-mentioned
genera are found in a fossil state in the cavern-deposits of New
South Wales and the Plistocene formations of Queensland, in
addition to which there are likewise several extinct representatives
of the genus *Macropus*, some of which considerably exceed the
largest living forms in point of size.   The same formations have
also yielded the remains of three extinct genera, namely *Palorchestes, Procoptodon*, and *Sthenurus*, all of which appear to have
been allied to the wallabies, although some of the species were
vastly larger than any existing kangaroo.   Another, but very imperfectly known genus *Triclis*, seems to have connected the
musk-kangaroo so closely with the phalangers, that it is scarcely
possible to draw any distinction between these two families.

In the family of the phalangers (*Phalangeridæ*), which
differs from the more typical representatives of the preceding
by the more generalised characters of the skull, teeth, and
limbs, there is an exclusively Australian form in the koala,
forming the sole representative of the genus of the same name.
Of the five species of cuscuses (*Phalanger*), one is, however,
common to northern Australia, New Guinea, and the Austro-
Malayan Islands, while the other four are restricted to the two
latter areas.   The two species of true phalanger (*Trichosurus*) are,
on the other hand, exclusively Australian ; while the ring-tailed

phalangers, constituting the genus *Pseudochirus*, are common to Australia and New Guinea. Another exclusively Australian type is to be found in the taguan flying phalanger (*Petauroides*); but of the non-volant striped phalangers (*Dactylopsila*) one species is common to Queensland, the Aru Islands, and New Guinea, while the second is exclusively Papuan. The true flying phalangers of the genus *Petaurus* include two Australian species, and a third, common to northern and eastern Australia and New Guinea and the adjacent islands. Leadbeater's phalanger (*Gymnobelideus*), which appears to be closely related to the ancestral stock from which were evolved the members of the last genus, is restricted to Victoria; but the dormouse-phalangers of the genus

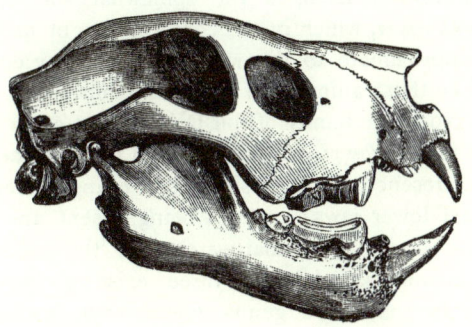

FIG. 3. SKULL OF EXTINCT PHALANGER (*Thylacoleo carnifex*).

Dromicia have both Australian and Papuan representatives, while the pen-tailed phalanger (*Distœchurus*) is exclusively from New Guinea, and of the two pigmy flying phalangers (*Acrobates*), one is Australian and the other Papuan. Lastly, the aberrant long-snouted phalanger (*Tarsipes*), representing a sub-family by itself, is confined to Western Australia. Remains of species belonging to some of the existing genera have been disinterred from the caves of New South Wales and the Plistocene deposits of Queensland; while several more or less imperfectly known extinct generic types have been described. Among the latter, by far the most remarkable is *Thylacoleo*, which was a gigantic phalanger comparable in size to a large leopard, and distinguished by the great development of the last premolar tooth in each jaw. The tooth in question has an elongated cutting-blade, adapted to work against its fellow in the opposite jaw with a scissor-like action, somewhat after the fashion of the carnassial teeth of a tiger ; but the other cheek-teeth were all relatively small, although the tusks were large. The giant among the marsupials was the extinct *Diprotodon* of the Australian Plistocene, a creature rivalling in size the extinct South American *Megalotherium*, and allied on the one hand to the kangaroos, and on the other to the phalangers. It was not, however, endowed with the leaping powers of the former, and doubtless walked on the ground in the ordinary manner, its toes having apparently been covered with structures intermediate between hoofs and nails. Nearly related, but likewise representing a family by itself, is the somewhat smaller, but still gigantic *Nototherium*, which in the conformation of its limb-bones appears to approximate to the wombats, and may consequently have been, like those animals, of fossorial habits.

The last Notogæic family of the Diprotodont section is that of the wombats (*Phascolomyidæ*), distinguished from all the preceding forms by the presence of only a single pair of incisor teeth in both the upper and lower jaws ; canines being absent, and the whole dentition thus curiously simulating that of the rodents among the higher mammals. All the three existing species, which are included in the single genus *Phascolomys*, are confined to Australia and Tasmania ; and, except certain extinct species belonging to the same genus, the only other member of the family is the extinct *Phascolonus* (*Sceparnodon*) from the Australian Plistocene, distinguished by the peculiarly flattened and chisel-like form of the upper incisor teeth. This, the only known species, attained much larger dimensions than either of the existing wombats.

The Polyprotodonts likewise include three existing families found within the limits of the Australian region, none of the members of which stray either into the Austro-Malayan or Polynesian regions, although the separate family of the opossums (*Didelphyidæ*) inhabits the New World. In the family of the bandicoots (*Peramelidæ*), the two species of rabbit-bandicoot (*Peragale*) are exclusively Australian, whereas the true bandicoots (*Perameles*) have both Papuan and Australian representatives ; the third genus

(*Chæropus*), which includes only the pig-footed bandicoot, being confined to Australia.

In the second family, or *Dasyuridæ*, the genus *Thylacinus* is now confined to Tasmania, but it was represented during the Plistocene period on the Australian mainland, where one species is stated to have been obtained from beds of Pliocene age. A similar distribution also obtains in the case of the genus *Sarcophilus*, now represented only by the well-known Tasmanian devil.

Fig. 4. BANDED ANTEATER. (*Myrmecobius fasciatus.*)

Although mainly Australian, the smaller animals known as dasyures (*Dasyurus*) have, however, a single Papuan representative; while the pouched mice (*Phascologale*) are likewise common to the two areas, one of the species ranging from New Guinea to eastern and southern Australia. On the other hand, both the narrow-footed pouched mice (*Sminthopsis*) and the jumping pouched mouse (*Antechinomys*) are exclusively Australian. The same is the case with the aberrant banded anteater (*Myrmecobius*), which although generally included in the *Dasyuridæ*, should perhaps form the type of a family by itself; this animal differing from all the foregoing in the number, and also in the structure of the cheek-teeth, and thereby making a marked approximation to certain Marsupials of the Jurassic epoch noticed in the sequel.

Before leaving this family, it should be mentioned that certain extinct Marsupials from the Tertiaries of Patagonia, referred to in the next chapter, seem to be inseparable from it, while there are strong reasons for regarding one of them (*Prothylacinus*) as very nearly allied to the existing genus *Thylacinus*.

The last of the Australian families (*Notoryctidæ*) of the sub-order is represented solely by the marsupial mole (*Notoryctes*), from the sandy deserts of central South Australia; this being the only member of the order which has taken to a subterranean mode of life. There are no extinct Australian genera of the sub-order.

Exclusive of the bats, the only other order of mammals well represented in the Australian region is that of the Rodentia, or Gnawing Mammals, which bear, however, a small proportion to the marsupials, and all of which belong to the mouse-family (*Muridæ*). And it is noteworthy that although several of these belong to generic types unknown elsewhere, the whole of them are animals of comparatively small size, so that it is possible that their ancestors may have been introduced without a direct land-connection with any other part of the world. A curious feature in connection with this group is that two of the Australian species, namely *Hydromys chrysogaster* and *Mus fuscipes*, are aquatic in their habits; whereas, as we have seen, none of the Australian marsupials are natatorial, although the duckbill is eminently so. The Australian water-rat (*Hydromys*), which is common to Australia and New Guinea, belongs to a sub-family typically distinguished from all other *Muridæ* by the reduction of the molar teeth to two pairs in each jaw. While this animal has partially webbed toes, and is strictly aquatic in its habits, the allied *Xeromys* from Queensland is terrestrial, and approximates to the more typical members of the family, although to which group is still uncertain. The only other representatives of the sub-family *Hydromyinæ* are met with in the mountains of Luzon,

Rodents.

in the Philippine group, where there is one genus allied to
Hydromys, while other species have been assigned to the genus
Xeromys. Whereas the typical Australian representative of the
latter has but two pairs of molar teeth, one of the Philippine
forms has three, thus approximating to more ordinary murines.
The occurrence of these Australian types of rats in the Philippines
is of the utmost importance in respect to Australia having received
its mammalian fauna from south-eastern Asia.

Of the typical genus *Mus*, whose geographical distributional
area includes the whole of the eastern hemisphere with the exception of Madagascar and many of the Polynesian islands[1],
Australia has upwards of twenty-six representatives, while two
species occur in the Duke of York group, and others probably on
the Papuan mainland. One of the Duke of York species (*M.
prætor*) ranges eastwards into the Solomons, where three other
kinds are also found. The jerboa-rats (*Conilurus*[2]) form a peculiar saltatorial group restricted to the sandy deserts of the
mainland of Australia, where they are represented by about a
dozen species ; while the broad-toothed rat (*Mastacomys*) is confined to Tasmania, although its fossilised remains, like those of
the other genus, are met with in the caverns of New South Wales.
More nearly allied to the true rats and mice, the mosaic-tailed rats
(*Uromys*) inhabit Queensland and the Aru Islands, one of the
species from the former area also occurring in the Solomons.
Lastly, the prehensile-tailed rat (*Chiruromys*), from the mountains
of south-eastern New Guinea, represents a genus distinguished
from all other placental mammals of the eastern hemisphere,
with the exception of the British harvest-mouse and the Oriental
binturong, by the prehensile nature of its tail.

In connection with these rodents it is important to observe
that fossil *Muridæ* are unknown from any part of the world earlier
than the Miocene epoch, so that it is evident the living Australian
representatives of the family are comparatively recent immigrants
into the region they inhabit.

---

[1] The Pacific rat (*Mus exulans*) appears to be widely distributed in these
islands, see note on p. 29.

[2] Commonly known by the preoccupied name *Hapalotis*.

Much discussion has taken place with regard to the date of
introduction of the native dog, or dingo (*Canis dingo*)
into Australia, and it was long considered that it
was imported by human agency. Seeing, however, that its remains
have been found in association with those of extinct kangaroos
and *Diprotodon*, there seems considerable probability of its being
an indigenous inhabitant of the country[1].

Carnivores.

The only other non-volant mammal found in the Australian
region is a species of pig (*Sus papuensis*). This
animal is, however, so closely allied to certain
Malayan species that it seems quite possible that its introduction
may be due to human agency.

Ungulates.

The Australian region contains representatives of all the families
of Bats with the exception of the Neogæic *Phyllostomatidæ*; some of the genera, such as the tube-
nosed bats (*Uronycteris*[2]), among the *Pteropodidæ*, being peculiar
to this and the Austro-Malayan region, while others are more or
less widely spread, or even cosmopolitan. It will be unnecessary
to mention the various genera by name; but the affinity of the
Notogæic Chiroptera to those of Eastern Arctogæa, as exemplified
by the abundance of fruit-bats (*Pteropodidæ*) and the absence of
the *Phyllostomatidæ*, is noteworthy.

Bats.

In the following synoptical list the higher groups and genera
(exclusive of Bats) peculiar to the Notogæic realm
are printed in italic type; the letters A, P, and M
following the names respectively indicate that the
groups in question occur in Australia (inclusive of
Tasmania), New Guinea (with the adjacent Papuan islands), or the
Austro-Malayan region; extinct groups have an asterisk prefixed.

List of Australian and Papuan Genera.

I. *MONOTREMATA.*

Ornithorhynchidæ, A.

Ornithorhynchus, A.

Echidnidæ, A. P.

Echidna, A. P.   (The living species common to
both areas.)

Proechidna, P.

[1] See Ogilby, "Catalogue of Australian Mammals," Sydney, 1891—92.
[2] This name replaces the preoccupied *Harpyia*.

II. **Marsupialia.**

i. DIPROTODONTIA, A. P. M. (Elsewhere represented only by an aberrant group in South America.)

1. *Macropodidæ*, A. P.
   Macropus, A. P.
   Petrogale, A.
   Onychogale, A.
   Lagorchestes, A.
   Dorcopsis, P.
   Dendrolagus, A. P.
   Lagostrophus, A.
   Potorous, A.
   Caloprymnus, A.
   Bettongia, A.
   Æpyprymnus, A.
   Hypsiprymnodon, A.
   *Palorchestes, A.
   *Procoptodon, A.
   *Sthenurus, A.
   *Triclis, A.

2. *Phalangeridæ*, A. P. M.
   Koala, A.
   Phalanger, A. P. M.
   Trichosurus, A.
   Pseudochirus, A. P.
   Petauroides, A.
   Dactylopsila, A. P.
   Petaurus, A. P. M.
   Gymnobelideus, A.
   Dromicia, A. P.
   Distæchurus, P.
   Acrobates, A. P.
   Tarsipes, A.
   *Thylacoleo, A.

3. *Diprotodontidæ, A.
   *Diprotodon, A.

4. *Nototheriidæ*, A.
   Nototherium, A.
5. *Phascolomyidæ*, A.
   Phascolomys, A.
   Phascolonus, A.

ii. POLYPROTODONTIA, A. P.

1. *Peramelidæ*, A. P.
   Peragale, A.
   Perameles, A. P.
   Chæropus, A.
2. DASYURIDÆ, A. P.
   Thylacinus, A.
   Sarcophilus, A.
   Dasyurus, A. P.
   Phascologale, A. P.
   Sminthopsis, A.
   Antechinomys, A.
   Myrmecobius, A.
3. *Notoryctidæ*, A.
   Notoryctes, A.

III. **Rodentia.**

1. MURIDÆ. Cosmopolitan.
   Hydromys, A. P.
   Xeromys, A. and Philippines.
   Mus, A. P. M.
   Conilurus, A.
   Mastacomys, A.
   Uromys, A. P.
   Chiruromys, P.

IV. **Carnivora.**

CANIDÆ, Cosmopolitan.
Canis, A. Cosmopolitan.

V. **Ungulata.**

SUIDÆ. Throughout Eastern Hemisphere, except Australia.
Sus, P. Elsewhere throughout greater part of Eastern Hemisphere.

VI. **Chiroptera.** All the families, with the exception of the Neogæic Phyllostomatidæ, well represented.

The Polynesian region, as already said, is characterised by the general absence of non-flying mammals, and therefore claims but little notice here. The only mar- Polynesian Region. supial occurring within the region is a variety of the widely-spread grey cuscus (*Phalanger orientalis*), which occurs in the Solomon Islands, where four species of *Mus* are likewise met with. As the cuscus, together with one of the rats, is also found in the Duke of York group, it may be inferred that the non-volant mammals of the Solomons have been derived from the latter area. In addition to members of widely-spread types, the Solomons possess two peculiar genera of bats.

New Zealand appears to be inhabited only by two peculiar generic types of bats, each represented by a single species, and a rat (*Mus maorium*), but whether the latter is indigenous or introduced appears doubtful.

Although a work devoted to mammals has little to do with an area where the sole member of the class is a bat of the genus *Atalapha*, brief mention must be made of Hawaiian Region. the Sandwich Islands, which from their bird-fauna are regarded as entitled to distinction from the Polynesian region. Of the birds of this area, Mr W. L. Sclater[1] writes that the greater number not only of the species, but even of the genera "are peculiar and wholly restricted to these islands. It is, of course, among the smaller land-birds (Passeres) that this individuality is most marked; but even in the other groups, where the distribution is generally wider, the Hawaiian birds are, in many cases, local."

Poverty, and an admixture of Australian and Malayan types, with a very marked preponderance of the latter, are the leading features in the mammalian fauna of the Austro-Malayan Region. Austro-Malayan region. This area includes the islands lying between Makassar Strait and the narrow channel separating Lombok from Bali on the west and the Australian region

---

[1] Appendix, No. 28.

on the east. The largest of these is Celebes, while those of the Moluccan group, such as Gilolo, Buru, Ceram, and Timor-Laut, together with Timor and Sumbawa, are of smaller size. Unfortunately no complete lists of the fauna of these islands, so far as I am aware, have yet been published.

Commencing with Timor and the Moluccas, we find several of the latter group of islands inhabited by four species of cuscus (*Phalanger*), two of which are common to the Australian region, while the third (*P. ornatus*) is peculiar, and the fourth (*P. celebensis*), which in this group is found only in Sanghir Island, is an inhabitant of Celebes, where the other three are unknown. The only other Austro-Malayan marsupial[1] is a variety of the Australian lesser flying-phalanger (*Petaurus breviceps*), this variety ranging eastwards from Gilolo to the New Britain group. With the possible exception of certain shrews, most of the few Moluccan species of eutherian mammals appear to be identical with those of Celebes, whence they were probably introduced. A deer from Timor has received a distinct name (*Cervus timoriensis*), and the same island is also inhabited by a common Malayan monkey (*Macacus cynomolgus*), a palm-civet (*Paradoxurus hermaphroditus*), and a true civet (*Viverra tangalunga*), the latter being common to the Moluccas. The common Javan porcupine (*Hystrix javanica*), which is widely spread in the Malayan islands, is also found in Timor. There is likewise a cat in the same island, which although described under the name of *Felis megalotis* as a distinct species, and regarded by Mr Jentink as such, has been identified by Mr W. L. Sclater in his "Catalogue of the Mammalia in the Indian Museum" as a mere variety of the domestic species. In regard to all the Timorese forms which are closely allied to, or identical with well-known Malayan species, it is necessary to take into consideration the well-known partiality of the Malays for taming wild animals and carrying them about during their voyages; and it is highly probable that all or most of such animals found in Timor have been thus introduced. From the small island of Flores Mr Jentink has described a rat (*Mus armandvillei*), which is the largest member of its genus.

---

[1] The Kei Islands, like the Aru group, may be best affiliated to Papua.

In addition to the above-mentioned cuscus, which appears to be its only marsupial, Celebes possesses several peculiar types of eutherian mammals.   Among these is a black and nearly tailless ape (*Cynopithecus niger*) representing a genus by itself; while there is also a species of macaque (*Macacus maurus*) peculiar to the southern portion of the island.   A species of the lemuroid tarsiers (*Tarsius fuscomanus*) is found both in Celebes and the neighbouring islands of Salayer and Sanghir, although represented by an

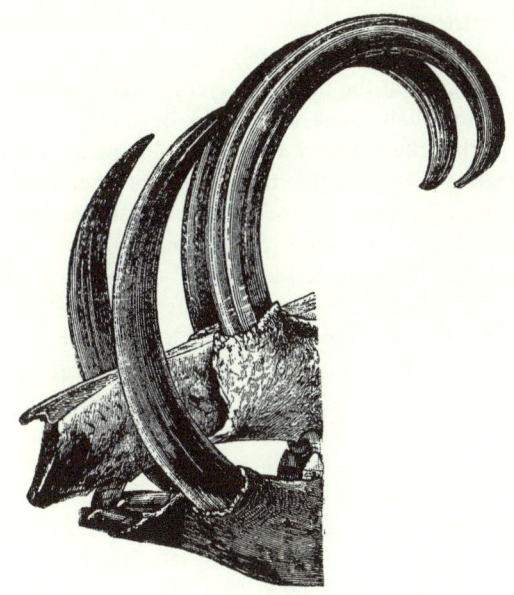

FIG. 5.   FORE PART OF SKULL OF BABIRUSA (*Babirusa alfurus*).

allied form in the Philippine group.   In the Carnivora there is a Malayan species of civet (*Viverra tangalunga*), and also a peculiar species of palm-civet (*Paradoxurus musschenbroecki*).   In the pig-tribe the babirusa (*Babirusa alfurus*), characterised by the extra-ordinary development of its tusks, is the sole representative of a genus confined to this island and Buru; while scarcely less peculiar is the small and somewhat antelope-like buffalo known as the anoa (*Bos depressicornis*), which although allied to the tamarao (*B. mindo-*

rensis)[1] of the Philippines, has its nearest relatives in certain extinct species from the Pliocene of Northern India. There is also a true pig (*Sus celebensis*), nearly allied to Malayan forms; as well as a deer forming a variety of the widely spread sambar (*Cervus unicolor*). Among the rodents, a rat with an extremely long muzzle constitutes a peculiar genus (*Echinothrix*), and there are also other *Muridæ*, as well as squirrels (*Sciuridæ*), in addition to numerous bats, mostly belonging to Oriental types; certain of the squirrels, such as *Sciurus prevosti*, being common to the Malayan countries.

Unfortunately there is absolutely no palæontological evidence to help us in regard to the past history of these islands; but from the living mammalian fauna we should be inclined to place the whole area within the limits of the Oriental region. On the other hand, a large number of the birds both of the Moluccas and Celebes are peculiar; and Australian affinities are displayed by the presence of a bird of paradise (*Semioptera*) in Gilolo and Batjan, and of a cassowary (*Casuarius*) in Ceram. Accordingly, it may be well to include not only the Moluccas, but likewise Celebes within the limits of the Notogæic realm, which will then embrace the whole of the countries where monotremes, typical diprotodont marsupials, birds-of-paradise, and cassowaries occur. Still it must be confessed that this is, after all, mainly a matter of convenience, seeing that since, as will be shown below, the diprotodont marsupials have in all probability originated in the Australian region, those inhabiting the Austro-Malayan region must apparently be regarded as comparatively late immigrants from the south-east[2]; the same being also true with regard to the single bird-of-paradise and the cassowary inhabiting this area. And it is noteworthy that both genera of Austro-Malayan marsupials are precisely such as, from their arboreal habits, would be likely to be transported on floating timber. Dr Wallace has suggested that Celebes has been separated from the Oriental region since the

---

[1] It has been suggested that this animal is a hybrid between the anoa and the Indian buffalo.

[2] In this view I am in accord with Dr Blanford, who (*Geol. Mag.* decade 3, vol. IX. p. 165, 1892) writes that the marsupials of Celebes "are probably of later introduction than the mammals with Oriental affinities."

Miocene; this, however, is obviously too early a date, since the only known allies of the anoa are met with (in common with the earliest of all the oxen) in the Siwalik Pliocene of northern India.

In regard to the Moluccas, Dr Wallace[1] observes that the absence of many characteristic groups of Papuan birds, and likewise of kangaroos and the smaller marsupials of New Guinea, leads to the conclusion that these islands "cannot be mere fragments of the old Papuan land, or they would certainly, in some one or other of their large and fertile islands, have preserved a more complete representation of the parent fauna. Most of the Moluccan birds are very distinct from the allied species of New Guinea; and this would imply that the entrance of the original forms took place at a remote period. The two peculiar genera with clearly Papuan affinities, show the same thing. The cassowary, found only in the large island of Ceram and distinct from any Papuan species, would however seem to have required a land connection for its introduction, almost as much as any of the larger mammalia."

In another work[2], summing up the general conclusions with regard to the fauna of Celebes, the same writer observes that "we are fully justified in classing it as an 'anomalous island,' since it possesses a small but very remarkable mammalian fauna, without ever having been directly united [during Tertiary times] with any continent or extensive land; and, both by what it has, and what it wants, occupies such an exactly intermediate position between the Oriental and Australian regions that it will perhaps ever remain a mere matter of opinion with which it should properly be associated. Forming, as it does, the western limit of such typical Australian groups as the marsupials among mammalia, and the *Trichoglossidæ*[3] and *Meliphagidæ*[4] among birds, and being so strongly deficient in all the more characteristic Oriental families and genera of both classes, I have always placed it in the Austra-

[1] *Geographical Distribution of Animals*, Vol. I. p. 419.
[2] *Island Life*, p. 432.
[3] Equal *Loriidæ;* includes the brush-tongued lories and loriquets.
[4] Honey-suckers.

lian region[1]; but it may perhaps with equal propriety be left out of both till a further knowledge of its geology enables us to determine its early history with more precision."

Having now briefly surveyed the leading features of the terres-

Palæonto-logical History of Marsupials.
trial mammalian fauna of the whole Notogæic realm, and discussed the relationship of the mammals of the Austro-Malayan to those of the Australian region (in the restricted sense of the term), we are in a position to enter upon the consideration of the probable past history of Australia and New Guinea, so far as the same group of animals is concerned. Before doing so, it is, however, essential to state what is known concerning marsupials from other regions of the world. Here it may be premised that in regard to Australia mammalian palæontological history is a total blank previously to the Pliocene epoch; while apparently but little is known even of that period, the great majority of the fossil mammals of that country belonging to the Plistocene epoch of the earth's history. As to the past history of the mammals of New Guinea, we know absolutely nothing; and, as already mentioned, the same is the case with regard to those of the Austro-Malayan islands. This, however, by no means exhibits the whole depth of our deficiency of information. Throughout the whole of eastern Asia, to say nothing of Alaska and western Canada, we have no information whatever as to mammalian life previous to the Pliocene era; while even in that period our sole knowledge relates to a portion of northern India and China. If, therefore, some of the modern types of marsupials originated in eastern Asia from the older forms, the blank in the palæontological history of the group relates to just the very countries where these animals might naturally be expected to occur during the Tertiary period. While there is no record of their existence in Asia, in Europe fossil Tertiary marsupials are unknown at a later date than the upper Oligocene, and in North America than the middle Oligocene, and the whole of those hitherto discovered belong to the existing American group of opossums. If, however, we were to infer from this that the whole order (with the exception of that group) never existed in conti-

---

[1] Equivalent to the Notogæic realm of this work.

nental Arctogæa after the Secondary epoch, a very serious error might be committed. And although it is improbable that any marsupials of an Australian type ever existed in Europe or North America, there is no reason why they should not have occurred in south-eastern Asia.

The extinct dasyurids of the Patagonian Miocene have been already mentioned, and these, together with another S. American group, are more fully noticed in the next chapter. With regard to the opossums, it will suffice to state that while they are unknown in the aforesaid Patagonian deposits, certain species occur in the middle Oligocene White River beds of the United States, and others in the lower, middle, and upper Oligocene[1] beds of Europe. Although the number of their incisor teeth is unknown, there is little doubt that the European Oligocene opossums[2] (which have been very generally separated as *Peratherium*), should be included in the existing genus *Didelphys*. Remains of these animals have been obtained from the upper Oligocene of Cournon in France, from the middle Oligocene beds of Hordwell in Hampshire, from the equivalent deposits of Débruge in Vaucluse, and of Montmartre near Paris, and likewise from the Quercy Phosphorites in the south of France. With the exception of a peculiar South American group of diprotodonts, the remaining fossil mammals which can be referred to the Marsupialia are mainly if not exclusively confined to the Secondary period ; all being of small dimensions, and many of them exceedingly minute. While many of them evidently died out without leaving any existing descendants, one group seems to have been the ancestral type from which the existing *Dasyuridæ* have originated. Among the former, we have the family *Triconodontidæ*, as represented by the genus *Triconodon* of the upper Jurassic of England and also by nearly allied forms from the corresponding rocks of the United States. In this family the

---

[1] It may be well to mention that the beds of St Gérand-le-Puy, in France, which many writers reckon as lower Miocene, are here classed as upper Oligocene. See Lydekker, *Cat. Foss. Mamm. Brit. Mus.* Pt. IV. p. xvii.

[2] The existing *Didelphyidæ* differ from the *Dasyuridæ* in the presence of four, in place of three, pairs of incisor teeth in the lower jaw, and of five pairs in the upper jaw instead of four.

molar teeth consist of three simple compressed trenchant cusps arranged in a longitudinal line; the upper ones biting on the outer side of those of the lower jaw. In the upper jaw the number of teeth is still unknown, but the lower jaw carries three pairs of incisors, four of premolars, and either three or four of molars, in addition to the tusks or canines, which are implanted by two distinct roots. In this respect the latter teeth present an approxi-

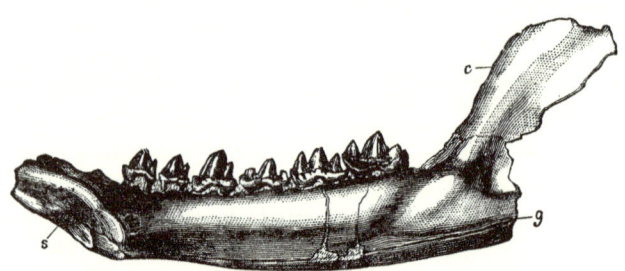

FIG. 6.   INNER SURFACE OF RIGHT HALF OF LOWER JAW OF *Triconodon*,
OR ALLIED FORM.

mation to those of the bandicoots, where the root of the canine is partially divided by a longitudinal groove. A second family (*Spalacotheriidæ*), likewise represented in the upper Jurassic rocks both of Europe and the United States, is distinguished by the cusps of the molars being arranged in a triangle, with the apex pointing inwards in the upper, and outwards in the lower jaw; these teeth being similar in structure to those of the marsupial mole.

Of more interest is the large family of the *Amphitheriidæ*, which may be taken to include a great number of forms apparently agreeing with the opossums in having four pairs of lower incisor teeth. The lower molars never consist solely of three simple cusps arranged in a straight line like those of the *Triconodontidæ*, or in a triangle like those of the *Spalacotheriidæ*. Among these forms the genus *Phascolotherium*, from the lower Jurassic Stonesfield Slate of Oxfordshire, appears to have had only seven pairs of cheek-teeth; the lower molars having three cusps arranged in a longitudinal line, of which the middle one is considerably larger

than the other two, while there are minute additional cusps at the
two extremities, and a distinct ledge at the base of the inner side

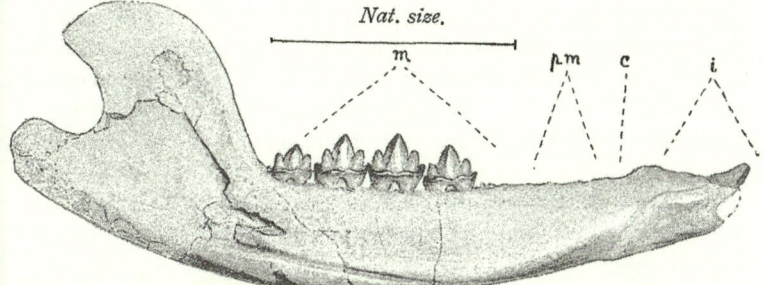

Fig. 7.   IMPERFECT RAMUS OF LOWER JAW OF *Phascolotherium.*

of each tooth.   In the allied *Amphilestes*, from the same formation, of which the imperfect lower jaw is shown in the annexed
figure, the molars are of the same general type as in the preceding,
but much more numerous, their total number being probably
nearly the same as in the next genus.

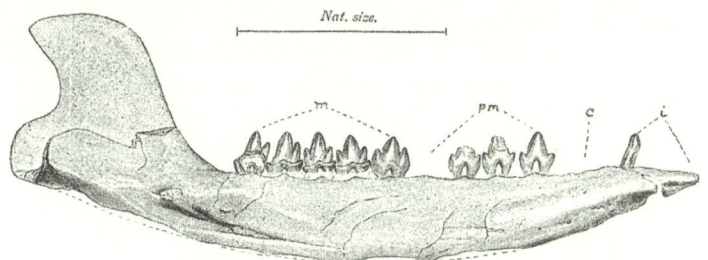

Fig. 8.   IMPERFECT RAMUS OF LOWER JAW OF *Amphilestes.*

Another type is represented by the genera *Amphitherium* of
the Stonesfield Slate and *Amblotherium* of the upper Jurassic of
Dorsetshire, in which, in addition to the canines, there are from
six to eight molar teeth, four premolars, and four incisors in each
half of the lower jaw.   The lower molars differ from those of the
preceding genera, and thereby resemble the corresponding teeth

of the opossums and bandicoots, in that they consist of an anterior portion carrying three cusps in a triangle, and of a posterior moiety or heel. Several more or less nearly related genera have left their remains in the upper Jurassic rocks of the United States, among which *Dryolestes* may be specially mentioned; and it appears that in North America the group survived till the succeeding Cretaceous epoch. The especial interest attaching to these marsupials is that they, and they alone, have molar teeth comparable in number, and to a certain extent in structure, with those of the living Australian *Myrmecobius*; and there can be but little hesitation in regarding the latter as the specially modified descendant of these ancient forms of mammalian life. It is, however, important to notice that all the Jurassic types have four pairs of lower incisor teeth, which are now retained by the opossums alone, having been reduced in all the Australian polyprotodonts to three.

Although a very low type of mammal (*Dromatherium*) occurs

How Australia received its fauna. in the Triassic rocks of North America, the foregoing include all the leading extinct forms that can be included among the marsupials. During the Jurassic epoch the group seems to have been widely distributed over Europe and North America; it is known to have existed in the latter area during the Cretaceous epoch, and it probably also survived to that date in some part of the northern half of the Old World. After that, our knowledge is a blank till we meet with the Oligocene opossums of Europe and North America, and the Miocene Patagonian marsupials; so that as regards the Eocene epoch there is absolutely no record whatever of the group.

That Australia received its original fauna of polyprotodont marsupials from the northward may be regarded as practically certain; and the question as regards the Notogæic realm accordingly narrows itself to the approximate date of the immigration. On this point Dr Wallace[1] writes that "it was probably far back in the Secondary period that some portion of the Australian region was in actual connection with the northern continent, and became stocked with ancestral forms of marsupials; but from that time till now there seems to have been no further land-connection,

---

[1] *Geographical Distribution of Animals*, Vol. i. p. 465.

and the Australian lands have thenceforward gone on developing the marsupial and monotremate types into the various living and extinct races we now find there."

Since this passage was written, the case has been somewhat materially altered by the discovery of the dasyuroid marsupials of the Patagonian Tertiaries; while recent researches have tended to show that the alliance between the *Dasyuridæ* and the *Didelphyidæ* is much more intimate than was formerly supposed to be the case[1]. This being so, it is a fairly safe assumption that both families are descended from a single common ancestral stock which, apart from any question of a connection between Australia and South America, can hardly have originated anywhere than in the northern hemisphere, seeing that the *Didelphyidæ* are totally unknown in Notogæa. There is, however, as already stated, no evidence of the existence of opossums before the Oligocene ; and it is in the highest degree improbable that the two families were differentiated as far back as the Jurassic, or even the Cretaceous epoch. Not improbably polyprotodont marsupials survived in south-eastern Asia till the early portion of the Eocene division of the Tertiary epoch, and in this region both *Dasyuridæ* and *Didelphyidæ* were differentiated. Representatives of the former family soon afterwards found their way into Australia and New Guinea, while the opossums would appear to have dispersed in one direction into Europe and in the other into North America, eventually making their way from the latter country at a late epoch in the Tertiary period into South America.

Assuming that the Patagonian dasyurids are more or less closely allied to the Australian forms (and this certainly appears to be the case), it may be taken for granted that they have not originated independently. From considerations advanced in the next chapter, it is almost impossible to believe that they travelled by way of North America ; and if this be so, their only mode of migration would be by means of a land-bridge between South America and Australia by way of the Antarctic continent, or

---

[1] In the British Museum "Catalogue of Marsupials and Monotremes," p. 315 (1888), Mr O. Thomas writes that the family *Didelphyidæ* "is, on the whole, very closely allied to the *Dasyuridæ*, from which, were it not for its isolated geographical position, it would be very doubtfully separable."

possibly in a zone nearer the equator[1].   Assuming such a connection to have existed in Tertiary times (and there is no reason why
it should not have existed), it must either have taken place
before the development of the diprotodonts in Australia, or must
have been in such high latitudes, or so transitory, as to permit of
the passage of only a few forms.   It is true that there is no
definite evidence that land mammals ever existed on the Antarctic
continent, but during a recent expedition certain seals were killed
bearing on their hides marks which appeared to have been inflicted
by the claws of a land carnivore.   If this be substantiated by
future discoveries, it would be not only probable, but essential
that there should have been a Tertiary connection between
'Antarctica' and other lands.   With regard to the probability that
'Antarctica' is of continental origin, in summarising what is known
with regard to the geology of 'Antarctica,' Messrs David and
Smeeth observe that whether a continent, or an archipelago the
islands of which are united by thick sheets of ice, the southern land
is considered to have a superficial area of 4,000,000 square miles,
being, therefore, larger than Australia.   A great chain of volcanoes
has been described, which in Victorialand rise .over 15,000 ft.
above the sea.   On the South American side of Antarctica may
be specially noticed the active volcano of Bridgman, and the
large and partially-submerged volcano of Deception Island, with
its crater over five miles in diameter, the wall of which, built up of
alternating layers of ice and volcanic scoriæ, rises to 1,800 ft.
above the sea.   Sedimentary rocks of Eocene age, with fossil trees,
were discovered in 1893 at Seymour Island; and the French ship
Talisman many years previously dredged off the Antarctic continent fragments of rock containing *Gyroporella*, a fossil plant very
characteristic of the Triassic rocks of Europe.   Near Laurie Island,
in the South Orkneys, limestone occurs.   The rocks collected by
Mr Borchgrevink are of especial interest as confirming the theory
that Antarctica is a continent rather than an archipelago, for the
microline-granite with garnet and tourmaline, and the mica-schists
must have had a continental origin, such rocks being almost
unknown in oceanic islands, but being of frequent occurrence in
continental areas.

[1] See Chapter III.

With regard to the presumed survival of marsupials in south-eastern Asia till the Tertiary epoch, it may be mentioned that although there is a total lack of knowledge of the early Tertiary mammals of Asia, yet there are not wanting indications of an affinity between the fauna of the eastern portion of the latter continent and that of North America which points to a migration from a common centre along the two sides of the Pacific;—a migration which in the early Tertiary period received on the American side an abrupt check by the sea then dividing North and South America. There is, for instance, living in Central Asia a species of deer so closely allied to the North American wapiti, that it is a question whether the two are really entitled to specific distinction; while the Chinese alligator has its nearest living ally in the species inhabiting the Mississippi. Another piece of evidence is furnished by the occurrence in the Tertiaries of the Balkan Peninsula of remains of the perissodactyle genus *Titanotherium*, belonging to a family only known elsewhere in North America. Quite recently remains of the North American mastodon (*Mastodon americanus*) have been discovered in eastern Russia[1].

The existence of such essentially American types in Eastern Europe and Central Asia clearly seems to point to a more intimate connection between the faunas of those regions than exists between those of Western Europe and North America; and thus lends countenance to the idea that marsupials may have lingered on in Eastern Asia till long after the earlier forms had disappeared from Western Europe. On this view, it is quite probable that the opossums of the Oligocene of Europe may have been immigrants into that area from the south-east; this being confirmed by the absence of the group from the Ethiopian and Malagasy regions. As already said, the existence of Australian types of *Muridæ* in the Philippines affords a pretty clear proof that Notogæa received its fauna from south-eastern Asia, where types that had died out elsewhere at an earlier epoch appear to have survived. Doubtless, however, the *Muridæ* effected their entrance into Australia at a later epoch than the marsupials.

[1] See Pavlow, *Mém. Ac. St Pétersbourg*, ser. 8, vol. i. pt. 3 (1894).

The case of the ratite, or flightless struthious birds of Notogæa, as represented by the extinct moas (*Dinornithidæ*) and the living kiwis (*Apterygidæ*) of New Zealand, and the cassowaries and emeus (*Casuariidæ*) of Papua and Australia, seems to confirm the conclusions drawn from the evidence of the marsupials as to the isolation of the Notogæic realm not being so ancient as supposed by Dr Wallace. It is to be presumed that all will agree that more or less complete land-connections must have been necessary for the migration of these birds; and if it can be shown that the group is a comparatively modern one, it cannot but support the marsupial evidence. Before proceeding to do so, allusion must, however, be made to the views of other writers as to the relationships of the different Notogæic lands.

In *Island Life*[1], Dr Wallace considers that during the Cretaceous, and probably also for a considerable portion of Tertiary times, Western Australia was cut off by a deep sea from the eastern margin of the continent, which was united with Tasmania, and possibly also with New Guinea. The eastern and western islands, he writes, "would then differ considerably in their vegetation and animal life. The western and more ancient land already possessed, in its main features, the peculiar Australian flora, and also the ancestral forms of its strange marsupial fauna, both of which it had probably received at some earlier epoch by a temporary union with the Asiatic continent over what is now the Java Sea. Eastern Australia, on the other hand, possessed only the rudiments of its existing mixed flora derived from three distinct sources.......The Marsupial fauna had not yet reached the eastern land, which was, however, occupied in the north by some ancestral struthious birds, which had entered it by way of New Guinea through some very ancient continental extension, and of which the emeu, the cassowaries, the extinct *Dromornis* of Queensland, and the moas and kiwis of New Zealand, are the modified descendants." He further concludes that a large area of what is now the Tasman Sea was upheaved, and nearly, or quite, connected New Zealand with Australia, whereby the fauna and flora then existing in Eastern Australia were enabled to colonise New

[1] Pages 465 *et seq.*

Zealand. Finally, this hypothetical bridge sank, isolating such forms as had reached New Zealand, and, soon after, Eastern and Western Australia became connected by land, and thus assumed a homogeneous fauna. From this and other passages, we are led to conclude that the author believes that the Notogæic flightless birds immigrated, if not in Cretaceous, at least in early Tertiary times[1], from more northern regions.

Dr Wallace's conclusions are, however, challenged by Mr C. Hedley[2], who, from a study of the floras of these regions, supplemented by the distribution of land molluscs, and recent geological observations, refuses to admit that Western Australia ever possessed a monopoly of characteristic Australian animals or plants. Although he considers the separation of the western and eastern portions of the continent by a Cretaceous sea may be granted, yet the land representing Western Australia was much smaller than Wallace supposes. " The shallow inland Cretaceous sea was studded with islands, large and small, which served the fauna and flora as stepping-stones in their migrations from west to east and from east to west." During a late era in the Tertiary epoch he believes New Guinea to have been in connection with Australia ; and further urges " that an ancient continent, separated on the west from Australia by the abysses of the Coral and of the Tasman Sea, is represented by the Solomons, the Fijis, the New Hebrides, New Caledonia, Lord Howe Island, and New Zealand, with its outlying islands....In conclusion, I would contend that New Zealand is associated with the Solomons and the New Hebrides, firstly, as a member of their volcanic system ; secondly, by community of fauna and flora ; whereas to Australia it is related not at all physically, and to a foreign and intrusive element biologically ; and that a theory which derives the fauna and flora of New Zealand primarily from these archipelagoes and remotely from New Guinea, necessitates fewer unproved assumptions than that which derives them from Australia."

To return to the ratite birds, it may be observed in the first place

[1] If the immigration into Eastern Australia was Tertiary, what becomes of the author's statement that Australia has been isolated since the Secondary epoch ?

[2] Appendix, No. 16.

that the giant flightless species such as *Phororhachis* and *Brontornis* of the Patagonian Tertiaries have been recently shown to form a totally distinct group—the Stereornithes,—and it is quite probable that the same may prove to be the case with *Gastornis*, *Dasornis*, and *Diatryma* of the lower Eocene of the northern hemisphere. Apart from these, the earliest known ratites are *Hypselornis* of the Pliocene of India—which appears to be allied to the emeus and cassowaries—and the Australian *Dromornis*, one species of which is likewise of Pliocene age; all the other forms being Plistocene. Moreover, it is now tolerably certain that the true ratites have originated from flying birds, and it is therefore highly probable that the group is an essentially modern one[1]. Accordingly, there is a strong presumption that the ancestors of these birds did not enter Notogæa till comparatively late in the Tertiary period; and that, in fact, their southern migration was not far removed in time from that of the giant land-tortoises, noticed in the next chapter. Possibly they may have entered Australia by way of New Guinea during the connection which Mr Hedley believes to have existed between those two countries late in the Tertiary epoch; while the New Zealand forms may have made their way by means of the presumed land-connection between these islands and the Solomons, New Hebrides, etc.

Of course there is the difficulty as to why mammals did not enter the Polynesian region at the same time; but it is conceivable that even at this date the mammalian fauna of South-eastern Asia may have been very poor in Eutherians, while it is quite possible that the ancestral forms of these birds may not have required the complete land-connection necessary for the passage of the higher mammals. Such connection as served for these birds, however, may have well sufficed for the transit of the ancestors of the Australian murine rodents, which almost certainly entered the country at a later date than the original marsupials and monotremes.

Assuming, then, that the marsupials and monotremes of the Australian region did not reach their present home till the early part of the Tertiary epoch, we must make the further assumption

[1] Captain Hutton is of opinion that the moas originated directly from flying birds in New Zealand, but the evidence in favour of this view appears insufficient.

that at this period South-eastern Asia was entirely, or to a great extent, devoid of higher mammals. Nor is this unlikely, seeing that the ungulates and carnivores of the lower Eocene of the northern hemisphere would clearly have required time to spread themselves to the southward. Hence it may be suggested that, towards the close of the Cretaceous epoch there was first a migration towards the south-east of the ancestral marsupials (and monotremes) inhabiting the northern hemisphere during the Secondary epoch ; and that similar migrations of the higher mammals took place during Tertiary times.

When once the ancestral polyprotodont marsupials obtained a footing in New Guinea and Australia, where they have since been isolated from any serious competition with the higher mammals, they flourished and developed to a degree which they could not possibly have attained in any other part of the world under existing conditions. And it is doubtless within this region that the more specialised diprotodont types were evolved. Remarkable as it undoubtedly is, the present state of development of the Australian marsupials is nothing to what it was during the Plistocene period, when there lived the giant kangaroos, phalangers, wombats, diprotodons, and nototheres already alluded to, by the side of which the largest existing species would appear almost dwarfs. The cause of this universal extinction (for universal it is) of all the larger types of mammalian life throughout the world soon after the appearance of man, is one of those problems which at present is not capable of being satisfactorily solved, as not even a glacial period could have made a clean sweep of the whole globe. It may be added that the evolution of the diprotodont marsupials within the limits of the Australian region, points to the conclusion that the outlying cuscuses of the Austro-Malayan regions are immigrants from the south-east.

With regard to the monotremes, it has already been mentioned that there is no record of their past history beyond the limits of the Australian region. It can, however, scarcely be doubted that their ancestors came from the north with the primitive marsupials ; and if, as is not improbable, the Secondary and early Tertiary Multituberculata of the northern hemisphere are an allied type, there can be no doubt whatever as to this having been the case.

That Notogæa, as typified by the Australian region, is entitled to form one of the three primary zoological divisions of the globe, the distinctness of its mammalian fauna from that of any other area, not only at the present day, but likewise during the Plisto-cene, and probably also the Pliocene epoch, amply demonstrates. The inclusion within the same realm of the Polynesian region,— which evidently never had such a close connection with south-eastern Asia during the time that area was mainly populated with marsupials,—is justified partly on account of its containing more or less similar types of birds, and partly by the practical absence of terrestrial mammals. On the other hand, the Austro-Malayan region, which is really a kind of zoological No-man's-land, is placed within the limits of the same great realm more as a matter of convenience than anything else, although it is undoubtedly sharply differentiated from the Oriental region by Wallace's line.

In conclusion, a few lines may be devoted to showing that certain other groups indicate that the vertebrate fauna of Notogæa, as a whole, has had a northern origin. Among the lizards, the family of iguanas (*Iguanidæ*), which in this realm occurs only in the Fiji and Friendly Islands, is represented in a fossil state in the Oligocene beds of France; while the gigantic extinct monitor (*Varanus priscus*) of the Australian Plistocene appears to have its nearest ally in the smaller *V. sivalensis* of the Pliocene of northern India. The Notogæic Chelonians, which are confined to Australia and New Guinea, all belong to the side-necked group (Pleurodira) of the order, and are represented by the families *Chelyidæ* and *Carettochelyidæ*, the latter containing only a single species from the Fly River. Now, although none of the Australian genera have been detected in the northern hemisphere, the side-necked chelo-nians, as shown in the next chapter, were abundantly represented there during the early Tertiary and Secondary epochs; and it is a remarkable fact that an extinct genus believed to be allied to *Carettochelys* occurs in the Eocene of northern India. Although from their aquatic, and sometimes partially marine habits, the crocodiles are of less importance than some other groups from a distributional point of view, yet it is noteworthy that the single representative of that group (*Crocodilus porosus*) inhabiting Noto-gæa (where it is found in North Australia, the Solomons, and

Fiji) is spread over India, Ceylon, and the south of China; while the absence of caimans and jacaras from Notogæa affords, so far as it goes, an additional argument that any land connection which may have existed in Tertiary times between Australia and South America must either have been very transitory, or must have been situated in such latitudes that tropical forms could not have used it as a means of transit.

Of more importance than all is the tuatera lizard (*Sphenodon*) of New Zealand—the sole existing representative of the order Rhynchocephalia,—since this curious creature is closely allied to the extinct *Rhynchosaurus* and *Hyperodapedon* of Triassic strata of the northern portion of the Old World. Finally, the Port Jackson shark (*Cestracion philippi*) belongs to a genus which was living in the seas of Europe during the Jurassic and Cretaceous periods; and the sole living survivor of the swarms of species of the Molluscan genus *Trigonia* inhabiting the same seas occurs in Australian waters.

CHAPTER III.

THE NEOGÆIC REALM.

Extent and Characters—Mammaliferous Deposits—Monkeys—Bats—Insectivora — Carnivores — Ungulates — Horses — Litopterna — Astrapotheria —
Toxodonts — Pyrotheria — Proboscideans — Rodents — Edentates — Armadillos and Glyptodonts—Sloths—Anteaters—Ground-sloths—Marsupials—
Cetaceans—Early Distinction of the Neogæic Fauna—Early Separation
of N. and S. America—Incursion of Northern Mammals—Distinctness of
the existing Fauna—Origin of the Santa Cruz Fauna—Antarctica and the
South American element in the Ethiopean Fauna—Conclusion — Sub-
regions.

THE second primary zoological division of the globe may be

Extent and
characters. known as Neogæa[1], or the Neogæic realm. It
includes only the Neotropical region. Comprising
not only the whole of South and Central America,
as well as the West Indian Islands, this area also embraces the
lowlands lying on either side of the Mexican plateau—the so-
called *tierras calientes*—thus running up in a fork-like manner to
the lower extremity of North America. While, therefore, the
greater part of this vast area is sharply delineated by its coast-
boundary, to the north it has a kind of No-man's-land connecting
it with the Sonoran region of Arctogæa, and, as will be shown
later, through this transitional area there has been a certain
amount of intermixture of the proper faunas of the Neotropical
and Sonoran regions. Neogæa, as a whole, may be characterised
as a country of extensive tropical forests or open grassy plains ;
deserts occupying only a few scattered areas in the upper Argentine (Tucuman, etc.), and certain parts of the coasts of Chili

---

[1] This term was originally proposed by Dr Sclater to include the whole of
the New World, but has been used by an anonymous writer (Appendix, No. 4)
in the present sense. Dr Sclater's term Dendrogæa (Appendix, No. 27,
p. 214) is open to considerable objection, as the greater part of Argentina is
woodless.

and Peru; the whole of the rest of the area, with the exception of the higher regions of the Andes, being thus admirably adapted for the support of animal life. At least one half of the whole area is occupied by a dense tropical forest, attaining its richest development in the hot steamy tracts of Brazil and Paraguay, and being unequalled in extent in any other part of the globe. With a width of some three thousand miles from the Atlantic seaboard at Pernambuco to the foot of the Andes, this forest extends north and south for nearly thirty degrees of latitude; while not only does it clothe the lowlands and valleys, but extends high up the mountain-sides, as may be seen in the exquisitely lovely harbour of Rio de Janeiro, where the forest-vegetation commences immediately above the wash of the waves, and thence extends in one continuous leafy mass to the summits of mountain-ranges at an elevation of eight or nine thousand feet. In the northern part of the area open grass-lands, like the "campos" of Brazil and the savannas of Venezuela, alternate with the forest; while in the neighbourhood of Buenos Aires the open pampas[1] forms one extensive sea of grass. The Andes, constituting the backbone of the country, run in one continuous chain from north to south on the Pacific seaboard, and present the usual varieties of climate and physical conditions common to other elevated mountain-ranges. Such climatic variations are, however, only an epitome of those met with in travelling from the northern to the southern extremity of the area; the steamy valley of the Amazons having a tropical climate, whereas when we reach the southern point of Patagonia and Tierra del Fuego we are in the midst of snows and glaciers. To the hot forest-regions are restricted the monkeys, marmosets, sloths, ant-eaters, and tree-porcupines; while the open plains of the south are tenanted by guanaco, deer, viscachas, and rheas. Moreover, in the forests, the variety of mammalian life, especially as regards the larger forms, is in marked contrast to its comparative paucity in the open plains; not but that, till civilised man made his appearance on the scene, the number of individuals may have been nearly, if not quite as large in the latter area as in an equal extent of the former.

[1] Although in Spanish the term 'pampas' is plural, in English it seems preferable to use it as singular.

L.   5

At the present day the mammalian fauna of Neogæa is markedly distinct not only from that of Notogæa but likewise from that of the whole of the rest of the globe (Arctogæa), although the distinction is now, owing to free communication with the north, much less marked than it was in Tertiary times, and it is accordingly essential to enter at once into the consideration of the extinct forms in order to show why this part of the world is entitled to rank as one of three primary zoological regions.

There are several districts in South America where fossil remains of mammals have been found ; most of these being remarkable for the extraordinary profusion in which the bones occur. The first that may be mentioned are the celebrated caves of Lagôa Santa, in the province of Minas Geraes, to the northward of Rio, which have yielded remains of a great variety of Plistocene genera and species, inclusive of those of man. Probably contemporaneous with these are the sand-dunes on the coast of Buenos Aires, which likewise contain human remains in association with those of extinct mammals ; while the so-called Pampean beds of the Argentine pampas are apparently somewhat older, although still pertaining to the Plistocene period. As these Pampean deposits are exceedingly rich in fossil mammals, they may be described in some detail. They form the great level tract of country extending southwards from the Rio de la Plata and the Paranà to the Rio Colorado, south of Bahia Blanca, and westwards from the Atlantic seaboard about half the distance to the Andes; thus occupying some 200,000 square miles of country. The pampas is an almost level grass-covered plain, intersected by water-courses, and penetrated near its margins by small mountain-ranges, while it is almost entirely barren of trees. It is composed of a rich black alluvial mud, mingled with beds of sand, and underlain by, or in some places interstratified with layers of a hard white calcareous deposit known as *tosca*; but in certain spots it contains beds of marine shells belonging to species still living in the adjacent seas. Except in those spots where the *tosca* comes to the surface, there is not a stone or a pebble to be seen in the whole deposit, and near Buenos Aires the formation has been bored to a depth of ninety feet without touching bottom. From its composition it is evident

Mammaliferous Deposits.

that the deposit has been carried down from the interior of the north by the Paranà, Paraguay, and other tributaries of what is now the Rio de la Plata; but since there is no splitting of the latter river at its estuary, it is evident that the formation cannot properly be called a delta. That it is mainly of freshwater origin seems evident not only from its intrinsic character, but likewise from the vast number of entire skeletons of mammals buried within it, since these creatures must certainly have lived very near to the places where their bones are now entombed. In the more southern part of its area the pampas is, however, probably to a large extent of estuarine origin; and the presence of layers of marine shells in its uppermost horizon near Buenos Aires proves that at least a portion was submerged beneath the sea before its final upheaval. Whereas the Rio de la Plata now flows in a single channel in a south-easterly direction near the northern limit of the coast-portion of the pampas, it would seem probable that the Paranà and Paraguay rivers may have originally continued their southerly course across the southern pampas, through which they may have flowed in a number of streams. Most likely the Pampean formation was laid down in a slowly subsiding area, in which the rate of deposition approximately counterbalanced the sinking, so that the greater part of it has been always land until the period of the great submergence. After the latter, the entire area was upheaved to a small degree above the sea-level, when the rivers assumed their present approximate courses. Whereas in certain localities the deposit is barren of mammalian remains, in other spots it appears absolutely crowded with them, and the number of entire skeletons that are entombed in it must doubtless be counted by thousands.

Somewhat older than the Pampean is a mammaliferous deposit occurring on the coast near Bahia Blanca in a small hill known as Monte Hermoso; and beds of approximately equivalent age occur in Catamarca at the foot of the Andes. Probably these beds should be regarded as of Pliocene age; and it may be mentioned that equivalents of these deposits are met with in other parts of Argentina, while representatives of the Pampean occur in Patagonia, Chili, Bolivia, etc. Still older than the Monte Hermoso deposits are the Santa Cruz beds of Patagonia, occurring not only on the river of that name, but likewise further north in the Chubat district. These Santa Cruz beds are exceedingly rich in mammalian remains, which are stained of a deep black colour; and while they contain many groups common to the Pampean formation, they lack those forms found in the latter which have an Arctogæic facies, thus indicating that we have reached an epoch when the fauna of South America was far more completely isolated from that of the rest of the world than is at present the case. By Dr Ameghino the Santa Cruz beds were at first correlated with the lower Eocene of Europe, although he subsequently admitted that they must occupy a somewhat higher position in that period[1]. From the nature of their entombed mammals[2] it is, however, certain that they must be still newer, and they cannot be regarded as older than lower Miocene. Indeed, they are underlain by the so-called Patagonian beds[3], which are of marine origin, and contain numerous cetaceans, among which are whalebone whales (Mystacoceti). From equivalent beds in the Chubat district of Patagonia, a large number of such cetaceans have been described by the present writer[4], and it is quite evident that the oldest age that can be assigned to these beds is upper Oligocene, seeing that in Europe whalebone whales are unknown till a considerably later epoch. Probably the freshwater beds containing the peculiar mammal alluded to below under the name of *Pyrotherium* are the freshwater equivalents of the Patagonian horizon[5].

Additional evidence in support of the comparatively late age of the Patagonian beds is afforded by the researches of Prof. Cope[6] on the cetaceans of the Miocene (or Upper Oligocene?) of the United States. From these beds have been obtained remains of *Hypocetus* (*Paracetus*)—a genus elsewhere known only from the Patagonian beds—together with *Cetotherium* (also occurring in the latter deposits), *Balænoptera*, and a species of *Balæna* identified with one from the European Pliocene; the North and South American species of *Hypocetus* being closely allied.

[1] *Bol. Ac. Cordoba*, Vol. XIII. p. 260 (1894).
[2] Remains of the existing genus *Dasypus* occur in these beds.
[3] Ameghino, *loc. cit.*
[4] *Ann. Mus. La Plata,—Pal. Argent.* Pt. II. (1893).
[5] Ameghino, *op. cit.* p. 262.
[6] *Proc. Amer. Phil. Soc.* Vol. XXXIV. pp. 135—155 (1895).

With these preliminary considerations, we pass to a critical examination of the recent and fossil mammalian fauna of Neogæa, which will show how intimately investigations of this nature are connected with the present and past configuration of the land-areas of the globe.

Like as are many of them superficially to their cousins of the Old World, the monkeys of the New World, which are now confined to the tropical forest-regions of **Monkeys.** the Neogæic realm, are structurally quite different to the former; and this circumstance, coupled with their isolated distribution, indicates that their relationship is, at most, but a very distant one. Indeed it is not improbable that the Old and New World monkeys may have originated independently from the group of lemurs, which were formerly very widely distributed over Arctogæa. Of the two families of Neotropical monkeys, the first is represented by the beautiful little marmosets, constituting the family *Hapalidæ*. Although having the same number (32) of teeth as the Old World monkeys, the marmosets differ in that the number of premolars on each side of both jaws is three, and that of the molars two; these numbers being transposed in the other group. The marmosets are further distinguished by the broad septum between the nostrils, the absence both of pouches in the cheeks and of callosities on the buttocks, and the want of any prehensile power in the tail; in addition to which it may be noticed that the thumb is not opposable to the other fingers, while all the digits, with the exception of the first on the hind foot, are provided with long claws. None of these animals are known from the Santa Cruz

FIG. 9. PART OF RIGHT LOWER JAW OF *Homunculus.*

deposits. In the second family or *Cebidæ*, which includes much larger species than the diminutive marmosets, the total number of teeth is 36, owing to the retention of the three pairs of molars found in the Old World group ; these monkeys being further distinguished from the marmosets by the presence of nails on all the fingers, and by the tail being frequently prehensile. The Santa Cruz beds have yielded remains of monkeys (*Homunculus*) referable to this family, showing that they belong to the original South American fauna, but beyond this nothing is known as to the palæontological history or origin of the group. Lemurs, both in the past and the present, are quite unknown in the realm under consideration.

Although bats might be thought of comparatively little importance from a distributional point of view, yet the Neogæic realm presents some very remarkable features in this respect. In the first place the two Old World families of fruit-bats (*Pteropodidæ*) and horse-shoe bats (*Rhinolophidæ*) are entirely wanting, whereas the great family of vampire-bats (*Phyllostomatidæ*) is mainly restricted to it, although a few representatives straggle northwards along the Pacific coast of North America. The family of *Emballonuridæ* is also more strongly represented here than in any other part of the world. Probably on account of their small size, there is no palæontological record of the South American bats from the earlier Tertiary deposits.

Bats.

The Insectivora are almost unrepresented in the continental portion of the realm, although a shrew (*Sorex*) reaches Guatemala and Costa Rica, and a member of the allied N. American genus *Blarina* is also found in the last-named country; both these being doubtless very recent immigrants from the north. Very remarkable is the occurrence in the West Indian Islands of the two species of *Solenodon*, constituting a distinct family (*Solenodontidæ*) by themselves. These have generally been considered as very nearly allied to the tenrecs (*Centetidæ*) of Madagascar, but Mr O. Thomas[1] is of opinion that the relationship is not really very close, the similarity in the structure of their molars being merely a generalised character. Both, however, probably indicate an ancient group, which has

Insectivora.

[1] *Proc. Zool. Soc.* 1892, p. 500.

migrated from the higher latitudes of the northern hemisphere to find a refuge in these two far distant localities. It may be mentioned that the *Solenodontidæ* and *Centetidæ*, together with the *Potamogalidæ* of western, and the *Chrysochloridæ*, or golden moles, of southern Africa, constitute a section of the order distinguished from all the other forms by having the cusps on their upper molar teeth arranged in the form of the letter V, instead of in that of a W; and it will not fail to be noticed that the whole group is now exclusively a southern one. Whether it is represented in the Santa Cruz beds is not quite certain, although one lower jaw has been referred to it by Dr Ameghino[1] under the name of *Necrolestes*. Be this as it may, it is clear that, with the exception of the aforesaid shrews, insectivores with W-shaped molars are entirely wanting in the realm; and that such V-shaped types as still exist, or formerly occurred, evidently indicate an ancient northern group. Other instances of the survival in South America and Madagascar of allied forms will be noticed in the sequel, and admit of a somewhat similar explanation.

Although fairly well represented at the present day, the carnivores of this realm have but few absolutely characteristic forms, and as no remains of the true Carnivora occur in the Santa Cruz beds, the whole of them may be regarded as comparatively late immigrants from the north. The civet family (*Viverridæ*) and hyænas (*Hyænidæ*) are totally wanting at all epochs, as indeed they are throughout the whole of the New World; but the weasel tribe (*Mustelidæ*) have a comparatively small number of representatives. Cats (*Felidæ*) on the other hand are numerous, although several species, and more especially the puma (*Felis concolor*), range to a greater or less extent into North America. Such a cosmopolitan genus is, however, of no importance whatever from a distributional point of view; and much the same may be said of the extinct sabre-tooths (*Machærodus*), which were distributed in Tertiary times throughout the entire northern hemisphere. Unknown in the deposits of Monte Hermoso and Santa Cruz, this genus is represented by a gigantic species in the Pampean, which undoubtedly reached the country

Carnivora.

[1] *Bol. Ac. Cordoba*, Vol. XIII. p. 364 (1894).

from the north in company with the other late immigrants. The dog tribe (*Canidæ*) likewise comes under the category of cosmopolitan groups, and has numerous South American species, although only one from the Monte Hermoso beds dates earlier than the Pampean. It may be mentioned that while true wolves[1] are absent from this realm, all the continental living members of the genus *Canis* found in it form a group by themselves, quite distinct from the true foxes of other regions. Very remarkable is the occurrence in the Pampean of a large species (*C. moreni*) perfectly distinct from all those now inhabiting the country, and presenting some curious approximations in the structure of the skull to domestic dogs. Peculiar to the realm is the bush-dog (*Icticyon*), of Guiana and Brazil,—a small short-haired and short-legged species differing from all other members of the family by the small size and reduced number of the molar teeth. Its remains occur in the Brazilian caves, but are unknown from earlier deposits; and it may thus be regarded as a comparatively late immigrant from the north, which perhaps developed its special characters after its arrival in the country. Of the *Ursidæ* there is but a single existing species in Neogæa, namely the spectacled bear (*Ursus ornatus*) of the highlands of Chili and Peru; but there occur in the Pampean formation of the Argentine remains of an allied extinct genus known as *Arctotherium*, another species of the same genus being recorded from the superficial deposits of California. Far more characteristic of the realm are the raccoons and coatis (*Procyonidæ*), although several of these are common to North America. Till recently it was considered that this family was entirely restricted to the New World, but the Oriental cat-bear or panda (*Ælurus*) is now included; and since fossil remains of the latter genus have been discovered in the Pliocene of England, while those of raccoons and of the extinct genus *Leptarctus* occur in that of the United States, it is practically certain that the group was formerly widely distributed over the northern hemisphere. Among this family the raccoons (*Procyon*) extend over most parts of both North and South America; the coatis (*Nasua*) range from Mexico to Paraguay; the kinkajou (*Cercoleptes*) is found from

[1] The Falkland Island wolf (*Canis antarcticus*) forms a remarkable exception.

Mexico to Peru and Brazil; while the two species of the genus
Bassariscus are inhabitants of the southern United States, Mexico,
and Central America. From the Tertiaries of Catamarca and
Parana there has been described the extinct genus *Cyonasua*,
differing from the coatis in the form of its teeth; this genus
showing that the family had obtained an entrance into South
America as early as the Pliocene period, although it is quite
unknown in the Santa Cruz epoch. In the *Mustelidæ*—where all
reference to the cosmopolitan otters, of which there is one Brazilian
species, may be omitted—the southern skunks (*Conepatus*), which
have mostly but thirty-two teeth, are now practically characteristic of this realm—although the common species ranges into
Texas—and their remains occur fossil in the caverns of Brazil.
The true skunks (*Mephitis*), on the other hand, which have
thirty-four teeth, are North American, although one species
ranges as far south as Guatemala; and since the whole group is
unknown in the earlier Tertiary deposits of the Argentine, it may
likewise be considered a recent immigrant from the north. The
same is true of the genus (*Galictis*) now represented by the
South American grison and tayra, since remains of this group of
mustelines occur in the Plistocene of the United States, as well
as in the Brazilian caves.

Of far more importance than either of the preceding orders
are the hoofed mammals or ungulates, since here we
meet with certainly three, and probably four extinct
Ungulates.
subordinal groups exclusively confined to this realm, while a large
number of the existing families are unrepresented even in the
Pampean formation. Whereas this order is entirely absent from the
West India Islands, South America at the present day is singularly
poor in ungulates. The only existing forms are the peccaries
(*Dicotyles*), which are peculiar to the New World; certain deer
belonging to a genus (*Cariacus*) which is likewise confined to the
western hemisphere, and a Chilian form constituting a genus
(*Pudua*) by itself; the exclusively South American guanacos and
vicuñas, which, together with their domesticated allies constitute
the genus *Lama*; and tapirs (*Tapirus*). The first three of these
belonging to the Artiodactyla, while the last alone represents the
Perissodactyla, or odd-toed division of the order. At all periods

of its history true pigs (*Sus*), hippopotami (*Hippopotamus*), camels (*Camelus*), chevrotains (*Tragulidæ*), giraffes (*Giraffidæ*), true deer (*Cervus*), antelopes, sheep, goats, and oxen, together with a number of extinct forms connecting the ruminants with the pigs, have been conspicuous for their absence. The same is true, among the perissodactyles, of the rhinoceroses (*Rhinocerotidæ*) and the extinct palæotheres (*Palæotheriidæ*) and lophiodons (*Lophiodontidæ*) of Europe, and the uintatheres (*Uintatheriidæ*) and titanotheres (*Titanotheriidæ*) of the United States ; while the true elephants (*Elephas*) among the Proboscidea have likewise been always absent.

Of the existing South American ungulates the peccaries belong to a family (*Dicotylidæ*) which is abundantly represented in the Tertiary formations of the United States ; while in those of Europe and Asia there occur allied forms apparently connecting the peccaries with the true pigs. On the other hand, in South America their remains occur only in the superficial and cavern-deposits, so that there can be no doubt as to their late intrusion into the country from the north. The vicuñas and guanacos are the western representatives of a family (*Camelidæ*) whose other members are Asiatic and African, and of which the past history seems to have been very similar to that of the last group. Fossil camels occur in the Pliocene of India, while a host of extinct genera more or less closely allied to the living South American forms occur in the Tertiaries of the United States; and since in Argentina and Brazil remains of *Lama* and the related types occur only in the Monte Hermoso, Pampean, and cavern deposits, there can be no hesitation in regarding the group as comparatively recently immigrant into the country. The deer of the genus *Cariacus* are likewise only known in South America from the Pampean and some other of the later Tertiaries, as well as the Brazilian caves, while in the Pliocene of the United States they appear to be represented by the ancestral *Blastomeryx*; and these accordingly come under the category of intruders from the north. As regards the tapirs, the genus and family now presents a remarkable instance of discontinuous distribution, one species being confined to the Malayan countries, while all the others are South American. Whereas in the realm under consideration remains of these animals occur only in the superficial deposits, they are met with abundantly in the Miocene and Pliocene deposits of Europe and Asia, as well as in the United States; and it is accordingly clear that the group was once widely distributed over the northern hemisphere, whence its surviving members have wandered southwards to Malaysia in the east and South America in the west.

Till introduced by the early Spanish settlers, horses, which are now so abundant in the pampas, were totally un- **Horses.** known in South America in a living state, although their fossilised remains occur commonly in the Pampean, as well as in the somewhat older deposits of Paranà, and Monte Hermoso. They are, however, entirely unknown in the Santa Cruz beds. Some of these fossil Argentine horses belong to the typical genus *Equus*; while others, on account of the simpler structure of their molar teeth and the great length of the slits in the skull beneath the nasal bones, are referred to a separate genus, under the name of *Hippidium*. A third genus (*Onohippidium*) is distinguished from the latter by a large lachrymal depression in the bones of the sides of the face, corresponding to the so-called larmier of the deer. Fossil species of *Equus* occur in the Plistocene deposits of the whole northern hemisphere, while those of the United States yield remains of a genus (*Pliohippus*) nearly allied to the South American *Hippidium*; and as both these genera are the descendants of extinct forms whose remains occur in the older Tertiaries of the same hemisphere, the extinct South American horses must likewise be classed with the groups that have entered the country from the north. Why these Plistocene South American *Equidæ* became extinct in a country so admirably suited to their existence as Argentina, is a question to which it is impossible to find a satisfactory answer. With the horses we reach the last of the Neogæic representatives of the more typical ungulates, that is to say the Artiodactyla and Perissodactyla; and we pass on to the other extinct subordinal groups, at least three of which, as already said, are peculiar to it.

From the preceding paragraphs it will be apparent that, if we except the deer, horses, and the guanacos and their **Litopterna.** allies, South America, so far as the Artiodactyla and

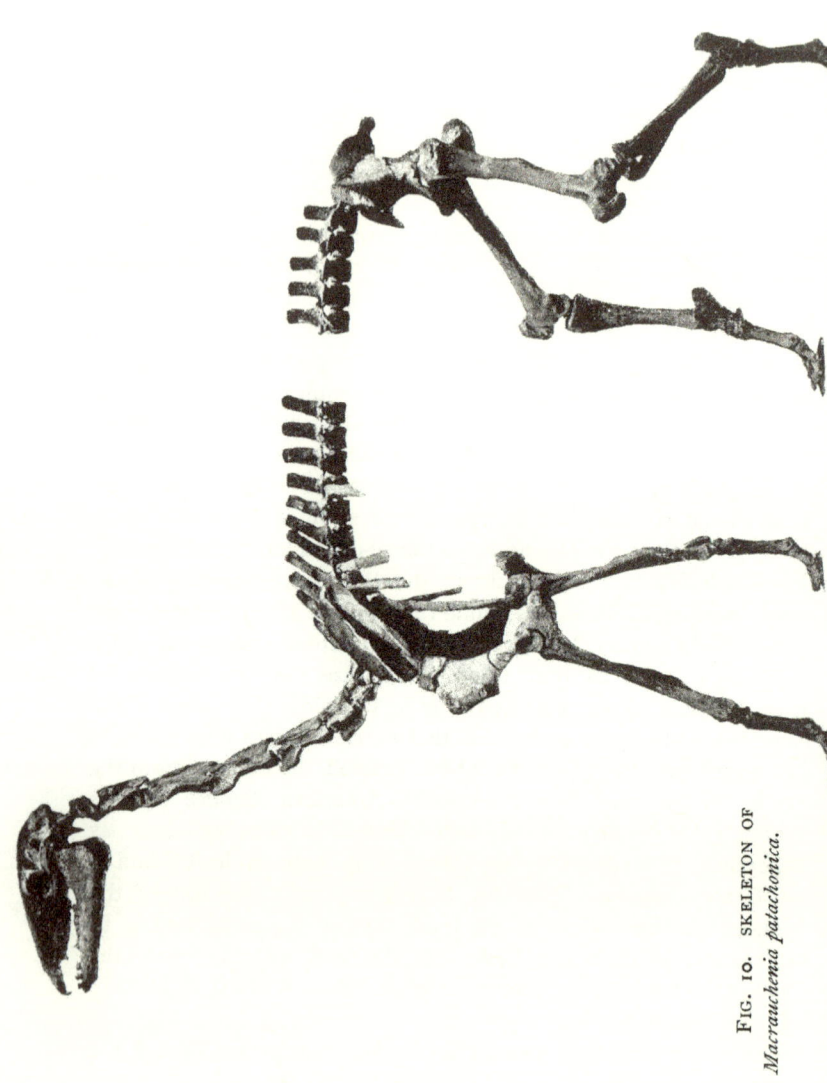

FIG. 10. SKELETON OF
Macrauchenia patachonica.

Perissodactyla are concerned, was exceedingly poorly off for ungulates during the Plistocene epoch. Nevertheless, the country was very rich in hoofed mammals, not only during the Pampean, but likewise during the Santa Crucian epoch, and in this respect it was quite as peculiar as it is in its edentates. It is not a little remarkable that three of the extinct subordinal groups of the order which are confined to this realm exhibit a more or less decided approximation, more especially in the structure of their molar teeth, to the earlier northern Tertiary representatives of the Perissodactyla. Hence it is probable that all the four suborders in question have originated from a common ancestral stock, although apparently before the perissodactyles were differentiated from an earlier group known as the Condylarthra. The date when the ancestors of the South American forms reached their present home is, however, enveloped in mystery, and although it is fairly certain that such ancestors had a northern origin, yet it is highly improbable that they entered South America from that direction.

The first of the three extinct subordinal groups in question, for which the name of Litopterna has been proposed, is the one showing the nearest parallelism with the Perissodactyla, and is typified by the genus *Macrauchenia*, of which the skeleton is figured in the accompanying illustration. In this group the cheek-teeth approximate in general structure to those of the well-known European Oligocene genus *Palæotherium*, although in the typical genus *Macrauchenia* they have been so modified as to render the resemblance obscure. An essential characteristic of the upper molar teeth (fig. 13) is to be found in the presence of two distinct lobes to their outer walls. The toes of both fore and hind feet are elongated, and constructed on the same general plan as those of the Perissodactyla, never exceeding three in number on each foot, and the middle one being symmetrical in itself. Moreover the astragalus of the tarsus or heel-joint resembles the corresponding bone of the latter group in having a deep pulley-like groove on its upper surface for the articulation of the tibia, although inferiorly it is unlike. The calcaneum, or heel-bone, on the other hand, resembles that of the Artiodactyla in bearing a small facet for the articulation of the fibula, or small bone of the leg. A more

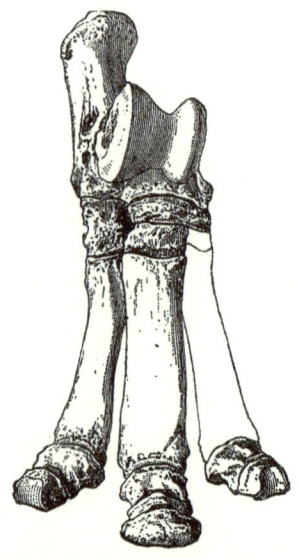

FIG. 11. SKELETON OF HIND FOOT OF AN EXTINCT SPECIES OF RHINOCEROS.
(*To exhibit Perissodactyle type of structure.*)

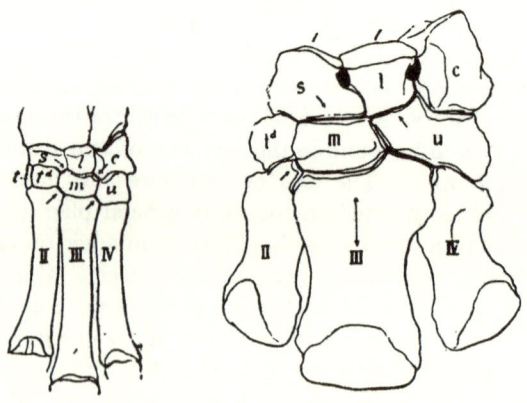

FIG. 12.   LEFT CARPUS AND METACARPUS OF MACRAUCHENIA AND
EXTINCT RHINOCEROS (*Aphelops*)
(*Showing linear and alternating arrangements of carpal bones.*)

important difference, from both the Perissodactyla and Artiodactyla, occurs in the structure of the carpus (wrist) and tarsus
(ankle), in both of which the two rows of small component bones
are arranged in vertical series one above another, instead of
overlapping and interlocking; this so-called linear arrangement
being a more primitive type than the interlocking or alternating
arrangement characterising the two existing suborders. The
vertebræ of the neck are elongated, with flat terminal faces, and
have the vertebral artery piercing the sides of the neural arch
in a manner elsewhere found only in the camel family and the
great anteater. In the femur, or thigh-bone, the projection known
as the third trochanter is much less developed than in the Perissodactyla. All the members of the group were long-limbed, long-
necked, and slenderly built animals, *Macrauchenia* itself being
as large as a camel, although the genera from the Santa Cruz
beds of Patagonia were represented by species of much smaller
dimensions.

Of the typical genus *Macrauchenia* fossil remains occur not
only in the Pampean formation, but likewise in the superficial
deposits of both Patagonia and Bolivia. The most curious feature
about this remarkable animal is the position in the skull of the
aperture for the nostrils, which instead of being situated at the
extremity of the muzzle, is placed in the middle of the forehead,
between the eyes. Otherwise the skull is not unlike that of a
horse in general contour. The teeth, which are forty-four in
number, and form a continuous uninterrupted series, are a specialised modification of the type of those of the undermentioned
genus; their crowns being taller, with a more complicated arrangement of folds. In nearly all points of its structure, especially
in the number of the teeth, and the absence of large tusks, as well
as in the structure of the wrist- and ankle-joints, *Macrauchenia*
is a very primitive kind of animal. In the Santa Cruz beds of
Patagonia the family to which this genus belongs is represented
by several smaller animals, such as those named *Oxyodontotherium*, in which the aperture for the nostrils occupied a more
normal position in the skull, and the crowns of the molar teeth
were shorter and of a more simple structure. The crown-surfaces
of worn upper molar teeth from the right side of the jaw of both

genera are shown in the accompanying figures (13 and 14); the letters indicating the corresponding elements in each.

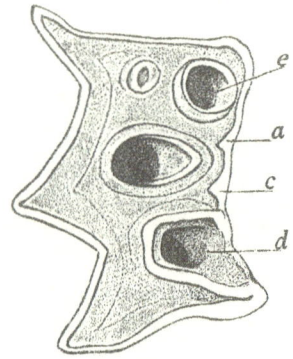

FIG. 13. GRINDING SURFACE OF RIGHT UPPER MOLAR OF *Macrauchenia*.

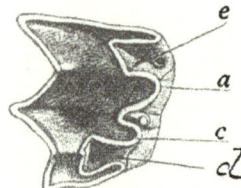

FIG. 14. GRINDING SURFACE OF RIGHT UPPER MOLAR OF
Oxyodontotherium.

The second family, or *Proterotheriidæ*, is confined to the Monte Hermoso, Paranà, and Santa Cruz beds of Patagonia, in the last of which it is represented by the genera *Proterotherium* and *Diadiaphorus*. In these animals, none of which much exceeded a sheep in size, the upper molar teeth are much more like those of the European *Palæotherium*; and, in place of forming a continuous series, the teeth are interrupted; one pair, both in the upper and lower jaw, being developed into tusks. In structure the feet presented a general resemblance to those of the extinct three-toed horses (*Hipparion*) of the northern hemisphere, but in some instances the toes were reduced to a single one, thus showing a curious parallelism in the development of this group to that of the horses.

A second sub-order of ungulates, entirely confined to the Santa Cruz and Patagonian beds of Patagonia, is repre- **Astrapotheria.** sented by the two generic types known as *Astrapo-therium* and *Homalodontotherium*, each of which constitutes a family by itself. Rivalling the rhinoceros in bulk, both of these extraordinary creatures differ from the last group by the structure of their molar teeth, which approximate to those of the *Rhino-cerotidæ*, those of the upper jaw having a continuous outer wall, not divided into lobes. In the ankle-joint the astragalus differs from that of the preceding group in having its upper surface flat, thus resembling that of the elephants. In both the wrist and ankle

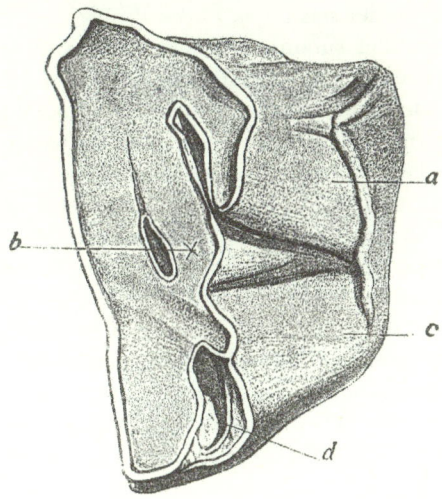

FIG. 15. GRINDING SURFACE OF RIGHT UPPER MOLAR OF *Astrapotherium.*

the component bones were arranged on the linear plan, and not improbably each foot had five toes. The vertebræ of the neck were short, with flat terminal faces; and in correlation with this shortness of the neck, the limbs and feet were more or less abbreviated. In the first-named of the two genera each jaw was provided with an enormous pair of tusks, wearing obliquely against one another like those of a wild boar; and although there were no small teeth in the upper jaw, the lower jaw carried three pairs of spatulate incisors quite unlike those of any other known animal. On the other hand, *Homalodontotherium* takes its name from having its forty-four teeth arranged in a continuous even series, unbroken either by tusks or by gaps. Whereas the molar teeth of *Astrapotherium* present a marked resemblance to those of the rhinoceroses, it is noteworthy that those of the allied genus make a certain approximation to the corresponding teeth of an extinct ungulate from the European Oligocene known as *Cadurcotherium*, apparently more or less closely allied to the rhinoceroses. It is, however, quite improbable that it is to these allied forms we have to look for the origin of the peculiar South American ungulates, seeing that they must have branched off from the primitive stock at an earlier stage.

The third extinct subordinal group, or Toxodontia, takes its
Toxodonts.   name from a gigantic species from the Pampean
formation to which Owen, on account of the peculiarly curved form of the molar teeth, gave the name of *Toxodon*. Rivalling the largest existing rhinoceros in point of size, the *Toxodon* has at first sight very much the general appearance of one of those animals, having a massive skull, short limbs and neck, and three-toed feet. The middle toe is, however, scarcely larger than the lateral ones; and while the bones of the wrist are arranged in the alternating manner, those of the tarsus are placed in a linear series, and the astragalus has a nearly flat, instead of a grooved, superior surface; the terminal faces of the vertebræ of the neck being also flat, instead of articulating together by ball-and-socket joints. Omitting mention of other more or less strongly pronounced peculiarities in the structure of the limbs, attention may be specially directed to the teeth, which differ from those of all existing ungulates in that the whole of them grow continuously throughout the life of their owner, without ever forming roots. The front or incisor teeth are chisel-like, and thus resemble in form, although not in number, those of a beaver or rabbit; and, indeed, in the general conformation of the teeth, as well as in the non-development of roots, the whole dentition of this animal presents a most curious similarity to that of the rodents. Some

idea of the huge size of these animals may be gathered from the fact that the skull may measure as much as a yard in length. Whereas *Toxodon* itself is confined to the Pampean beds, it is represented in the somewhat older deposits of Monte Hermoso and Catamarca by allied forms known as *Toxodontotherium* and *Xotodon*.

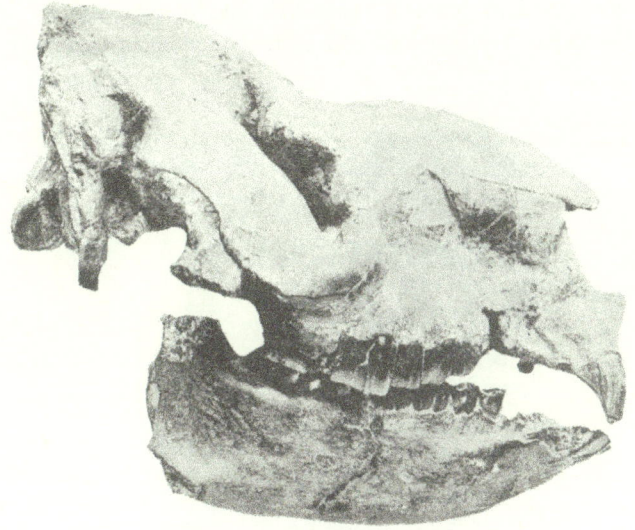

FIG. 16. SKULL OF *Nesodon*. Much reduced.

In the Santa Cruz beds of Patagonia the place of the fore-going is taken by the allied genus *Nesodon*, likewise belonging to the same family *Toxodontidæ*. In these animals, of which the skull is shown in the figure, the molars retain marked resem-blances to those of the rhinoceroses, which have been lost in the more specialised *Toxodon*; these teeth growing for a considerable period, yet late in life developing roots in the ordinary manner. The front teeth are very peculiar, the canines remaining small throughout life, but the second pair of incisors in the upper, and the third in the lower jaw growing beyond the others to form large tusks in the adult, and never developing roots. In their rooted molars the Nesodons depart less widely from the primitive Peris-

6—2

sodactyle type than do their more specialised descendants the Toxodons, although there are no forms known from other parts of the world to which they present any direct relationship or show any marked resemblance.

A second family (*Typotheriidæ*) of the sub-order is represented in the Pampean by the typical genus *Typotherium* and in the Santa Cruz, or Patagonian, beds by the allied *Trachytherus*. Both these have a dentition still more like that of the rodents; the front, or incisor teeth being reduced to a single chisel-like pair in

FIG. 17.   IMPERFECT PALATE OF *Typotherium*.   Reduced.

each jaw, while the molars are narrow, rootless, and of compara- tively simple structure and relatively small size.   Moreover, in- stead of hoofs, these highly modified ungulates appear, like many members of the rodent order, to have had their toes protected by nails. Like as are these animals to rodents, the resemblance to that group is carried to a still greater extent in certain small forms from the Santa Cruz beds which have been named *Pachyruchus* and *Hegetotherium*; these constituting a third family,—the *Pachyruchidæ*. From the features referred to, the reader might be disposed to consider that there is some direct genetic affinity between rodents and the group under consideration, but this is clearly not the case, such resemblances as exist being solely the result of that parallelism in development which appears to have been such an important factor in the evolution of mammals.

It may be well to mention here that in a recent paper[1] Dr Noack suggests an intimate affinity between the toxodonts and the existing hyraces (Hyracoidea) of Africa and Syria; adding that the presumed affinity affords evidence of a land-connection between Africa and South America. The two groups are, however, essentially different in regard to the structure of the fore-foot, in which the bones of the carpus or wrist are arranged, as already shown, in the alternating manner among the toxodonts, whereas in the Hyracoidea they form a linear series. Moreover, the molar teeth of the two groups are markedly distinct, although in both they approximate to the perissodactyle type. Still there is a possibility that the Hyracoidea may represent a less specialised branch which has originated from the primitive toxodont stock, but has retained the linear type of carpus; and if this really be the case, it would afford very strong confirmation of the view that South America received its earliest ungulates by way of Africa and the Antarctic Continent.

A remarkable ungulate from the Patagonian beds, with molar teeth very like those of the extinct European genus *Dinotherium*, and bearing a pair of large tusks in   *Pyrotheria.* the lower jaw at least, has been tentatively assigned by the present writer to the Proboscidea; but judging from the distinction of the other ungulates of the older Argentine Tertiaries from those of the rest of the world, it is more probable that it represents a sub-order by itself. *Pyrotherium*, as the animal is called, was

---

[1] *Zool. Jahrb.-Abtheil fur Systemat.* vol. VII. pp. 540—542.

associated with nesodons, astrapotheres, and homalodontotheres, which have been assigned by Dr Ameghino to genera distinct from those of the Santa Cruz beds. But from my own observations on a series of remains of these animals in the La Plata Museum obtained in association with those of *Pyrotherium*, they appear to be generically inseparable from their Santa Crucian representatives.

Now represented only by the living Indian and African elephants the Proboscidea were formerly a somewhat extensive subordinal group of the ungulates, easily characterised by their peculiar molar teeth and tusks (the latter of which may be present either in the upper or lower jaw alone, or in both); their five-toed feet, in which the bones of the ankle and wrist are arranged on the linear plan, and the astragalus has a flat upper surface; and the presence of a trunk. While true elephants (*Elephas*) are totally unknown in South America, the genus *Mastodon*, distinguished by the low crowns and simpler structure of the molar teeth, is represented by two species in the Pampean of Buenos Aires, and is also stated to occur in the Monte Hermoso beds; but (assuming the distinctness of *Pyrotherium*) the sub-order is unknown in the older deposits.

Proboscideans.

Although at the present day Neogæa contains no existing families of either carnivores or ungulates absolutely peculiar to it, the case is widely different with regard to the rodents, or gnawing mammals. Existing rodents are divided into seventeen families, arranged under four sectional groups; nine out of these seventeen families occurring in the Neogæic realm, among which four are absolutely peculiar to it. A fifth (*Octodontidæ*) is mainly South American and West Indian although possessing a few representatives in Africa south of the Sahara, and a sixth (*Hystricidæ*) has two genera which are practically only South American. The significance of these facts will be apparent when it is stated that of the other zoological regions of the globe only two have any families of the order peculiar to them, and neither of these has more than two such families. In the Ethiopian region, for instance, the only peculiar family is that of the African flying-squirrels (*Anomaluridæ*), represented by two genera; while the western division of the Holarctic region has the sewel-

Rodents.

lels (*Haplodontidæ*), with one genus; and the Sonoran region the pouched rats (*Geomyidæ*) with two. It is further highly significant that all of these families—which are mainly or chiefly Neotropical— belong to one division of the order, the Hystricomorpha, and that certain of them are represented either in the Santa Cruz beds of Patagonia, or the Paranà beds, where the whole of the other sections and families are totally unknown. This, however, is by no means all, for throughout the Tertiaries of North America below the Pliocene there are no Hystricomorpha at all, and at the present day there is but a single species—the Canadian porcupine (*Erethizon*)—inhabiting the whole of that country. On the other hand, both at the present day and in the Tertiaries, North America abounds in Sciuromorpha and Myomorpha. These facts clearly point to the existence of an impassable barrier between North and South America during a large portion of the Tertiary period. Moreover, even at the present day the Sciuromorpha are scarcely represented at all in Neogæa, while the Myomorpha are not very numerous.

Of the squirrel-like rodents, constituting the section Sciuro- morpha, and including the four existing families of the African flying-squirrels (*Anomaluridæ*), the squirrels and marmots (*Sciur- idæ*), the sewellels (*Haplodontidæ*), and the beavers (*Castoridæ*), as well as the extinct American *Castoroididæ*, the only living Neogæic representatives are certain species of squirrels, none of which range south of Paraguay. The extinct *Castoroididæ* include large beaver-like rodents with complex molars like those of the viscacha, and are represented by the typical *Castoroides* from the Plistocene of the United States, and likewise by *Amblyrhiza* (*Loxomylus*) from caves in the West Indies, and, it is said, also from the later Tertiaries of Argentina. This family is accordingly a northern type.

The second or murine group of rodents (*Myomorpha*) contains five families, namely the dormice (*Myoxidæ*), the jumping-mice and jerboas (*Dipodidæ*), the mice and rats (*Muridæ*), the mole-rats (*Spalacidæ*), and the American pouched rats (*Geomyidæ*). Out of all these, practically the only one represented in the realm is the cosmopolitan *Muridæ*, although among the *Geomyidæ* the genus *Heteromys* enters the transitional Mexican sub-region. In the

Muridæ, however, as in North America, the true rats and mice (*Mus*) of the Old World are entirely wanting, except through human introduction; their place being taken by the white-footed mice (*Sitomys*) common to the whole of the New World, and nearly related to the European hamsters. Being essentially a northern type, they are certainly recent immigrants into South America from the north; and the same is doubtless true of certain genera of the family now peculiar to this realm, such as the fish-eating rats (*Ichthyomys*), of Peru and Venezuela, the groove-toothed mice (*Rhithrodon*), one of which ranges as far south as Tierra del Fuego, and the Brazilian genus *Holochilus*, which is another form allied to the hamsters. Although the cotton-rat (*Sigmodon*) occurs in Central America, and occasionally wanders still further south, the whole tribe of voles and their kindred are absent from the entire realm.

Represented by six families, all of which occur within its limits, the porcupine-like group, or Hystricomorpha, may be regarded as the characteristic rodents of Neogæa. From both the preceding sections of the order the members of this section may be readily distinguished by the angle (or hinder inferior projection) of the lower jaw taking its origin from a prominent ridge running along the side of the jaw, instead of from the inferior edge of the socket for the incisor teeth. Of the families confined to this realm, that of the cavies (*Caviidæ*), includes heavily-built rodents, with four front and three hind toes, rudimental or short tails, and the cheek-teeth divided by transverse folds of enamel into a number of thin parallel plates. The genera of this family include the true cavies (*Cavia*)—so well known through the domestic guinea-pig—all of which are small short-limbed forms; the larger and taller Patagonian cavy (*Dolichotis*); and the carpincho, or capivara (*Hydrochœrus*), which is the largest living member of the order, and is characterised by the large number of plates going to form the last molar tooth in each jaw. Although they do not appear to have been recorded from the Santa Cruz beds, remains of members of this family occur in the Paraná[1] horizon, and also in that of

[1] There is some confusion with regard to the age of the Paraná beds. They are overlain by marine strata which have been identified with the Patago-nian; but it is more probable that they are newer than the Santa Cruz beds,

Monte Hermoso. Of the former *Plexochœrus* differs from *Hydro-chœrus* only by the somewhat simpler structure of the last molar, *Hydrochœrus* itself occurring in the Monte Hermoso beds. Other forms from the Paranà stage are *Eucardiodon* and *Cardiotherium*, apparently more nearly allied to *Cavia*. Here it should be mentioned that certain European Oligocene genera (*Issiodoromys* and *Nesocerodon*) have the molars so like those of the cavies that they are regarded by Dr Schlosser as nearly allied to the family. By Prof. Zittel they are, however, included in the extinct family *Theridomyidæ*, which is classed with the dormice and certain other families in a group regarded as intermediate between the Sciuromorpha and Hystricomorpha. If, as would seem probable, they are really allied to the latter, they are of the utmost import-ance as indicating a connection between the middle Tertiary rodent fauna of Europe with that of South America[1].

Nearly allied to the cavies are the agutis (*Dasyprocta*) and pacas (*Cœlogenys*), collectively constituting the family *Dasy-proctidæ*, and differing from the former in that the folds of enamel merely form notches on the sides of the crowns of the cheek-teeth. In a fossil state the family seems to be only known by remains of the existing genera from the Brazilian caves. The third peculiar family (*Dinomyidæ*) is known merely by a single specimen from Peru. The only other family exclusively confined to the realm is that of the *Lagostomatidæ* (*Chinchillidæ*), which includes not only the true chinchillas (*Eriomys*) and Cuvier's chinchilla (*Lagidium*), but likewise the viscacha (*Lagostomus*) of the Argentine pampas. All the members of this family have long bushy tails, elongated hind limbs, and the cheek-teeth divided by complete transverse folds of enamel into thin plates. *Lagostomus* occurs fossil not only in the Pampean, but likewise in the older Tertiaries of Paranà; while it may be doubted whether *Pliolagostomus* and *Prolagostomus* of the Santa Cruz beds are really entitled to generic distinction. Other allied forms from the latter deposits are

---

although their lowest portion is older than the Monte Hermoso stage. (See Ameghino, *Bull. Ac. Cordoba*, Vol. XIII. pp. 260, 261, 1894.) They may be partly made up of the remains of pre-existing beds.

[1] Dr Schlosser writes me to the effect that he is fully assured these forms are the ancestors of the *Caviidæ*.

described under the names of *Perimys* and *Sphodromys*, and thus indicate that the family is essentially South American. The largest known rodent is the extinct *Megamys*, from the Paranà beds; the typical species of the genus being described as equal in size to an ox. The porcupines (*Hystricidæ*), which form a practically cosmopolitan family, sufficiently distinguished by their spiny covering, are represented in South America by the two arboreal genera *Synetheres* and *Chætomys*, differing from all their allies by their prehensile tails, although otherwise related to the North American *Erethizon*, with which they form a separate sub-family. Of *Chætomys* an extinct representative occurs in the cavern-deposits of Brazil; and in the Santa Cruz beds the family is represented by apparently extinct generic types described under the names of *Stiromys*, *Acaremys*, and *Sciamys*. Since fossil remains of *Hystrix* date from the European Miocene or Oligocene, there is distinct evidence of a connection between the early South American rodents and those of the Old World; while as *Erethizon* is first known from the Plistocene of North America, it may probably be regarded as a late immigrant into that country from the south.

The largest of all the families of the Hystricomorpha is that of the *Octodontidæ*[1], in which out of a total of nineteen genera upwards of fifteen belong to the realm under consideration, while the remaining four are African and mainly Ethiopian. In addition to other features, the rodents of this family have the crowns of the cheek-teeth marked by infoldings of enamel from both sides; there are usually five toes to each foot; and the general form is more or less rat-like. Of the Neogæic forms, the typical genus *Octodon* is represented by the degu of Chili and Peru, which is a large rat-like animal, with a brush-tipped tail; other species occurring in Bolivia. The latter country is likewise the home of the allied genus *Habrocoma*, the members of which vie with the chinchillas in the delicate softness of their fur. Nearly related are the burrowing South American tuco-tucos (*Ctenomys*), characterised

---

[1] By some the family is divided into three; viz. *Capromyidæ*, with the West Indian *Capromys*, the S. American *Myopotamus*, and the African *Triaulacodus; Ctenodactylidæ*, including the remaining African forms; and *Octodontidæ*, comprising all the other American types.

by the comb-like bristles on the hind feet, and their bell-like cry; two Chilian species, constituting the genus *Spalacopus*, being distinguished by their rudimental ears. *Schizodon*, on the other hand, of which there is a single species from the Southern Andes, has the ears larger than in the tuco-tucos. The South African *Petromys* has been regarded as very closely allied to *Spalacopus*, but it is more probable it should be classed with the other two African genera *Ctenodactylus* and *Pectinator*, to constitute a separate sub-family. The third sub-family includes all the other genera, one of which (*Triaulacodus*[1]), as represented by the cane-rats, is Ethiopian, while the whole of the remainder are Neogæic. Many of the species are of large size, some being arboreal and others aquatic. Of the latter the best known and largest is the coypu (*Myopotamus*), widely spread over South America; while in the West Indies the group is represented by the equally large arboreal hutias (*Capromys* and *Plagiodon*). The other seven genera are South American, and include smaller rat-like forms, which in the case of the two genera *Loncheres* and *Echinomys* have flattened spines mingled with the fur; the others being known as *Mesomys*, *Dactylomys*, *Cannabateomys*, *Cercomys*, and *Cartero-don*. Several of the existing genera occur fossil in the caves of Brazil and the Pampean; *Myopotamus* also occurring in the infra-pampean beds of Paranà, together with the reputedly extinct forms described as *Orthomys* and *Morenia*. Other extinct genera, such as *Neoremys*, *Scleromys*, and *Adelphomys* occur in the Santa Cruz deposits, and appear to be very closely allied to *Myopotamus*. It is noteworthy that an extinct Octodont (*Pellegrinia*) allied to *Ctenodactylus* occurs in the Plistocene or Pliocene of Sicily; while *Ruscinomys* from the Pliocene of France is believed to belong to the same group. Finally, the genus *Eocardia*, together with certain other allied forms from the Santa Cruz beds, are regarded as indicating a separate family (*Eocardiidæ*) of the section.

With regard to the extinct family *Theridomyidæ* from the middle and upper Oligocene of Europe, which includes not only *Theridomys* and *Archæomys*, but probably also the above-men-

[1] The name *Aulacodus* being preoccupied, that of *Triaulacodus* is proposed in substitution.

tioned *Nesocerodon* and *Issiodoromys* (p. 89), it seems highly probable that these are really the ancestral forms of the modern Hystricomorpha, although their lower jaws approximate to the general type of those of the more generalised Sciuromorpha and Myomorpha.

The last section of the order (Lagomorpha), which includes the hares and picas, and is essentially a northern one, is but poorly represented in Neogæa; the picas (*Lagomyidæ*) being unknown there, while in the whole of the realm there are only two species of *Leporidæ*, one of which is Brazilian.

It has been usual in zoological systems to include under the title of Edentata not only the armadillos, anteaters, and sloths of South America, but likewise the Old World pangolins (*Manidæ*) and aard-varks (*Orycteropodidæ*). There can, however, be no doubt that there is little or no connection between the two groups, and the latter may accordingly be separated as a distinct order under the title of Effodientia. In this restricted sense the edentates at the present day are, perhaps, the most characteristic and remarkable of all the Neogæic mammals. Whereas, however, the sloths and anteaters are entirely Neogæic, a few of the armadillos have wandered at a comparatively modern date as far north as Texas; but this does not detract from their essentially southern character, seeing that they are well represented in the Santa Cruz beds, and, if we exclude certain remains of doubtful affinity from the Oligocene of France, are quite unknown elsewhere. This, however, is by no means all, since there are two extinct families of the order, dating from the Santa Cruz beds, which were extremely abundant during the Pliocene and Plistocene epochs; some few of these having managed to obtain an entrance into North America about the Miocene epoch. Central and South America may accordingly be considered as essentially the home of the edentates; and are thus broadly demarcated from all other parts of the world. It would be superfluous to point out all the distinctive features of the order, but it may be mentioned that none of the living forms have teeth in the front of the jaws; while in all those genera in which teeth are present, these are of comparatively simple structure, being unprovided with a coating of enamel, growing

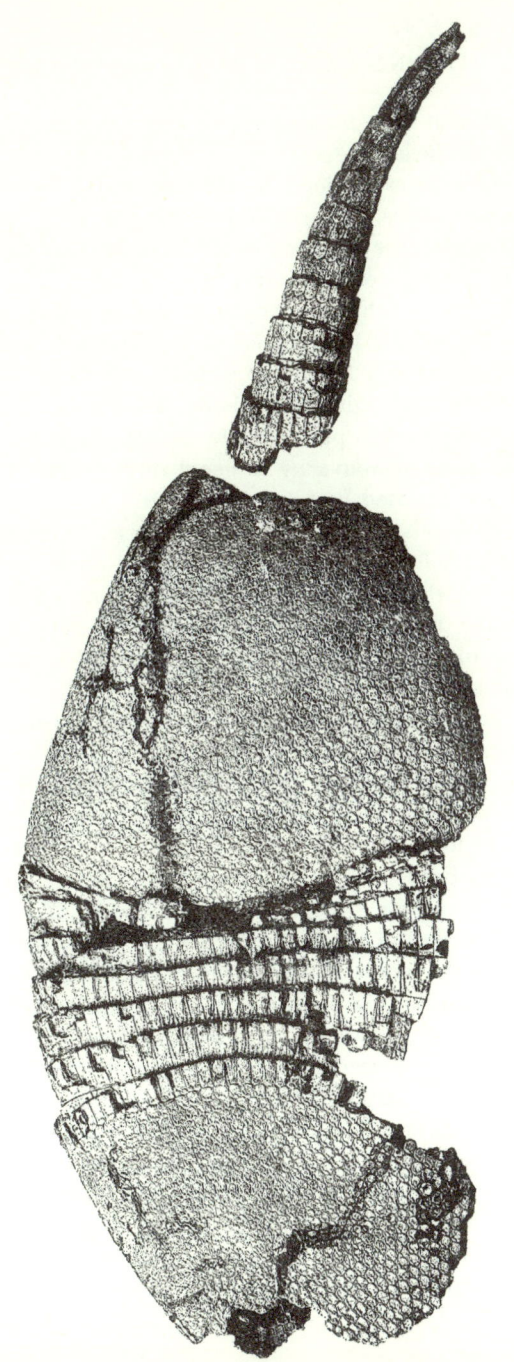

FIG. 18. EXTERNAL SKELETON OF AN ARMADILLO (*Tatusia gigantea*).

continuously without ever forming roots, and being mostly very similar throughout the series.

The mailed, or loricate edentates are represented by the two families of the armadillos (*Dasypodidæ*) and glyptodonts (*Glyptodontidæ*), the latter of which died out
at the close of the Plistocene or the commence-
ment of the Recent epoch, whereas the former is still abundant.

Armadillos and Glypto-donts.

With the exception of the two species of pichiciagos (*Chlamydophorus*) from Mendoza and Bolivia, in which a solid coat of mail is confined to the hinder region of the body, the members of both families have their bodies protected by a bony armour, while their heads are guarded above by a shield of the same nature, and their tails are enclosed in a tubular sheath. Covered externally with horny shields, like the shell of a tortoise, the carapaces of these animals are formed of a number of small plates of bone, either united everywhere by their edges into a continuous solid armour, or in the middle region of the body overlapping one another like the tiles on a roof. In the true armadillos (of which the living Argentine forms are all comparatively small creatures, although one Brazilian species reaches nearly a yard in length) the carapace consists of a nearly solid buckler in front and behind, while between the two are situated a variable number of movable overlapping bands, which in some instances admit of the body being rolled up into the form of a ball. They have all long snouts, and simple, subcylindrical teeth. The true existing armadillos may be divided into the genera *Dasypus, Xenurus, Priodon, Tolypeutes,* and *Tatusia*. The first of these, in which there are six or seven movable bands to the carapace, is found throughout the Argentine Tertiaries to the Santa Cruz beds, one of the fossil species from the higher beds of the series having a skull of nearly a foot in length, and thus vastly exceeding all its living congeners in size. *Tatusia*, in which the carapace has from seven to nine movable bands, does not appear to be known below the Pampean beds, where it is represented by the large species of which the external skeleton is shown in the figure on p. 93. A third genus, *Eutatus*, which likewise comprises species of large size, and ranges from the Pampean to the Santa Cruz beds, is distinguished by having over thirty movable bands in the carapace. More

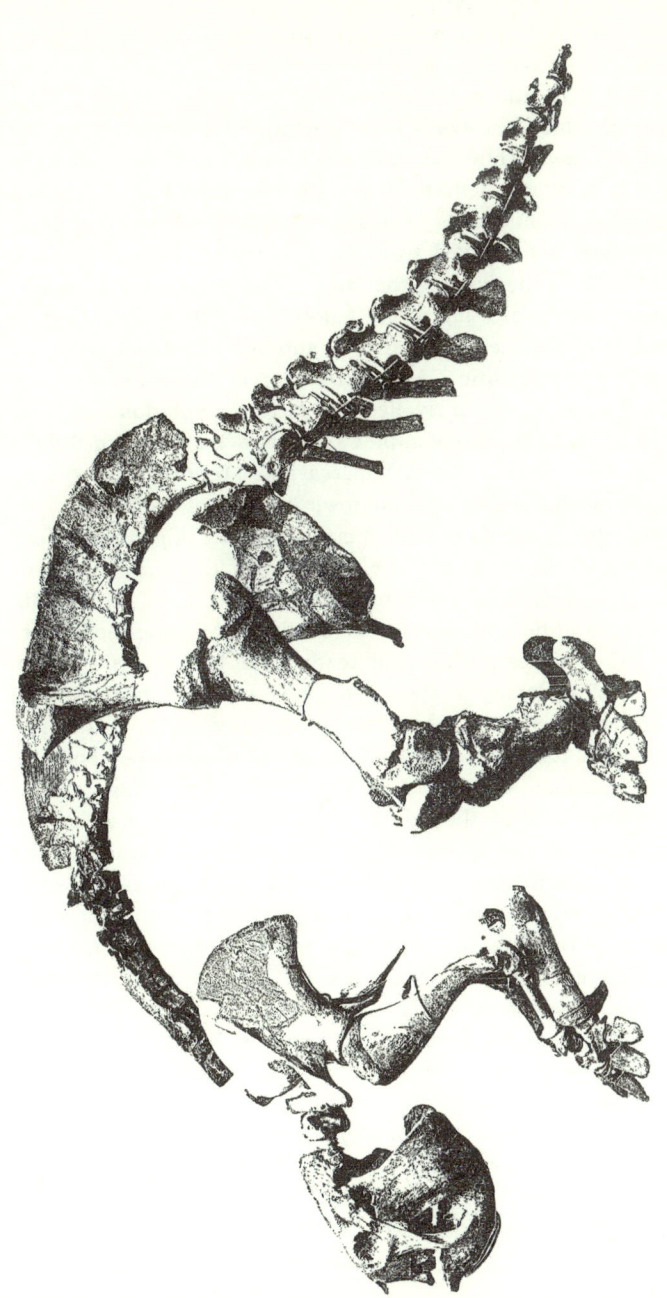

Fig. 19. INTERNAL SKELETON OF A GLYPTODONT.

remarkable than any is the extinct *Peltephilus* of the Santa Cruz deposits, in which the teeth form a continuous series up to the front of the jaws, while the skull has a very broad snout, and the humerus is of such a remarkable shape that it has been described as that of a monotreme. Indeed this genus seems to suggest that edentates are derived from animals with a fully developed series of teeth in the front of the jaws. The pichiciagos (*Chlamydophorus*), which are unknown before the Plistocene, form a sub-family by themselves; and yet another sub-family group is indicated by the gigantic *Chlamydotherium*, of the Brazilian caves and the Pampean, which rivalled the largest glyptodonts in size, and had teeth of a more complex type than the true armadillos. Other species occur in the Catamarca and Monte Hermoso Tertiaries, although the genus is unknown in those of Patagonia.

From the armadillos and their immediate kin the extinct glyptodonts differ in having the carapace in the form of a continuous solid shield, without any movable bands in the middle region; in addition to which the skull is characterised by its depth and shortness, while the teeth form long fluted prisms. The internal skeleton, as shown in figure 19, is characterised by the union of nearly the whole of the vertebræ of the back into a solid girder for the support of the massive carapace; and the feet are furnished with much shorter claws than those of the burrowing armadillos, the hinder ones being almost nail-like in form. Most of the species from the Pampean formation are animals of gigantic dimensions, the length of the carapace being not unfrequently from six to eight feet; and they are unquestionably some of the most extraordinary creatures that ever trod the earth. Although the majority are South American, some members of the genus wandered as far north as Texas, while from the upper division of the Loup Fork beds, corresponding to the lowest Pliocene, a North American form has been described under the name of *Carioderma*.

In the typical genus *Glyptodon*, which ranges from the sand-dunes and Pampean formation to the Monte Hermoso beds, the tail-sheath, as shown in the annexed figure, is composed of a number of spiny rings, gradually decreasing in size from the root to the tip, and the polygonal plates of the carapace are each

FIG. 20. EXTERNAL SKELETON OF *Glyptodon clavipes*.
(Greatly reduced.)

ornamented with a distinct rosette-like sculpture. The allied genus *Plohophorus*, of which the remains occur alike in the Brazilian caves and the Catamarca and Monte Hermoso beds of Argentina, while agreeing with the last in the characters of the skull, has a carapace more like that of the undermentioned *Panochthus*, and a tail-sheath resembling that of the next genus. In addition to well-marked distinctive features of the skull, *Lomaphorus* is characterised by the great elongation and slender form of the carapace, which is produced on either side of the neck in the same manner as in the armadillos, while its margins lack the large bosses exhibited by the typical genus. The tail-sheath consists of a small number of rings at the base, followed by a long terminal tube ornamented with smooth, oval, bony plates, of which those along the sides and at the tip are larger than the rest. The genus, of which the species are much inferior in point of size to those of *Glyptodon*, has the same geological range as *Plohophorus*.

Another type of the family is represented by the animals constituting the genus *Panochthus*, in which the hexagonal bony plates of the carapace are arranged in more distinct rows on the sides of the body; those of the back being ornamented either with a number of small granular tubercles (as in the species here figured), or with one circular central disc surrounded with several rows of much smaller discs. A more striking difference is displayed in the structure of the sheath of the tail, which consists at the base of six or seven large smooth rings diminishing very rapidly in diameter, and terminates in a long and massive depressed tube, the sides of which are ornamented with large roughened bosses, probably surmounted during life with horny knobs, while the intervening spaces are covered with small bony ossicles. The species are of very large or medium size, and range from the Pampean to the Monte Hermoso beds in Argentina, while some occur in the Brazilian caves. Still more extraordinary is the gigantic *Dædicurus* of the Pampean, represented by a somewhat smaller form from the Monte Hermoso beds. Having a total length in a straight line of close upon twelve feet, five of which are taken up by the ponderous tail, the Pampean representative of this genus has the outer surface of the plates of the

FIG. 21. EXTERNAL SKELETON OF *Panochthus tuberculatus*.

(Much reduced.)

7—2

carapace smooth; each being perforated by three or four large holes for the passage of blood-vessels, and the whole being probably invested with a continuous leathery skin, instead of each disc bearing a separate horny shield.   Commencing with a small series of enormous, narrow, hoop-like rings, the tail-sheath terminates in a long, massive, depressed, and nearly smooth club-shaped tube, at the extremity of which are a number of rough disc-like surfaces, apparently for the attachment of large horns.

None of the preceding forms, which include all the largest members of the family, range below the horizon of the Monte Hermoso, Catamarca, and Paranà beds, but the group is also represented in the Santa Cruz deposits of Patagonia, although the single species found there is of much smaller dimensions than any of its later cousins, the whole length of the carapace not exceeding about two feet.   It is noteworthy that this earliest known glyptodont, on which the unwieldy name of *Propalæohoplophorus* has been conferred, presents certain indications of affinity to the armadillos in the structure of its carapace, in which incipient movable bands may be detected on the margins of the middle region.   In this small size of their earliest definitely known representative, the glyptodonts resemble the under-mentioned ground-sloths.

Unless the aforesaid remains from the Oligocene Phosphorites of France should prove to belong to the group, we are at present totally in the dark as to whence both the glyptodonts and the armadillos originally came; and it is, indeed, quite probable that, like the other members of the order, they may have originated in South America (if not in an Antarctic continent) from some at present quite unknown mammalian type.   How such creatures, which seem absolutely unassailable, came to be exterminated, is one of those questions which it appears quite impossible to answer.

Although they have not hitherto been discovered in a fossil state, the sloths, constituting the family *Bradypodidæ*, are just as characteristic of Neogæa as the two preceding groups.   Their habits, however, necessarily restricting them to the tropical forest-districts, their absence in a fossil state[1] must

Sloths.

---

[1] A presumed fossil sloth was described from the Argentine, but the jaw on which it was founded proves to belong to the *Megalotheriidæ*.

not be taken as an indication that they did not exist during the Pampean epoch, since their remains are not likely to occur in the Argentine, although they might with more probability be looked for in Brazil. On the other hand, their specialised structure makes it highly probable that they had not come into being at the date of deposition of the Santa Cruz beds. Of the dimensions of medium-sized monkeys, sloths are characterised by their short, rounded heads, and extremely long limbs, armed with very elongated curved claws; in the genus *Bradypus* the latter being three in number on each foot, but in *Cholæpus* reduced to two in the fore feet. Their bodies are coated with very coarse ragged hair, and the tail is wanting. The teeth are oval prisms, somewhat cupped in the middle of their grinding surfaces; but in the last-named of the two genera the first pair in each jaw are larger than the rest, from which they are separated by an interval, and form tusks wearing against one another to oblique facets. Usually there are five upper, and four lower teeth on a side. The range of sloths extends from Mexico throughout the greater part of the forest-districts, although they do not appear to reach as far south as Paraguay.

Likewise unknown in a fossil condition, the true anteaters, or *Myrmecophagidæ*, constitute another exclusively Neogæic family, with nearly the same geographical range

Anteaters.

as the sloths, but represented in Paraguay. So unlike are these animals to sloths, that it is at first difficult to believe that there is any close relationship between the two, and it is largely due to the evidence of the ground-sloths referred to below that it has been possible to discover how close the connection really is. In place of being rounded and shortened, the skull in the present family is more or less elongated and slender, with the jaws entirely devoid of all vestiges of teeth, and the tongue long, cylindrical, and extensile. An equally striking difference obtains in regard to the structure of the limbs, the fore foot of the great anteater having five toes, of which the middle one is much more powerful than the others, while all except the fifth are furnished with strong claws. In walking, the outer side and part of the upper surface of the fore foot is applied to the ground; but in the hind limbs the sole forms the support in the ordinary manner. Whereas sloths are

highly specialised as regards the structure of their limbs, in the
present group the greatest degree of specialisation shows itself in
the skull, in the majority of the species.   There are but three

FIG. 22.   TAMANDUA ANTEATER.

members of the family, each representing a genus by itself, namely
the great anteater (*Myrmecophaga*), the tamandua (*Tamandua*), and
the two-toed or little anteater (*Cycloturus*); the latter being exclusively arboreal in its habits.

The foregoing remarks on some of the structural features of
the sloths and anteaters will the more easily enable
the reader to understand the peculiarities of the
extinct group of ground-sloths.   They have been
divided into a large number of genera and several families; but
the former may be considerably reduced, and the whole of them
included in the single family *Megalotheriidæ*.   Ranging in the
Argentine from the Santa Crucian to the Pampean epoch and the
overlying sand-dunes, the family has a geographical distributional
area including North America as far as Kentucky.   The South
American forms vastly exceed those of N. America in point of
number; and whereas the latter are found only in deposits of
upper Pliocene and Plistocene age, the former, as we have seen,
extend downwards to the Miocene.   The members of this family
may be defined as edentates without a carapace, the skull and
dentition of the general type of those of the sloths, and the

Ground-
sloths.

limb-bones and vertebræ like those of the anteaters. The skull is, however, somewhat more elongate than in the former, and in the case of the genus *Scelidotherium* approximates to that of the latter. The Plistocene forms include by far the largest representatives of the order, the *Megalotherium*[1] attaining a total length of about eighteen feet, with a bodily bulk fully as great as that of an elephant. Whereas all the members of the family whose remains occur in the Plistocene walked on the outer sides of their feet, in the small ancestral Patagonian forms this specialised character seems to have been less developed.

The typical genus *Megalotherium*—which includes several species, ranging in time from the Monte Hermoso and Cordoba beds to the Pampean, and in space from Argentina and Chili to South Carolina and Texas—is sufficiently distinguished by having the teeth in the form of large quadrangular prisms, sometimes measuring as much as a foot in length, and wearing on their summits into a pair of transverse ridges, owing to the presence of layers of unequal hardness. The allied genus *Mylodon*, including smaller forms which may be compared in size to rhinoceroses, differs from the preceding in the structure of the teeth, which are similar to those of the sloths; the skull, as shown in the figure on p. 104, being comparatively short, with the teeth extending nearly up to the extremities of the jaws. In the skeleton of this genus the limbs are of moderate length and very powerful. The two outer toes of the fore feet are rudimental and clawless, but the three innermost provided with claws, of which the third is much larger than either of the others, this discrepancy being carried to a still greater extent in *Megalotherium*. It will be observed that the creature walked on the outer sides and part of the upper surfaces of the fore feet after the manner of a sloth; but, unlike the latter, only the outer sides of the hind feet were applied to the ground; the great middle toe, which in *Megalotherium* carried a gigantic claw, not touching the ground at all. In the structure of their feet these animals are thus more like anteaters than sloths, although the hinder pair are of a somewhat more specialised structure than in the latter. It may be mentioned that the

---

[1] The name *Megatherium* clearly requires amendment to *Megalotherium*.

conformation of their teeth indicates that the ground-sloths were
vegetable-feeders, and it is probable that they subsisted largely
upon the young twigs and leaves of trees, which may have been

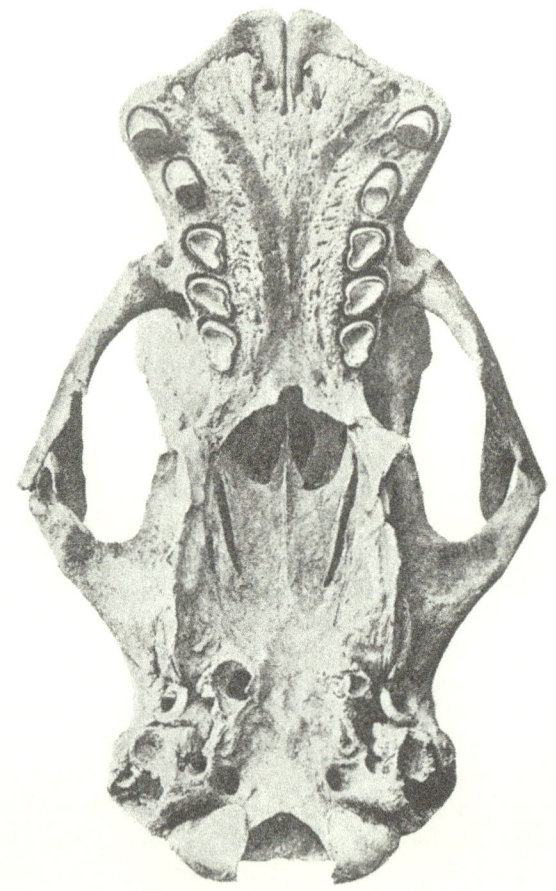

FIG. 23.   UNDER-SURFACE OF SKULL OF *Mylodon*.   (Reduced.)

brought within reach by the animals rearing themselves up
against the trunks and pulling down the boughs with their fore
paws. The present treeless condition of the Argentine pampas
suggests that the ground-sloths were grazing rather than browsing

animals, but their structure is not in favour of this view, and it must be remembered that their remains are likewise met with in Brazil, which was probably always as well wooded as at the present day. The disappearance of forests from the pampas cannot, indeed, be regarded as more marvellous than the extinction of its Plistocene mammals. In the sand-dunes near the coast at Buenos Aires bones of some of the ground-sloths, as well as of glyptodonts, have been found in association with human remains, so that their extinction is an event of comparatively recent date. The genus is typically represented by *Mylodon harlani* of the Plistocene of Kentucky and other parts of North America, but is nevertheless essentially South American, ranging in Argentina from the Pampean beds to those of Paranà and Monte Hermoso. The allied genus *Megalonyx* is exclusively restricted to the North American Plistocene and Upper Pliocene; and here may be repeated the observation that the absence of remains of these ground-sloths from the Miocene of North America, coupled with their presence in the Santa Cruz beds of Patagonia, clearly indicates that they are late immigrants from the south into the northern half of the continent.

Nearly allied to *Mylodon*, the genus *Glossotherium* from the Plistocene of Argentina and Uruguay serves to connect it with another generic representative of the family known as *Scelidotherium*. In place of the comparatively short skulls of the mylodons, the species of this genus have the muzzle of the skull greatly elongated, so that there is a long toothless space in advance of the dental series; and whereas the skulls of the species of *Mylodon* are essentially sloth-like, those of *Scelidotherium* show a marked approximation to the anteater-type. The species of *Scelidotherium* are of medium or rather small size; and in space the genus ranges from Patagonia, through the Argentine, to Brazil, Bolivia, and Chili; while in time it extends from the superficial sand-dunes and Pampean deposits to the lower Tertiaries of Paranà, Monte Hermoso, Catamarca, and Santa Cruz, with a gradual decrease in the size of the species as we descend in the geological scale[1]. Nearly allied is the genus *Catonyx*, from the

[1] The Santa Cruz form has been quite unnecessarily separated under the name of *Analcitherium*.

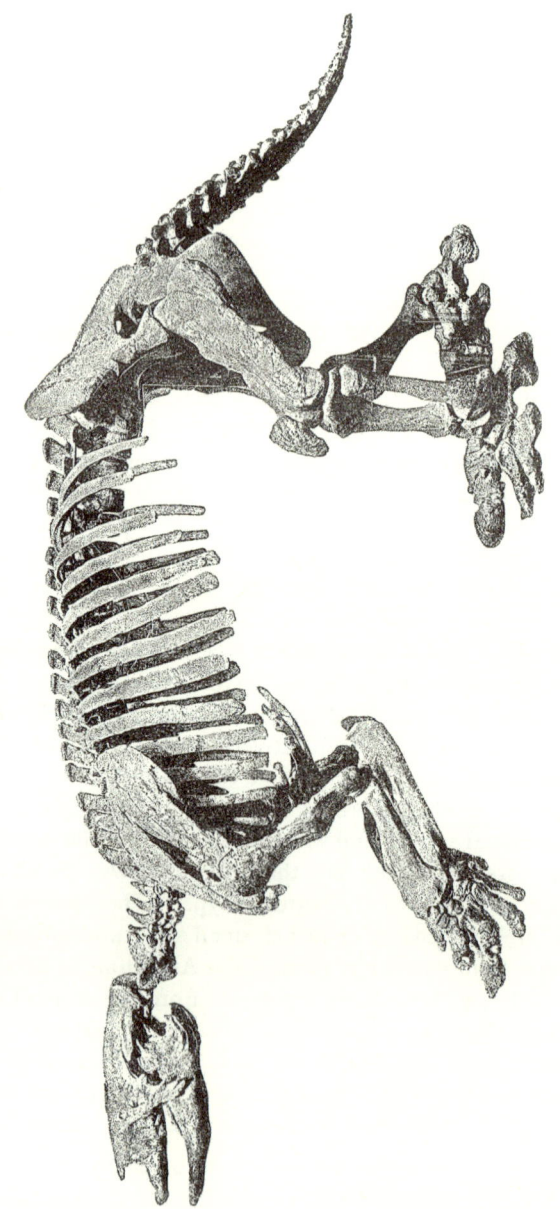

FIG. 24. SKELETON OF *Scelidotherium leptocephalum*.
(Much reduced.)

Brazilian caverns; while *Nothrotherium* of the same deposits seems to have been another nearly related form with teeth of the *Megalotherium* type. The imperfectly known *Nothropus* of the Pampean and *Ortotherium* of the Paraná beds seem, on the other hand, to be late survivals of another group typically represented by the genera *Eucholæops* and *Pseudhapalops* of the Santa Crucian epoch of Patagonia. These forms differ from all those noticed above in that the terminal joints of some of the toes have a median cleft as in the great anteater, and likewise in the elongation of the metatarsal bones; and it seems probable that the hind foot was not so much everted as in the later representatives of the family. The skull is of the general type of that of *Mylodon*; most of the molars being prismatic in form, and surmounted by a pair of transverse ridges, more or less closely connected at their extremities so as to produce an oval cavity on the grinding surface. The first tooth is, however, tusk-like, and separated by a gap from the others. In some of these early ground-sloths the skull did not exceed three inches in length; but other species were considerably larger. They are evidently generalised types, and were probably nearly allied to the ancestral stock which gave rise to *Mylodon* and *Megalonyx*, if indeed they be not the actual progenitors of both.

The last group for consideration is that of the marsupials, or pouched mammals, among which the family of the opossums (*Didelphyidæ*), with the three genera *Didelphys*[1], *Dromiciops*, and *Chironectes*, is now confined to the New World, the great majority of the numerous species being Neogæic, although the common opossum (*Didelphys marsupialis*) ranges from Chili and Brazil to the United States. Although certain forms from the Santa Cruz beds described under the names of *Eodidelphys* and *Prodidelphys* were originally assigned to the present family, these have been subsequently identified by Dr Ameghino[2] with the under-mentioned family of the *Microbiotheriidæ*. True opossums occur, however, in the Monte Hermoso

Marsupials.

---

[1] This genus is divided into several sub-genera, regarded by some writers as entitled to generic rank.

[2] *Bol. Ac. Cordoba*, Vol. XIII. p. 363 (1894).

beds; while, as mentioned in the last chapter, they were widely distributed in the northern hemisphere during the Oligocene. If the conclusions of Dr Ameghino are right as to the absence of these marsupials from the Santa Cruz beds, it is evident that opossums only reached South America from the north at the close of the Miocene or commencement of the Pliocene, and that they do not belong to the indigenous fauna. It has been generally considered that the common opossum of the United States is a direct survivor from the Oligocene forms of North America, but it is more probable that it is really a very recent immigrant from the south, seeing that fossil representatives of the genus are unknown from the North American Miocene and Pliocene. During the Miocene the group perhaps survived in the extreme south of North America.

Although opossums are apparently wanting, the Santa Cruz beds have yielded remains of undoubted marsupials, but several

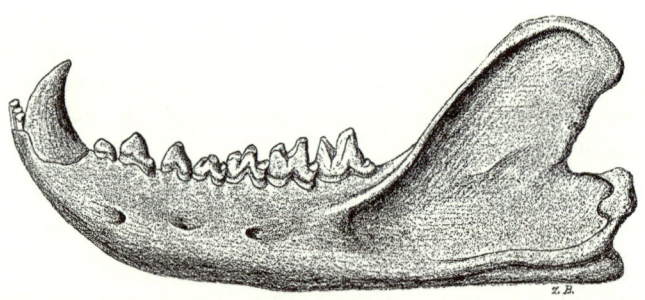

Fig. 25. LEFT HALF OF LOWER JAW OF *Prothylacinus.* (½ nat. size.)

of them are assigned by Dr Ameghino to a distinct group under the name of Sparassodonta. Foremost of these is the genus *Prothylacinus*, already mentioned in the last chapter, which may be provisionally assigned to the Australian *Dasyuridæ.* In having only three in place of four lower pairs of incisors this genus agrees with the latter family, and differs from the opossums; while the whole character of the lower jaw and dentition is very similar to that of the Tasmanian *Thylacinus*, with the exception that the premolars are closer together. As in the *Dasyuridæ* generally,

there are four pairs of upper and three of lower incisor teeth in
Prothylacinus, and the same is the case with the smaller Santa
Crucian form described as *Amphiproviverra*, which appears to be
of a distinctly dasyurid type, although not coming very near to
any Australian genus.

   With regard to the *Microbiotheriidæ*, as typified by the genus
Microbiotherium, these, although they are not included by Dr
Ameghino in the order, appear to be undoubted marsupials, since
they have a dentition numerically the same as that of the opossums,
vacuities in the palate, and an inflected angle to the lower jaw.
From the opossums they differ by the non-production of the palate
behind the last molars, and in the form of the lower jaw, in which
the extremity is produced to a greater extent in advance of the
canine.   In all these respects they approximate to the Dasyurid
genus *Phascologale*, from which they differ in having one pair less
of incisors in each jaw.   The ancestors of the Australian *Dasyuridæ*
must, however, have originally had five pairs of upper and four of
lower incisor teeth, as the former are retained in many of the bandicoots (*Peramelidæ*), while *Myrmecobius* occasionally develops four
pairs of these teeth in the lower jaw.   It seems therefore probable
that the *Microbiotheriidæ* were minute polyprotodont marsupials
of an Australian type.

   There is more difficulty in arriving at any satisfactory conclusions as to the serial position of certain carnivorous mammals
from the Santa Cruz beds, of which a large form described as
Borhyæna may be taken as an example.   In these animals the
dentition approximates to a certain extent to that of the primitive
or creodont Carnivora of the earlier Tertiaries of the northern
hemisphere, although retaining the marsupial feature of four pairs
of molars and only three of premolars.   The replacement of the
teeth is also fuller than in the marsupials.   Dr Ameghino has
suggested that these animals were transitional between marsupial
and eutherian carnivores, and that the latter group originated in
South America ; but this idea is obviously untenable.   A possible
suggestion is that they may be specialised offshoots from the
marsupial stock which died out without giving origin to any
descendants.

   A small mouse-like mammal first described in 1863 upon the

evidence of a single specimen from Ecuador under the name of
Hyracodon fuliginosus, but whose affinities were not determined
till 1895, when an example of a second and larger species was
procured from Bogota, belongs to the sole surviving genus of a
group of small marsupials which occur abundantly in the Santa
Cruz beds, and were till quite recently regarded as extinct. Unfortunately the name *Hyracodon* has been previously employed
to designate an extinct ungulate, and it has accordingly been
replaced by *Cænolestes*. The essential characteristic of this group
of marsupials, is that while their upper dentition is of a polyprotodont type, that of the lower jaw is very similar to the
diprotodont modification. In the living species, for instance,
there are four pairs of small upper incisors of a normal type,

FIG. 26.   FORE PART OF THE RIGHT HALF OF THE LOWER JAW OF
Acdestis oweni. (Much enlarged.)

*The first tooth on the right is the first incisor, and the one on the extreme
left the first molar.*

followed by a large canine, while in the lower jaw, as shown in
the accompanying figure of that of one of the extinct forms, there
is a large pair of horizontally projecting incisors, succeeded by
several minute functionless teeth, of which the first three represent
the second and third incisors and the canine. In all the forms
the molar teeth have quadrangular crowns, surmounted by four
blunt tubercles, somewhat resembling in structure those of certain
ungulates, and thus totally different from the triangular and
sharply-cusped molars of the opossums and other polyprotodonts.
In the living forms, as well as in certain fossil kinds (*Epanorthus,
Decastis* and *Acdestis*) from the Santa Cruz beds, the last premolar
tooth, as shown in the figure of the jaw of *Acdestis*, is of normal
dimensions : and these forms may consequently be grouped in a
single family, under the name of *Epanorthidæ*. In another group,

confined to the Santa Crucian horizon, where it is represented by
the family *Abderitidæ*, the last premolar in each jaw is much
larger and taller than the other teeth, and has its crown in the
form of a compressed cone, marked on the sides with vertical
grooves, as exhibited in the figure of *Abderites*. A third family is

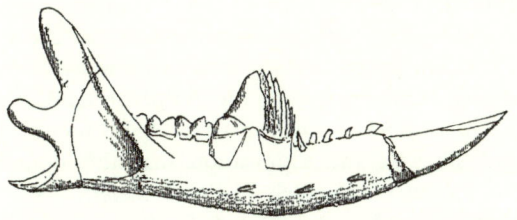

FIG. 27. RIGHT HALF OF LOWER JAW OF *Abderites*, MUCH ENLARGED.

known as the *Garzoniidæ*. In all the skull is of an elongated
form, with large vacuities both in the front and hinder part of the
palate, and presents a considerable general resemblance to those
of the Australian genera *Peragale* and *Perameles*. With the exception of the retention of four pairs of upper incisors and the small
size of all these teeth, the dentition exhibits, however, a remarkable approximation to that of the Australian diprotodont genus
Dromicia. On the other hand, the feet are of normal structure,
with five complete toes, none of which are united by integuments;
the thumb and great toe being apparently slightly opposable to
the other digits. Probably the rat-like tail is slightly prehensile
at the extremity ; and a small pouch is present in the female. In
the skeleton the lower jaw exhibits the usual inflection of the
angle; and the pelvis carries marsupial bones.

Probably these Patagonian marsupials, which may be known
as selvas, must be included in the diprotodont sub-order ; from
the Australian representatives of which they differ by the small
and numerous upper incisors and the non-syndactylous hind feet.
Both these being generalised features, it is evident that if the
selvas are true diprotodonts their ancestors must have originated
from the polyprotodonts in Notogæa, for if they are of exclusively
South American origin they must form a subordinal group by

themselves. Assuming their affinities with the Australian type to be rightly determined, they constitute a most important link in the chain connecting the faunas of South America and Australia.

In the last chapter it has been argued that, from the absence of allied forms in the Tertiaries of North America and Europe, as well as from their resemblance to the Australian dasyurids, it is difficult to come to any conclusion other than that the ancestors of the Santa Crucian polyprotodont marsupials reached the country from Australia, either by way of the Antarctic continent, or by a land-bridge in a more northern part of the Pacific. If this be correct, and likewise the supposition that the opossums originated from the ancestral stock in South-eastern Asia, it will be evident that *Didelphys* and *Cænolestes* met in South America after their ancestors had travelled half round the world in opposite hemispheres.

It may be added that the alleged occurrence of monotremes in the Santa Cruz beds is due to bones of aberrant armadillos (*Peltephilus*) having been mistaken for those of that group[1].

Although in this volume the writer avoids laying much stress upon aquatic mammals, it may be mentioned that there are two genera of dolphins belonging to the family *Platanistidæ*, each represented by a single species, which are peculiar to the Neogæic realm. These are *Stenodelphis* (*Pontoporia*) from the mouth of the Rio de la Plata, and *Inia* of the upper Amazon; the only other existing representative of the family being *Platanista* of the larger Indian rivers.

Cetaceans.

After the foregoing survey of the chief features of the recent and fossil mammalian fauna of the Neogæic realm, its general bearings on the relations of South America to other parts of the world may be taken into consideration. It will, however, facilitate matters to give a tabular view of the orders, suborders, and families of non-volant land mammals represented in the realm. In the following table such groups as are either confined to Neogæa, or have only reached North America at a comparatively recent epoch are

Early Distinction of the Neogæic Fauna.

[1] See Lydekker, *An. Mus. La Plata,—Pal. Argent.* Pt. III. p. 67 (1894).

printed in italics; the extinct types being indicated by the prefix of an *, while those dating from the Santa Crucian (or the earlier Patagonian) epoch are followed by the words Santa Cruz, and those from the Paranà beds by the word Paranà.

I.   PRIMATES.—Santa Cruz.

> *Cebidæ.*—Santa Cruz (* *Homunculus*).
> *Hapalidæ.*

II. CHIROPTERA.

> *Phyllostomatidæ.*—One genus ranging to California.
> Emballonuridæ.—Seven peculiar genera.
> Vespertilionidæ.—*Natalus, Thyroptera.*

III. INSECTIVORA.

> *Solenodontidæ.*—West Indies.

IV. CARNIVORA.

> Felidæ.—No peculiar genera.
> Canidæ.—In addition to the cosmopolitan Canis, represented in Brazil by the peculiar *Icticyon.*
> Ursidæ.—No peculiar genera; * Arctotherium, common to N. America.
> Procyonidæ—*Nasua, Cercoleptes, Bassaricyon*, and * *Cynonasua.*
> Mustelidæ.—*Galictis*, and *Conepatus*, the latter extending into Texas.

V.   UNGULATA.

  1.   ARTIODACTYLA.

> Cervidæ.—Cariacus, peculiar to New World.
> Dicotylidæ.—Peculiar to New World at present day.
> Camelidæ.—*Lama.*

L.   8

2. PERISSODACTYLA.

Tapiridæ.—Elsewhere only in Malaysia at the present day, but formerly widely distributed over the northern hemisphere.

Equidæ.—Now unknown, except through introduction. In addition to the cosmopolitan Equus (including Tertiary forms), the peculiar Pampean genera * *Hippidium* and * *Onohippidium.*

3. * *Litopterna.*—Santa Cruz.
     * *Macraucheniidæ.*—Santa Cruz.
     * *Proterotheriidæ.*—Santa Cruz.

4. * *Astrapotheria.*—Santa Cruz.
     * *Astrapotheriidæ.*—Santa Cruz.
     * *Homalodontotheriidæ.*—Santa Cruz.

5. * *Toxodontia.*—Santa Cruz.
     * *Toxodontidæ.*—Santa Cruz.
     * *Typotheriidæ.*—Santa Cruz.
     * *Pachyruchidæ.*—Santa Cruz.

6. * *Pyrotheria.*—Patagonian beds lying below those of Santa Cruz.
     * *Pyrotheriidæ.*—Patagonian beds.

7. PROBOSCIDEA.

Elephantidæ.—Represented by Mastodon in the Pampean and Monte Hermoso beds.

VI. RODENTIA.

1. SCIUROMORPHA.

Sciuridæ.—Represented by Sciurus as far south as Paraguay.
*Castoroididæ.—Peculiar to New World.

2 MYOMORPHA.

Muridæ.—Represented by species of the New World genus Sitomys, as well as by several peculiar types such as *Rhithrodon, Ichthyomys* of Peru, *Holochilus* of Brazil, etc.

3. HYSTRICOMORPHA.—Santa Cruz.
   Caviidæ.—Paranà.
   Dasyproctidæ.
   Dinomyidæ.
   Lagostomatidæ.—Santa Cruz.
   Hystricidæ.—Santa Cruz.
   Octodontidæ.—Mainly Neotropical, but also Ethi-
       opian. Santa Cruz.
   * *Eocardiidæ.*—Santa Cruz.

4. LAGOMORPHA.
   Leporidæ.—Represented by two species of
       Lepus.

VII. *EDENTATA.*—Santa Cruz.
   Dasypodidæ.—Santa Cruz. A few forms range
       as far as Texas.
   * *Glyptodontidæ.*—Santa Cruz. One Neogæic genus
       ranges as far north as Texas, and a peculiar
       one has been described from further north.
   * *Megalotheriidæ.*—Santa Cruz. Mainly Neogæic,
       but also ranging into North America.
   Myrmecophagidæ.
   Bradypodidæ.

VIII. MARSUPIALIA.—Santa Cruz.
   1. DIPROTODONTIA.—Santa Cruz. (Elsewhere only in
       Notogæa).
       Epanorthidæ. Santa Cruz. (One existing genus,
           Cænolestes).
       * *Abderitidæ.*—Santa Cruz.
       * *Garzoniidæ.*—Santa Cruz.
   2. POLYPROTODONTIA.—Santa Cruz.
       Didelphyidæ.—Now mainly Neogæic, where they
           date from the Monte Hermoso stage, but
           ranging into North America, and formerly
           widely spread over the northern hemisphere.
           Chironectes, several subgenera of Didelphys,
           and *Dromiciops* peculiar to the realm.

8—2

VIII.   MARSUPIALIA (*continued*).

> Dasyuridæ.—Santa Cruz.   Now confined to
> Notogæa, but apparently represented in the
> Santa Cruz beds by * *Prothylacinus* and
> * *Amphiproviverra*.
> * *Microbiotheriidæ*.—Santa Cruz.
> *Incertæ Sedis*.
> * *Borhyænidæ*.—Santa Cruz.

Although the addition of the names of all the peculiar genera,
both recent and extinct, would have rendered the distinctness of
the Neogæic mammalian fauna still more pronounced, yet the
foregoing table as it stands is amply sufficient to show that
Neogæa is entitled to form one of the three primary zoological
realms of the world.   Starting with the Santa Crucian epoch of
Patagonia and the somewhat older Patagonian stage, which form
the earliest date at which the history of the Tertiary land mammals
of Neogæa can be taken up, there is evidence that at least the
southern part of the area was populated by the following peculiar fauna.   Firstly, monkeys of a type quite different from those
of the Old World, but evidently allied to the existing Neogæic
forms, were abundant; while rodents, belonging to the same
groups as those now inhabiting the continent, several of which
were nearly allied to existing African forms, and more remotely to
certain Oligocene European types, attained a great development.
Probably Insectivora with V-shaped molars were also present.
More peculiar are the extinct subordinal groups of ungulates
described above, which appear to have been allied to the ancestors
of the perissodactyles of the northern hemisphere, and may
possibly be remotely connected with the African hyraces.   At the
same period flourished several families of edentates (a group
which in its restricted sense was originally peculiar to Neogæa),
such as armadillos, glyptodonts, and ground-sloths, most of the
members of which were of comparatively small size; but of the
ancestry of this group nothing can be said with certainty.
Among the marsupials, although opossums appear to have been
wanting, there were several types seemingly allied to Notogæic
forms, while others which may be included in the order were more

or less unlike any from other regions. In addition to true marsupials, the only carnivorous types were the problematical
Borhyænidæ.

Now, as stated in the earlier part of the chapter (p. 68),
this fauna cannot be older than the lowest Miocene or highest
Oligocene; and among its deficiencies may be noted lemuroids, true
carnivores, creodonts, artiodactyle and perissodactyle ungulates,
and opossums, all of which were in existence during the Oligocene or Miocene in North America and Europe. Moreover, at those
epochs the former country lacked the whole of the Neogæic types.

Clearly, then, there must have been a barrier between North
and South America during the Oligocene and a
portion or the whole of the Miocene; but before
entering into the consideration of other evidence

Early Separ-
ation of N. and
S. America.

showing the nature of that barrier, it may be well to give a table of
the mammaliferous Tertiary strata of North and South America,
with their approximate European equivalents[1]. In descending
order this runs as follows :

Age	*South America.*	*North America.*		*Europe.*
Plistocene	Pampean	Equus beds.		Cave-deposits etc.
Up. Pliocene		Blanco		? Crag.
Low. Pliocene	Monte Hermoso		Palo Duro.	Pikermi
	(?) Parana	Loup-Fork	Nebraska	
Miocene			Deep River	Sansan
	Santa Cruz		(Hiatus)	
Up. Oligocene	Patagonian	John Day		St Gérand-Le-Puy
			Protoceras Beds	
Mid. Oligocene		White River	Oreodon Beds	Ronzon.
			Titanother-ium Beds	
Low. Oligocene		Uinta		Montmartre
Up. Eocene			Washakii	
Mid. Eocene		Bridger	Bridger.	Parisian
			Wind River.	
Low. Eocene		Wahsatch		Suessonian
Lowest Eocene		Puerco		Cernaysian

[1] In compiling this table the writer is indebted to Prof. W. B. Scott.
Many American geologists (among them Dr Scott) include the whole of the

Regarding the geological evidence of a separation between the two Americas, Cretaceous marine strata occupy a large portion of Mexico; and in 1879 Dr Le Conte[1] wrote as follows. The shore line of the Gulf of Mexico "was much more extended both northward and westward than either now or in Tertiary times. From the Gulf there extended northwestward an immensely wide sea, covering the Plains region and the Rocky Mountain region as far westward as the Wahsatch range, and dividing the continent into two continents, an eastern or Appalachian, and a western or Basin-region continent. Probably also this sea connected across the region of Mexico with the Pacific, thus dividing the western continent into two, a northern and southern." Later observations have shown that the Cretaceous sea undoubtedly made a wide gap between North America and the southern portion of the continent[2]; while the existence of Oligocene or Miocene strata in the region of the isthmus of Paranà shows that the separation, which probably continued through the Eocene, was in existence during

---

Loup-Fork in the Miocene; while to the lower part of the same era they assign the John Day beds of America, and the St Gérand-le-Puy beds of Europe. Others (*e.g.* Prof. Osborn, *Studies Biol. Labor. Columbia Coll.* vol. I. pt. 2, p. 28, 1893) refer the Equus beds of North America to the Pliocene. The following quotation from a paper by Prof. Cope (*American Naturalist*, 1895, p. 599) conclusively proves their Plistocene age. There it is stated that "the Equus beds are found covering areas of various extent in Oregon, Nevada, California, the Staked Plains, Southern Texas, Chihuahua and the valley of Mexico. Their most eastern station is western Nebraska. They contain a fauna which includes one extinct species (*Equus major* Dek.) of the Megalonyx fauna, and the recent *Castor fiber*. They contain the extinct genus of sloths *Mylodon*, of a species different from that of the east, and four species of camels of the extinct genus *Holomeniscus*, and a peccary. Recent species of *Canis* and *Thomomys* occur, while two extinct horses (*Equus occidentalis* Leidy and *E. tau* Owen) are common. The hairy elephant (*E. primigenius*) is abundant, while the *Mastodon americanus* is rare, if occurring at all. The proportion of recent to extinct species and genera in the Equus bed fauna is very similar to that occurring in the Megalonyx fauna, while they differ as to details."

[1] *Elements of Geology*, pp. 451, 452, New York (1879).

[2] This separation also existed in the Jurassic era, when, as shown by Neumayr (*Erdgeschichte*, 2nd ed. vol. II., p. 263), South America was united across the Atlantic area with Africa and Madagascar.

the middle portion of the Tertiary epoch[1].   When the connection
between North and South America was completed is not precisely
fixed by geological evidence ; but the occurrence of a glyptodont
in the Nebraska stage of the Loup-Fork group, shows that it must
have been by the end of the Miocene[2].   The question of a connection between the two continents by way of the West Indies
is discussed later in this chapter, where it is concluded that if
such a connection existed at all, it must have been of a transient
nature.

Having thus shown, both from palæontological and geological
evidence, that the early mammalian fauna of
Neogæa appears to have been totally isolated from
that of North America up to about the end of the
Miocene, the question of the origin of that fauna may be deferred,
and the irruption of northern types after the connection between
North and South America had been established taken into consideration.   It may be mentioned, however, that it was not till
after this irruption of the northern forms that the essentially
Neogæic fauna attained its maximum development in respect to
the bodily size of its constituents ; since a gradual increase in this
respect may be traced from the small glyptodonts and sloths of the
Santa Cruz epoch, through the larger ones of the Monte Hermoso
horizon, to the gigantic forms characteristic of the Pampean and
the cavern deposits of Brazil.

The presence of a glyptodont in the Nebraska stage of the
Loup-Fork group in North America, and of northern forms in the
Monte Hermoso horizon of South America, marks, then, the first
commingling of the original faunas of the two halves of the New
World.   For the first time in the history of the southern continent
this connection allowed of the immigration from the north of the
true Carnivora, such as the existing cats (*Felis*) the extinct sabre-
toothed tigers (*Machærodus*), dogs and foxes (*Canidæ*), bears
(*Ursus* and *Arctotherium*), raccoons (*Procyonidæ*), skunks and
their allies (*Mustelidæ*), together with various ungulates belonging
to sub-orders previously unknown in the realm.   These latter
include the guanaco and vicuña (*Lama*)—of which ancestral forms

Incursion of Northern Mammals.

---

[1] See J. W. Gregory, *Quart. Journ. Geol. Soc.* Vol. LI. pp. 299, 300 (1895).
[2] As stated above, many refer the whole Loup-Fork group to the Miocene.

are abundant in the North American Tertiaries—New World deer (*Cariacus*), horses (*Equidæ*) of various genera, tapirs (*Tapiridæ*), peccaries (*Dicotylidæ*), and mastodons. Among the rodents, squirrels, the various genera of *Muridæ*, and the hares, likewise at this epoch made their first appearance on the scene. Opossums also at this time effected an entrance into the land which has now become their chief home. That this new fauna came in from North America, and not from any other part of the world, may be regarded as certain from the presence of such essentially New World types as raccoons and their allies, skunks, peccaries, *Cariacus*, and *Camelidæ* (exclusive of the Old World genus *Camelus*, which is of late origin), coupled with the absence of true deer (*Cervus*), pigs (*Sus*), Old World monkeys, and lemurs.

At the same time this union of the northern and southern halves of the New World allowed certain members of the original Neogæic fauna to make their escape into North America, glyptodonts, as already said, making their appearance in the Nebraska stage of the Loup-Fork group of the United States, while the ground-sloth *Megalonyx* occurs in the Blanco Beds.

Although it is not universally admitted[1], there is some evidence to indicate that this land connection was of comparatively brief duration, seeing that none of the characteristic extinct types of South American ungulates, nor any of the peculiar Neogæic rodents, reached the northern half of the continent.

During the whole time that the alluvial deposits of the Paranà and Paraguay rivers were being laid down, and well on into the human period, the mammalian fauna of the Pampean epoch, formed by an admixture of southern and northern types, continued to flourish, until the time when there came a complete sweep of all the larger forms, clearing off the whole of the ground-sloths, glyptodonts, mastodons, toxodonts, macrauchenias, horses, sabre-toothed tigers, and the larger members of the camel tribe, and in the Argentine leaving only armadillos, guanacos, a few deer, a number of rodents, various cats and foxes, as well as skunks and certain other members of the weasel family, to represent the vast assemblage of strange and giant creatures that once roamed over

[1] See Gregory, *op. cit.* p. 300.

its plains. With regard to this extraordinary, and apparently sudden disappearance of almost all the larger forms of animal life from South America, it may be pretty confidently asserted, from the organisation of the creatures themselves, that at the time when the ground-sloths flourished, extensive portions of what is now the open pampas of Argentina were covered with forest ; and why the whole district should have become practically treeless, seeing that trees like the Australian eucalypti grow, when introduced, with more vigour than in their native home, is exceedingly hard to understand. That the country even when thus denuded was unsuited to the needs of the larger forms of mammalian life, is, however, negatived by the circumstance that in many parts the horses and cattle introduced from Europe have run wild and increased to an almost unprecedented extent. Neither does it appear that the extermination can be attributed to a period of extreme cold, since a glaciation of the pampas would assuredly have left unmistakable evidence of its presence. It is likewise practically certain that the clean sweep of the forests of Argentina and the larger mammals of the whole of South America is not due to the hand of man. It has, indeed, been suggested that the vast herds of guanaco which formerly roamed the pampas may have cleared the forests by preventing the growth of the seedlings ; but when we recall the fact that numbers of this group of animals flourished during the period when the alluvial formation was in the course of being deposited, it scarcely looks as though this could have been a *vera causa*. Moreover, if the forests were by some means or other actually destroyed, and the extermination of their animal denizens thus encompassed, there would still remain the disappearance of plain-dwelling forms, like horses, to be accounted for. Some have thought that pumas, by preying on the colts, were the active agents in this instance ; but even if such were really the case, it gives no help with regard to the disappearance of the ground-sloths and glyptodonts ;—the latter being such strongly armoured creatures that it is absolutely certain they were not killed off by any animal foes. The problem is further complicated by the circumstance that the fossil remains of nearly all the larger animals which formerly inhabited the pampas are also found in the caverns of Brazil, where the climate is now, and probably always has been, tropical. Up to the present, it is, accordingly, impossible to account satisfactorily for the disappearance of all the larger forms from among the mammalian fauna of South America.

Returning to the fauna itself, it may be asserted that before the great intrusion of northern forms the mammals of Neogæa were far more distinct from those of the rest of the world than is the case with the fauna of any other region, with the exception of Australasia; and that consequently there can be no hesitation in allowing this part of the earth's surface to take rank as a primary realm. At the time when the Santa Cruz fauna was so decidedly marked off there was a much more general similarity between the faunas of all the other regions of the world (exclusive of Notogæa) than is the case at the present day, and it is this antiquity of the differentiation of the Neogæic fauna that supports so strongly its claim to distinctness.

It has been suggested that the first land-connection between South and North America was probably of limited extent or short duration; and some evidence of a later separation between the two areas is afforded by the beds of marine shells already mentioned as occurring in the upper part of the Pampean deposits; these beds marking an epoch of submergence of a considerable portion of the area[1]. Subsequently to this final submergence of portions of Argentina and Uruguay there was a great upheaval of the country, indicated not only by the upraising of the aforesaid marine beds, but likewise by that of the sand-dunes fringing most parts of the Argentine coast. This upheaval, which has taken place within the human period, certainly resulted in the final union of the two Americas; and since its date there has probably continued to be a greater and greater admixture of the two originally distinct faunas, so far as climatic conditions have permitted. It is, however, not a little curious that some of the original northern types, such as the vicuñas and guanacos, have entirely died out in their original habitat, to flourish only in the southern half of the continent.

[1] There is some evidence to show that the isthmus of Panamà was never completely submerged after the Pliocene, but it may have been so narrow as not to allow of much migration of the fauna.

Hitherto especial stress has been laid on the fossil mammals of Neogæa as entitling the tract to form a primary realm, on account of the distinctness of its fauna from that of the rest of the world during a consider-

able portion of the Tertiary epoch. Even at the present day, however, when, as already shown, its mammalian fauna contains a very large admixture of types which have immigrated from the north at a comparatively recent epoch, it still stands widely apart from other regions. On this point may be quoted the admirable summary given in "Island Life"[1] by Dr Wallace, who writes that among the peculiar mammals we have "the prehensile-tailed monkeys and the marmosets, the blood-sucking bats, the coati-mundis, the peccaries, the llamas and alpacas [vicuñas and guanacos], chinchillas, the agutis, the sloths, the armadillos, and the ant-eaters; a series of types more varied and more distinct from those of the rest of the world than any other continent can boast of. Among birds we have the charming sugar-birds, forming the family *Cærebidæ*, the immense and wonderfully varied group of tanagers (*Tanagridæ*), the exquisite little manakins and the gorgeously-coloured chatterers (*Cotingidæ*); the host of tree-creepers of the family *Dendrocolaptidæ*, the wonderful toucans (*Rhamphastidæ*), the puff-birds (*Bucconidæ*), jacamars (*Galbulidæ*), todies (*Todidæ*), and motmots (*Momotidæ*); the marvellous assemblage of four hundred distinct kinds of humming-birds (*Trochilidæ*), the gorgeous macaws (*Ara*), the curassows (*Cracidæ*), the trumpeters (*Psophiidæ*), and the sun-bitterns (*Eurypygidæ*). Here again there is no other continent or region that can produce such an assemblage of remarkable and perfectly distinct groups of birds; and no less wonderful is its richness in species, since these fully equal, if they do not surpass, those of the two great tropical regions of the Eastern Hemisphere (the Ethiopian and the Oriental) combined." Not less noteworthy among the birds are the screamers (*Palamedeidæ*); the tinamus (*Tinamidæ*), which while outwardly resembling game-birds, agree with the struthious birds in the structure of their skulls; and the rheas (*Rheidæ*), or South American ostriches, whose nearest allies are the true

---

[1] Pages 50, 51. In this quotation the scientific names of some of the groups have been added.

ostriches of the Old World. The hoatzin (*Opisthocomus*), the oil-bird (*Steatornis*), and the boat-bill (*Cancroma*) are likewise peculiar Neogæic types. Still more remarkable is the solitary Andean survival (*Cænolestes*) of the diprotodont marsupials of the Santa Cruz epoch. A curious feature of the Neogæan forest-mammals—whether they belong to the old fauna or the new—is the frequency of prehensile power in the tail. Not only does this occur in several genera of monkeys, but also in *Cercoleptes*, *Synetheres*, *Chætomys*, *Capromys prehensilis*, *Cycloturus*, *Didelphys*, and *Cænolestes*. A parallel is to be found elsewhere only in Australia.

After referring to the deficiency of the many types of mammals alluded to in an earlier paragraph of the present chapter, the author adds that "Among birds we have to notice the absence of tits, true flycatchers, shrikes, sun-birds, starlings, larks (except a solitary species in the Andes), rollers, bee-eaters, and pheasants, while warblers are very scarce, and the almost cosmopolitan wagtails are represented by a single species of pipit......... Whether, therefore, we consider its richness in peculiar forms of animal life, its enormous variety of species, its numerous deficiencies as compared with other parts of the world, or the prevalence of a low type of organisation among its higher animals, the Neotropical region stands out as undoubtedly the most remarkable of the great zoological regions of the earth."

The distinctness is, however, by no means confined to mammals and birds. Of the land molluscs, Mr A. H. Cooke[1] writes that they present a marked contrast to those of North America. "Instead of being scanty, they are exceedingly abundant; instead of being small and obscure, they are among the largest in size, most brilliant in colour, and most singular in shape that are known to exist. At the same time they are, as a whole, isolated in type, and exhibit but little relation with the Mollusca of any other region."

Having arrived at the conclusion that the original Neogæic mammalian fauna, exemplified in the Santa Cruz beds, has not been derived from North America, it remains to endeavour to account for its origin.

Origin of the Santa Cruz Fauna.

[1] *The Cambridge Natural History*—Mollusca, p. 342 (1895).

This, however, is a difficult and perplexing subject which it is scarcely possible to explain fully in the present imperfect state of palæontological knowledge.

With regard to the marsupials of an Australian type, it has already been stated in the preceding chapter[1] that these appear to have been derived from Notogæa by means of a southern land-bridge. The hypothesis there suggested is that the immigration has taken place *viâ* the Antarctic continent, probably across the southern Pacific[2]. It is known that shallow water extends from southern Patagonia and Tierra del Fuego to South Shetland; while between Australia and the Antarctic land there are no depths exceeding 2000 fathoms. On the other hand it is just possible that the connection may have been by way of Polynesia.

In this place reference may be made to certain very remarkable resemblances existing between animals of groups other than mammals respectively inhabiting Neogæa and Notogæa. The first instance is that of two peculiar families of freshwater fishes, known as the *Haplochitonidæ* and *Galaxiidæ*. Of these the former has one Australian and a second South American genus, while the latter is represented by a single genus (*Galaxias*), common to New Zealand, Australia, and the extremity of South America. This, however, is by no means all, since one species of the last-mentioned genus (*G. attenuatus*) is found in regions as remote from one another as New Zealand and Tasmania on the one hand, and the Falkland Islands and the extremity of Patagonia on the other. Commenting on this, Dr Wallace remarks[3], it is impossible to believe that a land connection between South America and Notogæa could have existed "within the period of existence of this one species of fish, not only on account of what we know of the permanency of continents and deep oceans; but because such a connection must have led to much more numerous and important cases of similarity of natural productions than we actually find. Rather must we look to the transport of the ova across the

---

[1] *Supra*, p. 55.

[2] Possibly, as suggested below, the connection may have been nearer the tropics.

[3] *Geographical Distribution of Animals*, Vol. I. pp. 401, 402; see also Vol. II. pp. 82, 83.

southern seas, aided perhaps by the Antarctic ice, and a former greater extension of South America towards the pole." After remarking how such transmission might take place with but little extension of the present Antarctic lands, Dr Wallace adds that "there is evidently some means by which ova or young fishes are carried moderate distances, from the fact that remote Alpine lakes and distinct river-systems often have the same species. Glaciers and icebergs generally have pools of fresh water on their surfaces; and whatever cause transmits fish to an isolated pond might occasionally stock these pools, and by this means introduce the fishes of one southern island into another."

Allowing all due weight to these objections to a land connection between Notogæa and Neogæa, it seems almost impossible to believe that the transit has taken place in the manner suggested by Dr Wallace.

Another piece of evidence is afforded by some observations of Mr F. E. Beddard [1] in regard to the intimate relationship existing between the earth-worms of New Zealand and Eastern Australia on the one hand, and those of Patagonia on the other. Without committing himself to any theory as to how the communication took place, the author is content to say that " the facts seem to point to a more recent communication between Patagonia and New Zealand than between either of those countries and the Cape of Good Hope."

Assuming a land connection, earth-worms would suggest that it was probably not in very high latitudes. Now I have been informed on verbal authority that there is a curious similarity between the slugs of Patagonia and those of Polynesia. And it is probable that the latter tract indicates a subsiding area which was formerly connected with Patagonia. Possibly, therefore, there may have been a land connection between Patagonia and Australia *viâ* Polynesia; and this may have been the line through which Neogæa received the Notogæic elements in its fauna. Whether it could have existed at a date sufficiently late for the passage of the marsupials, it is impossible to say. If it existed, it probably allowed only a limited communication between the

---

[1] Appendix, No. 5, pp. 170, 171.

Notogæic and Neogæic mammals; and it is easy to imagine that the Polynesian mammals (if they existed) were drowned out by submergence, as has undoubtedly been the case with many of those of the West Indies. In dismissing this part of the subject, it may be observed that it appears impossible to adequately explain the presence of a Notogæic element in the fauna of Neogæa without the aid of some form of southern land connection; although there is not sufficient evidence to show in what latitude such connection (or connections) existed.

Attention must now be directed to the Santa Crucian mammals other than marsupials. With the exception of the edentates, which probably originated in Neogæa, or possibly in some still more southern land, all the evidence points to the whole of these being originally of northern derivation; the ungulates having affinities with the ancestral types from which the earlier Tertiary perissodactyles were descended, while the rodents have certain relationships with the early European members of the order. The monkeys again were probably descended from lemuroids; and the solenodons are evidently related to the Old World insectivores. It has been shown that this portion of the fauna did not come from North America; and it is certainly not derived from Notogæa. Accordingly, the only route by which it could have entered is by way of Africa. The only marked community between the Ethiopian and Neogæic faunas as regards mammals relates to the hystricomorphous rodents; but this community is a very marked one, and difficult to explain on any other hypothesis than that of a connection between the two areas. The possibility of a close community of origin between the toxodonts and the hyraces has already been mentioned; and if it be substantiated, it will be highly important. Of course, on the supposition of an African origin for the eutherian mammalian fauna of Tertiary Neogæa, it must be taken for granted that the ancestral types entered Africa long before the progenitors of its modern fauna; although probably not before the ancestors of the Malagasy fauna. It may be objected that we ought to find Neogæic Tertiary types of ungulates in Africa; but we are unacquainted with the Tertiary palæontology of that country, and it is quite probable that the peculiar subordinal Neogæic types of ungulates were only developed as such in America itself. Even if they ever existed in Africa there is no more reason why they should have survived there than in America. As the evidence for the presence of Insectivora in the Santa Cruz deposits is not very strong, the case of the West Indian solenodons must not be pressed too strongly, but their affinity to the tenrecs of Madagascar, and the absence of allied types in the North American Tertiaries, both point to their having reached Neogæa, with the other eutherians.

Regarding a possible connection between Africa and South America by way of the Antarctic continent, Dr Blanford[1] writes as follows:—"Singularly enough, so far as our present information as to the depths of the southern oceans goes, there would appear at first sight to be less difficulty in supposing a former extension of the southern continent to Australia and South America than to Africa, the depth as shown on the 'Challenger' charts south of the former continents nowhere exceeding 2000 fathoms, whereas to the south of Africa there is represented a considerable belt of greater depth. But on an Admiralty chart on which all the known deep soundings are marked, none are shown south of the southern extremity of Africa....So far as our present information goes, the ocean south of the Cape of Good Hope may be no deeper than the Mozambique channel, though probably the depth is greater in the former case."

Before discussing certain relationships between the Ethiopian fauna and that of Neogæa, it seems advisable to refer to some recent views as to the existence of a great southern circumpolar continent in Tertiary times, extending into comparatively low latitudes, and connected, at all events temporarily, with America, Africa, and Australia. For this continent the name Antarctica has been suggested by Dr H. O. Forbes[2], who urges that many of the types of animal life now confined to the southern hemisphere have originated there. It is chiefly to show the fallacy of these latter views that the subject is referred to here; palæontological evidence clearly proving that several of the groups of

"Antarctica" and the South American element in the Ethiopian Fauna.

---

[1] Appendix, No. 8, p. 100.

[2] Appendix, No. 15.

animals assumed to be essentially southern, really had a northern origin. It may be premised that, according to the view of Dr Forbes, "Antarctica" followed nearly the 2000 fathom line, extending northwards from a circumpolar area by broad expansions, one to join an old New Zealand continental island (including the Antipodes, Macquarries, New Zealand, and Chatham, Lord Howe, Norfolk, and the Kermadec and Fiji Islands); a second to East Australia and Tasmania; a third to the Mascarene and adjacent islands; perhaps one to South Africa; and finally one to South America.

As regards the marsupials, which are among those considered to be southern types, the evidence of the northern Jurassic and Cretaceous kinds alluded to in the preceding chapter, coupled with the presence of opossums in the Oligocene of the northern hemisphere, renders it practically certain that the group did not originate in the southern hemisphere.

Among other groups cited by Dr Forbes as being mainly or exclusively southern in their distribution and origin are the parrots (*Psittaci*) and trogons (*Trogonidæ*). But both these are represented in a fossil state in the Oligocene strata of France, and are thus shown to have been originally denizens of the temperate regions of the northern hemisphere. Take, again, the case of the struthious birds, or Ratitæ, which although cited as a southern group, are represented by an ostrich (*Struthio*) in the Pliocene of Northern India and the Crimea, while the former deposits have yielded remains of a three-toed genus allied probably to the emeus and cassowaries. Much the same may be said in regard to the giant land-tortoises (*Testudo*), which although now confined to the Galapagos and Mascarene islands, were abundant in Northern India, Greece, France, and the United States during the Pliocene, and also occur in the French Miocene and Oligocene, as well as in the Plistocene deposits of the Maltese caves. The group was thus evidently a northern one originally, and as it is unknown in the southern hemisphere before the Pampean epoch of Argentina and the superficial deposits of Madagascar, its southern migration probably did not take place till the Miocene, or even the Pliocene. Indeed, the separation of North and South America indicates that, if the Galapagos tortoises came from the former country, they could not have reached their present habitat till the end of the Miocene.

On the question of the southern or northern origin of some of the above-mentioned birds, Professor Huxley [1], so far back as .1868, wrote as follows :—"I watch the progress of M. Alphonse Milne-Edwards's researches with great interest, to know whether parrots, pigeons, *Dromæidæ* (*Casuariidæ*) and *Rhæidæ* occur in force, or at all, among the Miocene birds. If they are absent from the Miocene fauna of Arctogæa, it will be necessary to suppose that these groups of birds are of sufficiently ancient origin to have been separated, even before the Miocene epoch in Austro-Columbia (Neogæa) and Australasia, whence they have subsequently colonised part of Arctogæa ; while, on the other hand, their presence in European Miocene formations will render it possible that the colonisation has taken place the other way, and that these birds have attained their wonderful multiplicity and diversity of forms in Austro-Columbia and Australasia simply in consequence of the very favourable nature of the conditions to which they have been exposed in that country.

"I confess I incline to the latter supposition. The distribution of *Psittacula*, for instance, is quite unintelligible to me upon any other supposition than that this genus existed in the Miocene epoch, or earlier, in Northern Arctogæa, and has thence spread into Austro-Columbia, South Africa, India, and the Papuan Islands, where it is now found."

Although the term *Psittacula* has now been restricted so as to include only the Neogæic forms, this passage is almost prophetic ; both parrots and pigeons having, as already stated, been discovered in the French Oligocene, while the Australian and probably the South American ratite birds appear to have had an Indian forerunner. And here it may be mentioned that the South American ostriches (*Rhea*) which are primitive types allied to the ostrich, would seem to have made their way into Neogæa *viâ* Africa, as there are no traces of ancestral forms in the North American Tertiaries.

On the other hand, there is considerable probability that the

[1] Appendix, No. 18, p. 319.

penguins (*Spheniscidæ*), which present a relation to other birds
somewhat analogous to that exhibited by the edentates to other
mammals, having no apparent affinity with any group—may prove
to be an exception to the rule of the northern origin of most of the
existing southern types of terrestrial vertebrates, since they are
quite unknown in the north, and occur fossil both in New Zealand
and Patagonia.

Another marked instance of the northern origin of southern
types is afforded by the side-necked, or pleurodiran Chelonia,
which although now restricted to the more southern parts of the
globe, were abundant during Secondary and early Tertiary times
throughout the northern hemisphere.   A striking example of this
is shown in the family *Pelomedusidæ*, whose existing representatives
are confined to Africa, Madagascar, and South America.   Among
these, two out of three genera, namely *Sternothærus* and *Pelomedusa*
are found in Ethiopian Africa and Madagascar, one of them also
ranging into the Sinaitic peninsula ; while the third (*Podocnemis*)
has five species in South America and a sixth in Madagascar.
There occurs, however, in the upper Cretaceous of the United
States the allied extinct genus *Bothremys*, and *Podocnemis* itself is
represented in the London Clay and the Eocene of the Punjab.
Here the inference would seem to be that the latter genus
originated in the northern half of the Old World, passed by way of
India into Madagascar and Africa, and thence by a southern route
into Neogæa.   Even if this particular genus occurred in the early
Eocene of North America, which it does not, it could scarcely have
crossed the sea into South America ; and the migration can hardly
have taken place since the union of the two continents.   In
commenting on the distribution of *Podocnemis* Dr Blanford[1]
observes that as the incursion of more modern types into Africa
appears to have driven out many of the older, it is in Madagascar
that traces of the relationship of the modern fauna to that of
Neogæa should be looked for.   One such instance is the occurrence there of *Podocnemis*, and a second that of the *Centetidæ* in
that island.   Perhaps the occurrence of sucker-footed bats only in
Brazil where they are represented by the single species of

---

[1] Appendix, No. 8, p. 101.

Thyropoda, and in Madagascar, where there is the sole member of the allied genus *Myxopoda*, may be an analogous instance.

In the second family of the pleurodiran Chelonia, the *Chelyidæ*, which are now restricted to South America and Australia (the genera in the two areas being distinct), there is at present no evidence of the derivation of the Australian forms, but of the Neogæic types *Platemys* is represented in the Cretaceous of the United States, and *Hydraspis* in the Eocene of Bombay. Although the northern origin of the family is thus proved, the explanation of how the existing forms attained their present distribution is very difficult. Possibly *Hydraspis* may have reached South America by way of Africa; but it is difficult to believe, in the absence of its remains, that *Platemys* survived in North America till the late Miocene communication with Neogæa was established.

Among snakes, the boas of the genera *Corallus* and *Boa* are confined to South America and Madagascar, and thus have precisely the same distribution as *Podocnemis*. Now although true boas are unknown as fossils, the allied extinct genus *Paleryx* occurs in the European Oligocene, thus pointing to the northern origin of the group, which has probably reached South America by way of Madagascar and Africa.

Another remarkable case is afforded by the limbless lizards of the family *Amphisbænidæ*, which are now almost equally divided between South America and South Africa, although one genus extends into the Mediterranean area, and two are found in North America; the two genera *Amphisbæna* and *Anops* being common to South America and Africa, while the northern ones are different. The northern origin of the family is, however, indicated by the recent discovery of fossil forms[1] in the White River Oligocene of the United States. Here the evidence strongly points to a southern connection between Neogæa and Africa; Tertiary forms having probably existed in Europe or Asia as well as in North America. A second instance that may be cited among lizards is the family of the *Iguanidæ*, which while now mainly Neogæic, has representatives in the warmer parts of North America, and also includes

[1] Baur, *American Naturalist*, Vol. xxvii. p. 998 (1893).

two outlying genera in Madagascar, and a third in the Fiji and Friendly Islands. But fossil iguanas occur in the French Oligocene, and it may hence be suggested that the group may have reached Neogæa *viâ* Madagascar and Africa; while if the connection between Patagonia and Polynesia alluded to above were substantiated, the origin of the Polynesian forms could be accounted for.

During the middle portion of the Secondary period a very curious resemblance between the fauna of Ethiopia and Neogæa is exhibited by the occurrence in both of certain very peculiar reptiles known as *Mesosaurus* (*Stereosternum*), which have been referred to the Sauropterygia. Remains of these reptiles have been obtained at San Paolo in Brazil, and in Griqualand West and other parts of South Africa, but nowhere else; and, although the type may be of northern origin, this curious distribution apparently points strongly to a connection between Africa and South America as far back as the Secondary epoch. This connection, as pointed out by Neumayr[1], was, however, probably by way of the Atlantic.

Somewhat similar relationships to those of living reptiles are exhibited by fishes, among which the *Haplochitonidæ* and *Galaxiidæ* have been already mentioned. Very remarkable is the case of the lung-fishes *Lepidosirenidæ*, where there is a very close relationship between the West African *Protopterus* and *Lepidosiren* of Brazil and Paraguay; the Australian *Ceratodus* being markedly distinct from both. Although the two former are unknown as fossils, teeth of the latter are abundant in the Trias and Jurassic of Europe, India, South Africa, and the United States; while during the Palæozoic era extinct families of the subclass (Dipnoi) were abundant in the northern hemisphere. Clearly, then, the group was originally northern in origin; and *Ceratodus* apparently migrated south both into Africa and Australia. Taking into account the Cretaceous separation of North and South America, and the close alliance between *Lepidosiren* and *Protopterus*, it is, however, difficult to see how the latter reached its present habitat except by way of Africa. If this be so, and the connection between South

[1] *Vide supra*, p. 118, note.

Africa and South America in Tertiary times was only in high latitudes, a warm epoch in the southern hemisphere must have been necessary for the passage of such tropical forms. It might be urged that as *Ceratodus* dates from the Trias, the other two genera might have reached their present habitats at a very distant epoch; but their specialisation is against their antiquity. Another family which is essentially southern is that of the *Osteoglossidæ*, represented by *Arapaima* of the Brazilian rivers, and *Osteoglossum*, with one species from Brazil and the Guianas, a second from Sumatra, and two others from Australia. But the northern origin of the family is indicated by the occurrence of the extinct genus *Dapedoglossus* in the Eocene of Wyoming. Here there is a presumption that *Osteoglossum* originated in Asia, from which it passed in one direction by way of Malaysia to Australia, and in another through Africa to South America. Two other families of freshwater fishes have a somewhat similar distribution; the first being the *Chromididæ*, which includes spiny fishes mainly characteristic of tropical America and Africa, but extending eastwards into Syria, and sparingly represented in Southern India and Ceylon. In a fossil state they occur in the Cretaceous of Syria; and, although none of the genera are common to the two continents, they are highly suggestive of a connection between Africa and South America. The second family is that of the *Characiniidæ*, comprising fish more nearly allied to the carps, and now exclusively confined to tropical America and Africa. Although the palæontological record is a blank, this can scarcely be taken as a sufficient indication that the family has always been a southern one.

From the foregoing facts it may be considered that the assumption of an Antarctic continent is unnecessary to explain the origin of the many forms of vertebrate life which are now exclusively or mainly southern; nearly all of these, with the exception of the edentates and penguins, being of northern derivation, and thus apparently showing a southern migration of the older forms of life. The Cretaceous and Tertiary break between North and South America appears, however, to have prevented the occurrence of such migration in the western hemisphere till the close of the Miocene: and it is accordingly

Conclusion.

necessary to look elsewhere for the origin of the Neogæic fauna. That Africa has been the great feeder appears the most probable explanation; although in the case of the marsupials it seems necessary to look to Notogæa as the point of origin. Clearly, however, the presumed connections between Neogæa, Notogæa, and Africa have not been very continuous or very extensive in Tertiary times, or the faunas of these areas would have been more alike than they are; and this suggests that the northern extension of Antarctica has not been so great as has been supposed. Whether the presumed connection between Notogæa and Neogæa has taken place by way of Antarctica or Polynesia may be left an open question. With regard to Africa, the recent researches of Dr Gregory[1] on the West Indian corals, in the course of which it is urged that a shallow-water connection "extended across the Central Atlantic in—at latest—Miocene times," while the southward extension of the Atlantic is a comparatively recent feature, indicate the possibility that the land-connection which existed in Jurassic times between Brazil and Western Africa may have persisted till the Tertiary era.

As already mentioned, the Neogæic realm includes but a single region—the Neotropical; and in this four sub-regions have been defined, and are named as follows.

Sub-regions.

Firstly we have the Brazilian sub-region, which includes not only Brazil, but likewise the Guianas, Venezuela, Colombia, Ecuador, Paraguay, and those portions of Peru and Bolivia lying on the Brazilian side of the Andes, together with the eastern slopes of that portion of the great mountain-chain itself. This is essentially an area of dense tropical forests, locally interspersed with open pastures, or "campos." The second is the Chilian sub-region, comprising Chili, Argentina proper, Uruguay, Patagonia, and such portions of Peru and Bolivia as are not included in the preceding. It is chiefly an area of open plains and pampas, although including the high Andes. Thirdly, there is the Mexican sub-region, which embraces the isthmus of Panamà, Central America, and Southern Mexico, and may be regarded to a great extent as a transitional tract between the typical Neo-

---

[1] *Quart. Journ. Geol. Soc.* Vol. LI. pp. 306—307 (1895).

tropical and the Sonoran regions. Lastly, the Antillean sub-region includes the West Indian Islands, exclusive of Trinidad, which for zoological purposes may be regarded as part and parcel of the South American continent.

From the survey of the fossil forms it has been shown that during the Plistocene epoch the mammalian fauna of the Chilian and Brazilian sub-regions was similar, and it may consequently be inferred that the present differentiation of the two areas in this respect is a comparatively modern feature, probably due to the disappearance of the forests from the Argentine. At the present time the mammalian fauna of the Brazilian sub-region is essentially that of the Neotropical region as a whole, nearly all the characteristic groups being present within its limits, while several are almost or quite peculiar to it. Among the latter, the great ant-eater (*Myrmecophaga*) is practically confined to this sub-region [1], while most of the sloths and marmosets are limited to it, although a few extend northwards into or through the isthmus. The pacas (*Cœlogenys*), and the giant armadillo (*Priodon*)—the sole representative of its genus—are likewise restricted to this tract ; as is the bush-dog (*Icticyon*), and also one genus of tree-porcupines (*Chætomys*), while most of the spiny rats (*Echinomys* and *Loncheres*) are confined to it. The carpincho (*Hydrochœrus*)—the largest of living rodents—likewise chiefly pertains to the Brazilian region, although extending southwards into Uruguay. The American monkeys are also very abundantly represented here ; the genera *Lagothrix*, *Pithecia*, *Brachyurus*, *Brachyteles* (*Eriodes*), and *Callithrix* being restricted to it. Among the forms that are unrepresented, may be mentioned guanacos and vicuñas (*Lama*), viscachas (*Lagostomus*), the Patagonian cavy (*Dolichotis*), and chinchillas (*Eriomys* and *Lagidium*).

The Mexican or Central American sub-region differs chiefly from the last by the paucity of the essentially Neotropical forms, and the large mingling of Arctogæic types ; among the latter, the shrews (*Soricidæ*), a pouched rat (*Heteromys*), and the caxomistle (*Bassariscus*) being noticeable. In the Chilian sub-region marmosets, monkeys, sloths, tapirs, and peccaries are wanting ;

[1] According to Señor Figueira it just enters Uruguay.

while the carpincho, as already mentioned, only borders on it in Uruguay. Among the characteristic types are prominent the vicuñas of the Andes, the guanacos of the Argentine pampas and Patagonia, the spectacled bear of the Andes, the chinchillas of the same elevated regions, together with the aquatic coypu (*Myopotamus*), the burrowing viscacha, and the cursorial Patagonian cavy; all the three latter being plain-dwelling forms. Armadillos are abundant; and among these the sub-family represented by the beautiful little pichiciagos, or fairy-armadillos (*Chlamydophorus*) is peculiar, one of the two species inhabiting open plains near Mendoza, in the Argentine, while the second (regarded by some as a distinct genus) is found in the Bolivian highlands.

The Antillean, or West Indian sub-region, which comprises the West Indian Islands (exclusive of Trinidad, Tobago, and some of the adjacent islets, which are zoologically a part of continental South America), differs widely from the other three by the extreme poverty of its mammalian fauna; monkeys, marmosets, carnivores, and edentates being wanting, and the class mainly represented by bats, insectivores, and rodents, although a species of aguti (*Dasyprocta antilliensis*) is found in the islands of St Vincent and Santa Lucia, in the Lesser Antilles group, as also in Tobago. In addition to a single species of white-footed mouse (*Sitomys*) said to inhabit Hayti and Martinique, and which may also occur in some of the other islands, the West Indian sub-region is especially characterised by the large arboreal rodents known as hutias, which, while belonging to the family *Octodontidæ*, represent two genera totally unknown elsewhere. Of these, the genus *Plagiodon* has but a single species, confined to Hayti and Jamaica; although in the allied *Capromys* three existing species are found in Cuba, and the fourth in Jamaica, the extinct kind occurring in the former island[1]. The nearest relative of the hutias appears to be the South American coypu, but the group seems also to show affinities with the porcupines. From caves in the small island of Anguilla, at the northern extremity of the Lesser Antilles group have been obtained remains of a large extinct beaver-like rodent known as *Amblyrhiza* (*Loxomylus*), which has also been recorded

---

[1] Chapman, *Bull. Amer. Mus.* Vol. IV. p. 314 (1892).

from the Pliocene of Argentina—a fact of importance as serving
to connect the Antillean fauna with that of the mainland. As
mentioned above, this genus belongs to a family (*Castoroididæ*)—
typified by the extinct *Castoroides* of the Plistocene of Ohio and
Georgia—with species which rivalled a bear in point of size. The
other Antillean native mammals (exclusive of the bats, to which it
will be unnecessary to refer) are the two species of the genus
Solenodon respectively inhabiting Cuba and Hayti, and constituting by themselves a separate family among the Insectivora. It
has been already mentioned that the nearest allies of these strange
creatures are the tenrecs (*Centetidæ*) of Madagascar; and thus both
are probably derived from unknown extinct insectivores formerly
inhabiting the northern hemisphere. As Jamaica and probably
several other of the West Indian islands contain large masses of
sedimentary deposits of Tertiary age, it is probable that they
come under the denomination of continental islands; and there
seems little doubt, from the evidence of their mammals alone, that
they have been connected with the mainland[1]. Dr Wallace is of
opinion that "originally they probably formed part of Central
America, and may have been united with Yucatan and Honduras
in one extensive tropical land. But their separation from the
continent took place at a remote period, and they have since been
broken up into numerous islands, which have probably undergone
much submergence in recent times. This has led to that poverty
of the higher forms of life, combined with the remarkable similarity
which now characterises them; while their fauna still preserves a
sufficient resemblance to that of Central America to indicate its
origin." Recently, the connection of the West Indies with the
mainland has been worked out more fully by Mr J. W. Spencer[2],
who, from observations made on the buried river channels so
numerous in some of the islands, concludes that there have been
several epochs of connection with the continent, one of which was
so late in date as the Plistocene epoch. The extinction in the
islands of the great majority of the mammals of the continent
is attributed to drowning.

[1] This is not the opinion of Dr A. Agassiz, who regards them as oceanic
islands.

[2] *Geological Magazine*, 1894, pp. 448—451.

Here a presumed connection between North and South America by way of the West Indies must be referred to, the evidence in favour of which has been summarised by Dr J. W. Gregory[1] as follows:—" It is not at all certain that when the isthmus of Panama was submerged there was a free communication between the Atlantic and Pacific Oceans. The Caribbean Sea may then have been a gulf from the Pacific, separated from the Atlantic by the land area of the hypothetical 'Antillia.' That there was once a connection between North and South America along the chain of the Windward Islands, Cuba, the Bahamas, and Florida is not improbable. Evidence for this, either in whole or in part, has been advanced by De Castro and others. Further evidence could be adduced from the study of the land-shells, and also from the remarkable distribution of *Peripatus*. That Cuba was once connected with Yucatan and Florida is almost certain ; that this connection was in existence in the Pliocene, and probably also in the Plistocene, is shown by the evidence collected by De Castro. That the area of the Windward Islands was occupied by land in the lower Tertiary is also most probable. But this was all submerged at the period when the Oceanic [Miocene] deposits of Barbados were laid down. There is no adequate evidence to show that at any time after this was there more land in this region than there is at present." With regard to the land-shells, Mr A. H. Cooke[2] writes "that a certain number of the characteristic North American genera are found in the Antillean sub-region, indicating a former connection, more or less intimate, between the West Indies and the mainland....A small amount of South American influence is perceptible throughout the Antilles, chiefly in the occurrence of a few species of *Bulimulus* and *Simpulopsis*. The South American element may have strayed into the sub-region by three distinct routes : (1) by way of Trinidad, Tobago, and the islands northward ; (2) by a north-westerly extension of Honduras towards Jamaica, forming a series of islands, of which the Rosalind and Pedro banks are perhaps the remains ; (3) by a similar approximation of the peninsula of Yucatan and the western extremity of Cuba." This seems to indicate that such Antillean

[1] *Quart. Journ. Geol. Soc.* Vol. l.i. p. 305 (1895).
[2] *Cambridge Natural History*—Mollusca, pp. 345, 346 (1895).

connection as may have existed between North and South America was of a very incomplete and transitory nature; and that before the end of the Miocene there was never any route in this direction by which mammals passed from the one continent to the other.

In this connection it may be mentioned that Dr Hart Merriam[1] has proposed to unite Central America with the West Indies to form a separate zoological region—the Tropical—of equal rank with the Sonoran; but however much may be urged in favour of this view, the multiplication of regions is much to be deprecated.

The practical or entire absence of non-volant native mammals, both recent and fossil, from all the other South American islands, with the exception of Tierra del Fuego and the Falklands, properly excludes their consideration from this volume, although it is almost essential that a few words should be said with regard to the Galapagos group, more especially since conflicting views have been expressed concerning their relations to the mainland. In respect to Tierra del Fuego and some of the adjacent islets, these may really be regarded as a part of the continent, since the main dividing channel is extremely narrow, and species like the guanaco are common both to the islands and the mainland. And although the Falkland Islands lie about 350 miles to the eastward of southern Patagonia, yet they are separated by a comparatively shallow sea (less than a hundred fathoms in depth), and it is thus evident that they were connected at no very distant date with the mainland. Of the two indigenous mammals, the most remarkable is the Falkland Island wolf (*Canis antarcticus*), which differs markedly from all the *Canidæ* of the mainland, and is apparently closely allied to the North American coyote (*C. latrans*). The other is a species of groove-toothed mouse (*Rhithrodon*). With regard to Fernando Noronha, the poverty of its fauna induces Mr Beddard[2] to class it among oceanic islands, although there is some affinity between its fauna and flora and those of South America and the West Indies.

Of far more interest are the Galapagos islands, situated on the equator, at a distance of about five hundred miles westward of the

[1] Appendix, No. 19, p. 33.
[2] *Ibid.* No. 5, pp. 190, 207.

coast of Ecuador. Entirely volcanic in structure, they are surrounded by a sea of great depth ; and according to the view both
of Wallace and Darwin, they have never been connected with the
mainland, while the latter observer is also of opinion that for
countless ages they have been separated from one another. The
known mammals include a bat (*Atalapha*), a rat (*Mus*), doubtless
introduced, and a peculiar species of white-footed mouse (*Sitomys
bauri*). Of reptiles the islands contain two peculiar genera of
iguanoid reptiles, and no less than five species of giant tortoises
belonging to the genus *Testudo*. The iguanoids are nearly related
to South American types ; but there are no tortoises now living
on the mainland, although a large species flourished in Argentina
during the Pampean epoch, while, as already stated, others are
now found living in the Mascarene islands, and extinct species
occur in the middle and later Tertiary deposits of the United
States, Europe, and Northern India. It may accordingly be
taken for granted that both the iguanoid lizards and the giant
tortoises reached the island from the South American mainland ;
but the question is how did they arrive ? Dr Wallace is of opinion
that both were transported across the sea, although by what means
is unknown. This view is, however, disputed by Dr G. Baur[1],
who believes that the Galapagos islands were formerly connected
not only with one another, but likewise with the mainland. He
observes that if the Galapagos be oceanic islands, their inhabitants
could only have been introduced by accident from other regions ;
"but on such a supposition we are absolutely unable to explain
the harmonious distribution, we cannot explain why every, or
nearly every, island has its peculiar race or species, not represented on any other island. If some animals could be carried
over hundreds of miles to the islands, why are they not carried
from one island to the other? But besides that, how could we
make plain the presence of such peculiar forms as the gigantic
land-tortoises? According to the elevation theory, we can only
think of an accidental importation of these tortoises by some
current, because they are unable to swim. After the islands had
been elevated out of the sea, it happened once, by a peculiar

---

[1] *Proceedings American Antiquarian Society*, Oct. 1891.

accident, that a land-tortoise was carried over. Alone it could not propagate. This was only possible after a similar accident imported another specimen of the same species, of the other sex, to the same island. Or we could imagine that at the same time animals of both sexes were thus accidentally introduced. By this we could at least explain the population of a single island. But how did all the other islands become populated? To explain this we should have to invoke a thousand accidents. The most simple explanation is given by the theory of subsidence. All the islands were formerly connected with one another, forming a single large island; subsidence kept on, and the single island was divided up into several islands. Every island developed, in the course of long ages, its peculiar races, because the conditions on these different islands were not absolutely identical."

Further evidence in favour of the same view is adduced by Mr W. B. Hemsley[1], who draws attention to the marked similarity of the flora of the Galapagos Islands to that of the South American mainland.

The difficulty of accounting for the transport of reptiles like land-tortoises across five hundred miles of sea is undoubtedly very great; yet, in the face of their volcanic nature and the depth of the surrounding ocean, it is somewhat difficult to accept the view that the Galapagos Islands were connected with South America. In the paragraph quoted Dr Baur appears to forget that the first tortoise carried to these islands may have been a gravid female; and also that if an ancestral species were established on one of the islands, there would be nothing very wonderful in individuals being carried to some of the others, where they might eventually differentiate into distinct specific types. Moreover, it seems quite within the bounds of possibility that the original introduction may have been effected by means of eggs transported on natural rafts. That the Galapagos tortoises were derived originally from equally gigantic continental forms, may be taken for granted; and the existence at the present day of such creatures only in these and the Mascarene islands is one more instance of the survival in the southern hemisphere of ancient types which were formerly abun-

[1] *Nature*, vol. lii. p. 623 (1895).

dant on the opposite side of the equator. The practical absence of mammals from the Galapagos Islands is of little import one way or the other, as they might have been drowned out during the subsidence ; but perhaps, on the whole, a suspension of judgment as to the relation of these islands to the mainland is the wisest course to adopt at present.

Finally, whether the hypotheses that have been advanced in the present chapter to explain the origin of the peculiar mammalian fauna of Neogæa be substantiated or the reverse, there can be little doubt that Dr Wallace has been misled in his statement that this area, so "far as we can judge from the remarkable characteristics of its fauna and the vast depths of the ocean east and west of it, has not during Tertiary, and probably not even during Secondary times, been united with any other continent, except through the intervention of North America."

CHAPTER IV.

THE ARCTOGÆIC REALM.

Features of the Arctogæic Fauna—Community of earliest Fauna—Evidence of Secondary Reptiles—Puerco Fauna—Lemuroids—Insectivora—Carnivores —Rodents—Ungulates—Summary of the characteristics of the Mammalian Fauna of Arctogæa.

ARCTOGÆA, or the Arctogæic realm, includes the whole of the countries of the globe which do not come within the limits of either the Neogæic or Notogæic realms, and thus embraces by far the greater portion of the land-surface. Nearly the whole of this vast tract lies to the northward of the equator ; the only portions lying below that line being the southern half of Africa, together with Madagascar and some of the Malayan islands. As stated in the introductory chapter, the term Arctogæa was originally proposed by Professor Huxley[1], but, although subsequently used in one case by Dr Sclater[2], and at a still later date adopted by Dr Blanford[3], has failed to obtain general recognition. There can, however, be no doubt that, so far as mammals at least are concerned, the whole of this vast tract is entitled to hold only the same relative rank as each of the realms treated of in the two preceding chapters; and that if we regard each of the regions into which the area under consideration is divided by Sclater and Wallace as equivalent to each of those two realms, we have an exceedingly unequal series of divisions. Not only have the Neogæic and Notogæic realms no species of mammal common to one another, but if we eliminate the genus *Canis*, which is of comparatively recent introduction into these areas, we shall find

[1] Appendix, No. 18.
[2] *Ibid.*, No. 27, p. 214.
[3] *Ibid.*, No. 8, pp. 76, 77.

that all the genera, and likewise most of the families, together
with some of the subordinal or even ordinal groups of mammals
are likewise perfectly different.   Were it not also for the comparatively recent union between South and North America, to which
allusion has been made in the preceding chapter, we should
likewise find just as well marked a distinction between the
mammalian fauna of these two countries; and, as a matter of
fact, when we go back to the middle portion of the Tertiary
epoch, we find such a distinction actually existing.   Again, were
it not for the intermediate connecting Austro-Malayan region,
which forms, as we have said, a kind of zoological No-man's-land,
there would be an equally stringent line of division between the
Notogæic realm and Asia.   If, on the other hand, we take the
different regions of Arctogæa, we find not only a certain number
of species of mammals common to two or more regions, but
when we pass back into the Tertiary epoch, the whole faunas of
several of such regions merge more or less completely into one
another, instead of becoming more distinct than they are at the
present day.   The lion and the leopard, for instance, are common
to India and Ethiopian Africa, and during the Plistocene epoch
ranged over a considerable portion of Europe; while the range of
the tiger includes not only India and Ceylon, but likewise a
considerable portion of Central Asia and China.   The caracal
and the hunting-leopard are also common to India and Africa;
the British fox ranges not only over Europe and a large portion of
Asia north of the Himalaya, but likewise over a part of North
America; and the common otter is found alike in India and
Europe.   Still more numerous are the species of mammals common to Europe, Northern Asia, and North America.

Recapitulating some of the details given in the introductory
chapter, it may be observed that by Messrs Sclater and Wallace
the area here included in the Arctogæic realm was divided into
the Nearctic, Palæarctic, and Oriental regions.   Professor Newton,
who was subsequently followed by Dr Heilprin, proposed to
brigade the first two of these together under the name of the
Holarctic region.   At a still later date Dr Blanford[1], who as

---

[1] Appendix, No. 8, p. 76.

already stated, takes Arctogæa as one of the three primary divisions of the globe, proposed to subdivide it as follows, *viz.*:—

1. *Madagascar.*
2. *Africa*, south of the tropic of Cancer.
3. *Oriental*, South-eastern Asia, and Malayan islands to Wallace's line.
4. *Aquilonian*, Europe, Asia north of the Himalaya, Africa north of the tropic of Cancer, and America north of about 45°.
5. *Medio-Columbian*, America, between about 25° and 45° north latitude.

The importance of this division was, firstly, the recognition of the right of Madagascar and the adjacent islands to form a region by themselves; and, secondly, the separation of the Medio-Columbian region from the rest of North America. And it will be noted that, if we take away that area, the Aquilonian region corresponds to the Holarctic of Newton and Heilprin. A further modification was proposed in 1892 by Dr C. H. Merriam[1], who gave the name of Sonoran region to the area corresponding approximately with the Medio-Columbian of Dr Blanford, and suggested that the southern portion of the Eastern Holarctic region should form a region by itself; the name of Boreal region being adopted for what remained of the Holarctic after the subtraction of the Sonoran region and a corresponding area in the Eastern Hemisphere.

Although from many points of view the retention of such well-known terms as Palæarctic and Nearctic would be a great convenience, the close resemblance of the existing mammalian fauna of the whole of northern Arctogæa compels us to adopt the view that the area forms but a single zoological region. For this region the name Holarctic may be retained; while for the southern portion of the old Nearctic region, the term Sonoran is the most appropriate. In the Eastern hemisphere the whole of that portion of Arctogæa not included in either the Malagasy, Ethiopian, or Oriental regions is provisionally included in the Holarctic, although when our knowledge of distribution is less

[1] Appendix, No. 19.

imperfect it may subsequently be found practicable to separate a distinct Mediterranean region.

In an area of such vast extent as the Arctogæic realm, which embraces countries from the equator to the most northern habitable lands, it is, of course, perfectly unnecessary to say anything as regards climate and physical features, and we may accordingly proceed forthwith to discuss the leading features of the mammalian fauna as a whole.

Features of the Arctogæic fauna.

From the Notogæic realm, in its typical form, Arctogæa is distinguished by the absence of monotremes and diprotodont marsupials, not only at the present epoch, but so far as we know, in past times also. From the Neogæic realm it is equally well differentiated by the absence at the present day of all the peculiar Neogæic types of edentates, and likewise at all epochs of Neotropical monkeys and marmosets. Such of the former as are found in North America are indeed, as we have seen in the last chapter, only intruders from the south since the epoch of the earlier Pliocene; and in the Miocene the Arctogæic fauna was further distinguished from that of Neogæa by the absence of the peculiar subordinal groups of ungulates characteristic of the latter. The Insectivora, which with the exception of the solenodons are practically wanting in Neogæa, and are unknown in Notogæa proper, are abundant in all the regions of this realm.

We might almost go one step further than this, and say that Arctogæa previous to the Pliocene epoch was characterised by being the sole habitat of almost all the families of Eutherian terrestrial mammals, with the exception of those characteristic of the Santa Crucian epoch of Neogæa. But although this would be practically true, it would land us in the difficulty that the Ethiopian region would probably have to be excluded from Arctogæa, seeing that the higher mammals of the former region are but comparatively recent immigrants. Still, it may be stated that northern Arctogæa is the original habitat of all the modern types of the higher Eutherian mammals.

Another feature of Arctogæa is the absence at the present day of all marsupials except opossums, while these are only sparingly represented in its western half. Moreover, with the exception of the same family group, which are only known from strata of the Oligocene and Miocene epochs, marsupials appear to have been absent from a large part of the realm during the Tertiary period, although there is reason to believe that during the Eocene they must have survived in south-eastern Asia[1]. Among volant mammals, bats of the Neogæic family *Phyllostomatidæ*[2] are now absent from the whole of the realm, with the exception of a part of the Pacific side of North America. Again, to revert to the non-volant forms, the Lemuroid suborder of the Primates seems to have been absolutely restricted to Arctogæa, at least since the Miocene epoch, although it may turn out that the monkeys and marmosets of South America may be descended from ancestral lemuroids which inhabited that country at an epoch previous to the deposition of the Santa Crucian beds.

To take a comprehensive survey of the whole Secondary and Tertiary mammalian faunas of Arctogæa would entail such a mass of palæontological and anatomical detail that it would only weary the majority of our readers, and we must accordingly limit ourselves to noticing some of the most striking features in the earlier faunas, and then pass on to the consideration of some of the more widely spread modern groups.

Community of earliest fauna.

In the chapter devoted to the Notogæic realm, it has been already pointed out (p. 51) that during the Jurassic period Europe and North America were populated with a fauna of polyprotodont marsupials of small size, some of which appear to have been the ancestral types from which those now inhabiting the Notogæic and Neogæic realms were derived, while others have disappeared entirely. It will be unnecessary to recapitulate the names of the more representative of these forms, but it may be stated that while the fauna of the lower Jurassic Stonesfield Slate has no equivalent in North America, that of the upper Jurassic Purbeck beds of Dor- setshire is paralleled in the latter area. Although some difference of opinion prevails among palæontologists as to the identity of the American with the European genera, there can be no doubt that

[1] *Vide suprà*, p. 55.

[2] The genus *Necromantis*, from the French Oligocene, has been assigned to this family.

many of them are very closely allied indeed, while some are probably inseparable. Contemporary with these early marsupials were members of the group known as Multituberculata, which are probably more or less closely related to the existing monotremes, or egg-laying mammals, and form with them the subclass Proto-theria. An essential feature of these multituberculates is that the molar teeth were divided by one or more grooves into longi-tudinal ridges, covered with numerous blunt tubercles; such grooves being very generally two in number in the upper molars, while the lower teeth have but a single one. Apparently in all cases the extremities of the jaws were armed with a pair of chisel-like incisor teeth, behind which in the upper jaw there may have been a pair of smaller teeth. Very generally the last premolar tooth, as in the English Purbeck genus *Plagiaulax*, was com-pressed and trenchant in shape, with its upper edge regularly

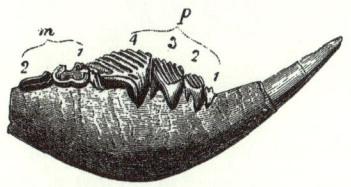

FIG. 28.   RIGHT SIDE OF LOWER JAW OF *Plagiaulax*.   Enlarged.
p. premolars,   *m*. molars.

convex, and its sides marked by oblique grooves; but in other forms (*Polymastodon*) this tooth was of a more tubercular type. Without going into disputed questions, it may be stated that this group was represented by closely allied forms in the Jurassic of both Europe and North America; while it is also known, from the evidence of a single genus (*Tritylodon*), to have extended its range to South Africa; this genus also occurring in the European Trias, and thus affording another instance of the wide range of the earlier faunas.

Although in Europe the only known traces of mammalian life during the succeeding Cretaceous period occur in the Wealden beds (which are the immediate successors of the Jurassic Pur-becks), in North America a well-developed fauna of polyprotodont marsupials and multituberculates is met with in rocks of Cretaceous age; and it is most probable that if suitable freshwater beds were

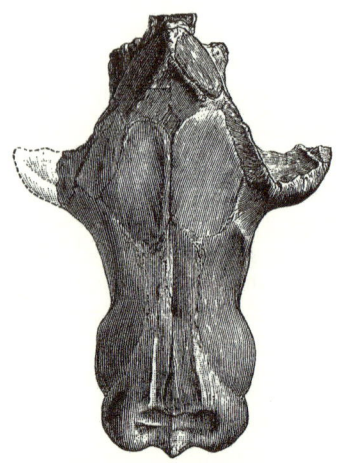

FIG. 29.   UPPER SURFACE OF SKULL OF *Tritylodon*.   Somewhat reduced.

extant, the same fauna would be found in Europe. By the commencement of the Tertiary epoch, most of this old fauna seems to have disappeared; but in the Puerco beds of the United States,

FIG. 30.   UPPER MOLAR OF A SMALLER SPECIES OF *Tritylodon*.
Natural size and enlarged.

and the equivalent deposits of Cernays, in the south of France, which seem to form a transition between the Secondary and Tertiary, the Multituberculata still persisted, and it is noteworthy that one genus (*Neoplagiaulax*) at least was common to the northern half of the eastern and western hemispheres.

It thus appears, so far as the available evidence permits of our forming a judgment, that during both the Jurassic and Cretaceous

epochs a single mammalian fauna was spread over Europe and
North America. This being so, it is a fair inference that a similar
fauna characterised a considerable portion of Asia ; while the
occurrence of the above-mentioned genus *Tritylodon* points to the
conclusion that it likewise ranged over Africa. Accordingly, it
would appear that not only did the whole of Arctogæa then form
a single zoological realm, but that this realm was indivisible into
regions.

The evidence for this unity is, however, by no means restricted
to mammals, but is supplemented and extended by
the extinct reptiles of the Secondary epoch of the
earth's history. During the Triassic and early
Jurassic periods there flourished an extinct ordinal group of reptiles known as the Anomodontia, remarkable for many structural
resemblances to mammals, and likewise for the peculiarities of
their dentition. As a well-known example of one section of this
group may be cited the dicynodonts, in which the teeth were
reduced, at most, to a single pair of tusks in the upper jaw, the
remainder of the jaws being ensheathed in horn to form a beak ;
whereas *Galesaurus* represents a second section in which the teeth
simulate those of the carnivorous mammals. These anomodonts
are known to have been spread over Europe, India, Africa, and
North America ; the dicynodont types from the three areas first
named being so alike that there is little question that some of
them were generically identical. The North American forms,
which mostly or exclusively come from beds assigned to the
Permian epoch, do not include dicynodonts, but are allied to
certain other African families, and are also closely related to their
European contemporaries.

If we turn to another order of the same class, namely the
Dinosauria, as represented by the *Iguanodon* of Europe and the
Atlantosaurus of the United States, we find not less well-
marked similarities in the Jurassic and Cretaceous fauna of the
whole of Arctogæa, this group being represented by closely allied,
and in many cases generically identical forms, not only in Europe,
India, and South Africa, but likewise in Madagascar and South
America. For instance, in that section of the order known as
the Sauropoda, which includes the most gigantic forms, and is

Evidence of Secondary Reptiles.

characterised by the presence of large chambers in the sides of the vertebræ of the neck and trunk, we find not only that several genera, such as *Morosaurus*, are common to the upper Jurassic and lower Cretaceous strata of Europe and the United States ; but we also find one genus (*Titanosaurus*) in India, Europe, and Patagonia, while a second (*Bothriospondylus*) occurs in countries as far apart as England and Madagascar[1]. Again, in the carnivorous or Theropodous section of the order, as typified by the English *Megalosaurus*, we find certain closely allied or identical generic types common to Europe and South Africa. Further evidence in the same direction is afforded by the discovery in the Jurassic of Madagascar of a genus of extinct crocodiles (*Steneosaurus*) which were abundantly represented during the same epoch in Europe. Among the class of fishes we have also the genus *Ceratodus*, now living in Queensland, represented in the Secondary rocks of Europe, India, Africa, and North America.

With regard to the land-fauna of Australia at the same epoch we have less evidence ; anomodonts, and, we believe, dinosaurs, being unknown from that country. Among the amphibians, however, we find in the extinct order of Labyrinthodontia certain genera, such as *Bothriceps* and *Micropholis*, common to Australia and South Africa, both of these being closely allied to the Indian *Brachyops*.

This reptilian evidence thus clearly points to the conclusion that during the greater part of the Secondary period not only had Arctogæa a single widely-spread fauna, but that the same fauna was represented in South America, and at least partially in Australia. Hence at this date no zoological realms can be distinguished, and it was probably not till late in Cretaceous times that Arctogæa was differentiated from the rest of the world as a realm. Needless to say, the great continents and islands during the epochs in question must have had free communication with one another, and it is highly probable, as Dr Blanford[2] suggests, that Madagascar then formed a line of connection between Africa and India. It is possible that even at the early part of the Secondary era,

[1] Possibly future discoveries may show differences worthy of generic distinction between these forms, but this would not affect the general question.

[2] Appendix, No. 8, pp. 88, et seq.

"when South Africa was united to India *viâ* Madagascar on one side, and to South America on the other, especially if the Indo-Malay continent was also connected with the Australian, there may have been a girdle of land, chiefly in low latitudes, round nearly three-quarters of the earth's circumference from Peru to New Zealand and the Fiji Islands." The vertebrate testimony does not, however, countenance the idea that such southern land was cut off from Europe and northern Asia by sea. It is true that the evidence in favour of such an isolation is afforded by the identity of the Carboniferous (Damuda-Talchir) floras of Australia, South Africa, Peninsular India, and Central Argentina[1], and their total dissimilarity from those of Europe, Northern Asia, and North America; and it is suggested that the same conditions may have prevailed during the Jurassic[2]. This, however, the vertebrate evidence certainly does not support; and hence, while admitting the isolation of a great southern (subtropical) continent during the Palæozoic era, it appears probable that since that epoch most of the southern lands have been from time to time more or less closely connected with those to the north[3].

Leaving these difficult problems with the foregoing remarks, we pass on to notice briefly the Puerco mammalian fauna of the United States, which, together with the approximately equivalent Cernaysian fauna of Europe, is of especial interest as showing a transition between the Cretaceous and Tertiary. As we have already said, this fauna includes several representatives of the multituberculates, which are essentially a Secondary group, and one of which is common to the Cernaysian fauna. In addition to these, four orders of eutherian mammals are represented, namely the Primates, the Carnivora, the Ungulata, and the extinct group known as the Tillodontia. It is, however, very noteworthy that in all these orders it is only the lowest sections that were in existence during the Puerco epoch. Thus

Puerco Fauna.

[1] See F. Kurtz, *Rev. Mus. La Plata*, Vol. VI. p. 117, and *Rec. Geol. Surv. India*, Vol. XXVIII. p. 111 (1895).

[2] Appendix, No. 8, p. 96.

[3] Dr Blanford writes to me that he believes the Palæozoic connection between South America and South Africa was tropical or subtropical, rather than antarctic, and hence that "the evidence for an antarctic continent in upper Mesozoic or Tertiary times is very slight indeed."

all the Primates belong to the lemuroid section, and include no monkeys or apes; and the carnivores are represented solely by the extinct creodont group, which differs from the existing members of the order by the simpler and more primitive structure of the limbs and teeth. The ungulates, again, belong exclusively to two extinct suborders, respectively termed the Condylarthra and the Amblypoda, both of which are very primitive types, with five-toed limbs of simple structure, the former still retaining evidences of affinity with the early carnivores. The tillodonts are quite unlike any existing forms, having a pair of incisor teeth similar to those of rodents in the front of the jaws, while their cheek-teeth recall those of the ungulates.

All the Puerco mammals are characterised by the lowness of the crowns of their molar teeth, which carry simple tubercles, generally arranged in a triangle; this type of tooth being known as the tritubercular, and occurring in all the orders found in the Puerco. It will be unnecessary to mention the names of the genera occurring in this horizon, and it will suffice to state that while peculiar to these beds, many of them belong to families characteristic of the overlying Tertiaries. Thus we have the *Anaptomorphidæ* among the lemuroids, the *Arctocyonidæ, Mesonychidæ, Proviverridæ,* and *Miacidæ* in the carnivores, and the *Phenacodontidæ* among the condylarthrous ungulates. The Cernaysian fauna is mostly represented by such fragmentary specimens that the determination of the affinities of its members is a matter of considerable difficulty; but the forms were all more or less nearly allied to those of the Puerco, and the creodont genus *Dissacus* is common to the two formations; while *Arctocyon,* which is met with in the Cernaysian, also occurs in higher horizons both in Europe and America.

A very remarkable fact connected with the Puerco fauna is that out of the 39 generic types by which it is represented, only eight are followed by analogous forms in the overlying Wahsatch beds, three of which became extinct in the still higher Bridger deposits. This leads Messrs Osborn and Earle to the conclusion that this early mammalian fauna was a kind of failure as regards development, and that only a few of its less specialised members persisted to give rise to the mammals of later periods.

With the Puerco and Cernaysian faunas we take leave of the
Secondary multituberculates, and as we ascend in
the Tertiary series we find a gradual and progressive   Lemuroids.
modification of the eutherian mammals towards the modern types.
In the ungulates especially the modification displays itself in the
more complex structure of the molar teeth, and in the reduction of
the number of toes ; the culmination of the latter line of development being reached in the modern horses among the perissodactyle
section of the order, and in the ruminants among the artiodactyles.
As regards the molar teeth, the chief features are a lengthening of
the crowns in the more specialised later forms, accompanied by
complex infoldings of the surface of the crown and sides, whereby
the short-crowned, or *brachydont* type, as exemplified in the tapirs,
has developed into the tall, or *hypsodont* type characteristic of the
horses.   Instead, however, of tracing the succession of the various
faunas, it will suit our present purpose better to refer to the
distribution of the more widely spread groups which are either
characteristic of Arctogæa as a whole, or which were common to
that realm together with Neogæa during the Plistocene period.

The lemuroids, which are at present unknown beyond the
limits of this realm, are first met with in the Puerco beds, where
they are represented by *Indrodon*, and it is probable that the
Cernaysian mammal described as *Plesiadapis* (of which the upper
cheek-teeth are shown in the annexed figure) belongs to the same

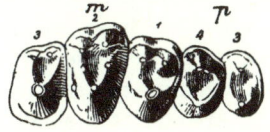

FIG. 31.   THE RIGHT UPPER CHEEK-TEETH OF *Plesiadapis*.
p. premolars, *m.* molars.

group.   It will be observed that in the latter genus the molars are
of the tritubercular type; the same being the case in *Anaptomorphus* of the lower or Wahsatch Eocene of America.   In other
forms, however, as in *Hyopsodus* and *Pelycodus* of the lower Eocene
of North America, and probably also of the European Eocene, as
well as in *Microchærus* of the Oligocene of France and England,
the upper molar teeth have squared crowns, and thus approximate

to those of modern lemurs. These early forms differ however from the latter in that the first of the three lower premolar teeth does not assume the form and functions of a canine. Another well-

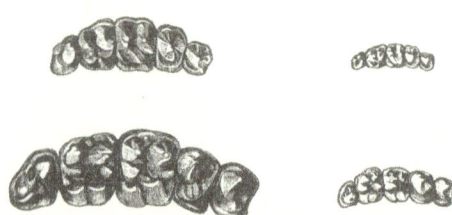

F IG. 32.   RIGHT UPPER CHEEK-TEETH OF TWO SPECIES OF THE LEMUROID
GENUS *Microchœrus*.   Nat. size and enlarged.

known European lemuroid is *Adapis*, of the European Oligocene, differing from all living forms in having four pairs of premolar teeth.

With the Oligocene, lemuroids seem to have disappeared from western Europe, and they apparently ceased to exist about the same date in North America, after which the entire order of the Primates is unrepresented in the latter country. At the present day, as we shall see, lemuroids are confined to the Malagasy, Ethiopian, and Oriental regions; but at what epoch the southern migration took place cannot yet be determined.

Omitting mention of the bats, we pass on to the Insectivora, among which we have the mole family (*Talpidæ*)
Insectivora.
distributed over the whole of the Holarctic as well as the Sonoran region, although all the genera but one are distinct on the two sides of the Atlantic; the single common type being the shrew-moles (*Urotrichus*), which have one species in Japan and another in the United States, thus affording an instance of the near affinity of the fauna of eastern Asia to that of North America. The earliest known fossil forms which have been assigned to the typical genus *Talpa* occur in the upper Oligocene strata of Europe, while in the middle Oligocene the family is represented by the allied *Amphidozotherium* (*Protalpa*). The shrews (*Soricidæ*), which likewise date from the Oligocene of the Continent, range over the whole realm, and also enter the Austro-Malayan region

as well as the Mexican subregion, although they are represented in Madagascar only by a single species of the widely spread genus *Crocidura*. Unknown in America, the latter genus, which includes the well-known musk-shrews, is widely spread over Europe, Asia, and Africa, extending as far east as Amurland; but the typical genus *Sorex* is practically confined to the Holarctic region[1], while other genera have a more local distribution.

With the exception of the civet tribe (*Viverridæ*) and hyænas (*Hyænidæ*), which are unknown in the New World, the majority of the existing families of the Carnivora

Carnivora.

have, if we except Notogæa, a cosmopolitan distribution, while in many cases this extensive distribution also holds good with regard to genera. In Europe the *Felidæ* and *Canidæ*, together with the *Mustelidæ* seem to have made their first appearance in the lower Oligocene, when they were accompanied by the extinct creodonts; while in America the two former are known from the John Day group, corresponding to the European Miocene. The bears (*Ursidæ*) are, however, a later group, being unknown before the Pliocene. Although the whole of the families mentioned above are represented in South America at the present day and during the Plistocene, they are, as we have seen in the last chapter, unknown in the presumably Miocene Tertiaries of Patagonia, and they are therefore originally of Arctogæic origin. Although most of the extinct American Tertiary genera of cats are distinct from those of Europe, the sabre-toothed tigers (*Machærodus*) were common to both areas, and likewise ranged into Notogæa; and the existing *Felis* has a similar cosmopolitan distribution. A more specialised sabre-toothed genus (*Eusmilus*) is likewise common to North America and Europe. As examples of extinct American cats, we may name *Nimravus* and *Archælurus*; while the Oligocene *Ælurictis* may be cited as an Old World form. The aforesaid distinction between the Oligocene and Miocene *Felidæ* of North America and Europe is, however, an indication that by this date the mammalian faunas of Western and Eastern Arctogæa had become differentiated to a certain extent, although, as now, there were many types common to the two areas.

---

[1] It has one species in the Sonoran.

Much the same story is told by the fossil dogs (*Canidæ*) of the two areas. In the Miocene of North America we meet with the genus *Temnocyon*, characterised by the cutting heel of the lower carnassial tooth; while the more civet-like *Cynodictis* appears to be confined to the Tertiaries of Europe. The bear-like genus *Amphicyon*, differing from modern dogs by its plantigrade feet, is confined to the European Oligocene and Miocene and lower Pliocene, but is represented in the Miocene of America by the nearly allied *Daphænus*. Through the intervention of the still larger *Dinocyon* of the European Miocene, the foregoing groups are intimately connected with the bears (*Ursidæ*) by means of the genus *Hyænarctus*, which is common to the Miocene and Pliocene of Europe and the Pliocene of India; and this suggests that the bears are originally an Old World group, which have subsequently migrated into America. As to whether true dogs (*Canis*) and cats (*Felis*) originated in America or Europe, we have no means of deciding[1]. The large weasel family (*Mustelidæ*) calls for no special mention here, although its comparative poverty in South America proclaims its Arctogæic origin.

The community between the mammalian faunas of Eastern and Western Arctogæa is perhaps better exemplified by the extinct creodont carnivores, since none of the families occurring there have been definitely recognised in the South American area. Differing from modern carnivores by the absence of a pair of differentiated carnassial teeth in each jaw, as well as by the scaphoid and lunar bones of the wrist generally remaining separate, and by the nearly flat upper surfaces of the astragalus in the ankle, these creodonts make their first appearance in the Puerco, and mostly died out in the Oligocene, although a few seem to have survived till the Miocene. In the typical family *Hyænodontidæ* we find the genus *Hyænodon* common to the Oligocene of both sides of the Atlantic, while the European *Pterodon* seems to have a transatlantic representative in the so-called *Hemipsalodon* of

---

[1] Scott, *Trans. Amer. Phil. Soc.* XVII. p. 75, concludes that the evolution of *Canis* took place in North America, the ancestry being traced through *Cynodesmus* of the John Day Beds to *Daphænus* of the White River, and thence to the creodont *Miacis* of the Bridger; *Cynodictis* forming a lateral branch.

Canada. *Oxyæna*, again, is found both in America and Europe, although in a lower horizon in the former than in the latter.   In a second family (*Proviverridæ*) the typical genus *Proviverra* is met

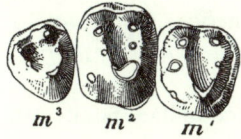

m^3 m^2 m^1

FIG. 33.   RIGHT UPPER MOLARS OF *Arctocyon*.

with in the Bridger Eocene of America, and the French Oligocene; while in the *Arctocyonidæ*, in which the upper molar teeth are bluntly tritubercular, *Arctocyon* of the lowest Eocene of Europe is represented by two allied genera in the American Puerco, one of which has been described as *Clænodon*.

In the rodents there are three more or less widely distributed existing families restricted to Arctogæa. Of these, the jerboas and their allies (*Dipodidæ*)[1] occur in all the regions with the exception of the Malagasy, Oriental, and Sonoran,—although the genera from the different areas are more or less markedly distinct.   The other two, namely the picas or tailless hares (*Lagomyidæ*) and the beavers (*Castoridæ*) are now severally represented by a single genus confined to the Holarctic region.   The picas date from the Oligocene of Europe, and the family not improbably originated in eastern Arctogæa; while the beavers have fossil representatives in both hemispheres, with the Miocene and Pliocene genus *Chalicomys* common to the two.

Although members of the typical genus range into the Neogæic realm as far south as Paraguay, the squirrel family (*Sciuridæ*) may be regarded as a typical Arctogæic one, the ground-squirrels (*Tamias*), marmots (*Arctomys*) and susliks (*Spermophilus*) being restricted to the Holarctic region, though others range over the

---

[1] It has been suggested by Dobson that the *Dipodidæ* are Hystricomorpha. This, however, is disproved by Dr Scott (*P. Ac. Philad.* 1895, pp. 269—286), who finds in the Uinta Oligocene genus *Protoptychus* an ancestral type of the family, which thus appears to be of N. American origin.   From this family are probably descended the *Geomyidæ*.

whole of the tropical and temperate parts of the realm with the exception of Madagascar. Remains of *Spermophilus* and *Sciurus* are met with in the later Tertiaries of Europe; and the extinct *Plesiarctomys*, which is common to the Oligocene and Miocene of Europe and North America, seems to be a connecting form between the squirrels and marmots, having upper molar teeth of the tritubercular type.

As regards the cosmopolitan family *Muridæ*, including the rats, voles, lemmings, etc., it will suffice to say that originally it was undoubtedly Arctogæic; the forms respectively inhabiting the Neogæic and Notogæic realms being comparatively recent immigrants. Both the subfamilies of the voles (*Microtinæ*) and the *Cricetinæ* are common to the entire Holarctic region; the latter being represented in the eastern half by the hamsters (*Cricetus*) and in the western by the white-footed mice (*Sitomys*), while they are the sole rodents inhabiting Madagascar, and have one species in Ethiopian Africa, where there is also the closely allied *Deomys*, forming a subfamily by itself. The cricetines are indeed evidently a primitive type, which in the Old World have been largely supplanted or driven south by the more specialised *Murinæ* (true rats and mice); but as these are represented in the Middle Tertiaries of both eastern and western Arctogæa, it is difficult to decide which was their original habitat. Little need be said in regard to the hares (*Leporidæ*), except that they range over the whole of Arctogæa, and have two outlying representatives in Neogæa, which are doubtless comparatively recent immigrants, although one is known to have inhabited Brazil since the Plistocene.

A not less marked feature of Arctogæa is the absence of most of the Neogæic rodent families noticed in the preceding chapter. The existence of members of one of these (*Octodontidæ*) in Africa is mentioned in the same place[1], where a reference to the occurrence of allied forms (*Theridomyidæ*) in the European Tertiaries will also be found[2].

One of the most important features in connection with the
Ungulates.   Arctogæic ungulates is the total absence of the
peculiar subordinal groups characteristic of the

---

[1] See also the chapter on the Ethiopian region.   [2] *Supra*, p. 86.

South American Tertiaries; although, as stated in the last chapter, there are indications of a distant affinity between these and some of the primitive early European perissodactyles. So far as our present knowledge goes, both the perissodactyle and artiodactyle suborders made their appearance in North America during the lower or Wahsatch Eocene ; the former group at least also dating from the same epoch in Europe, where the genus *Hyracotherium*, which is common to the Bridger and Wahsatch Eocene, and is one of the earliest ancestors of the horses, occurs in the London Clay.

Commencing with the Artiodactyla, or even-toed group—characterised by the toes corresponding to the third and fourth of the typical pentedactyle limb being symmetrical to a line drawn between them—and taking into consideration only such families as have a wide distribution, we have first to do with certain extinct types which serve to connect the pigs with the ruminants. The most pig-like of these are the animals forming the family *Chœropotamidæ*, characterised by their broad upper molar teeth carrying five blunt tubercles, three of which are on the front half of the crown. Although the typical genus *Chœropotamus* appears to have been confined to the lower Oligocene of Europe, the much larger animals known as *Elotherium* (fig. 35) were common to the upper Oligocene of both hemispheres. Nearly allied is the family of the *Anthracotheriidæ*, in which the low tubercles of the molars assume

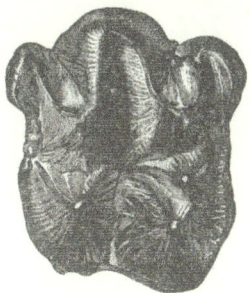

FIG. 34.   RIGHT UPPER MOLAR OF *Ancodus*.

a more or less crescentic, or selenodont structure, thus foreshadowing those of the ruminants. Here, again, the typical genus is restricted to the Old World, but the nearly allied *Ancodus* (*Hyo-*

L   11

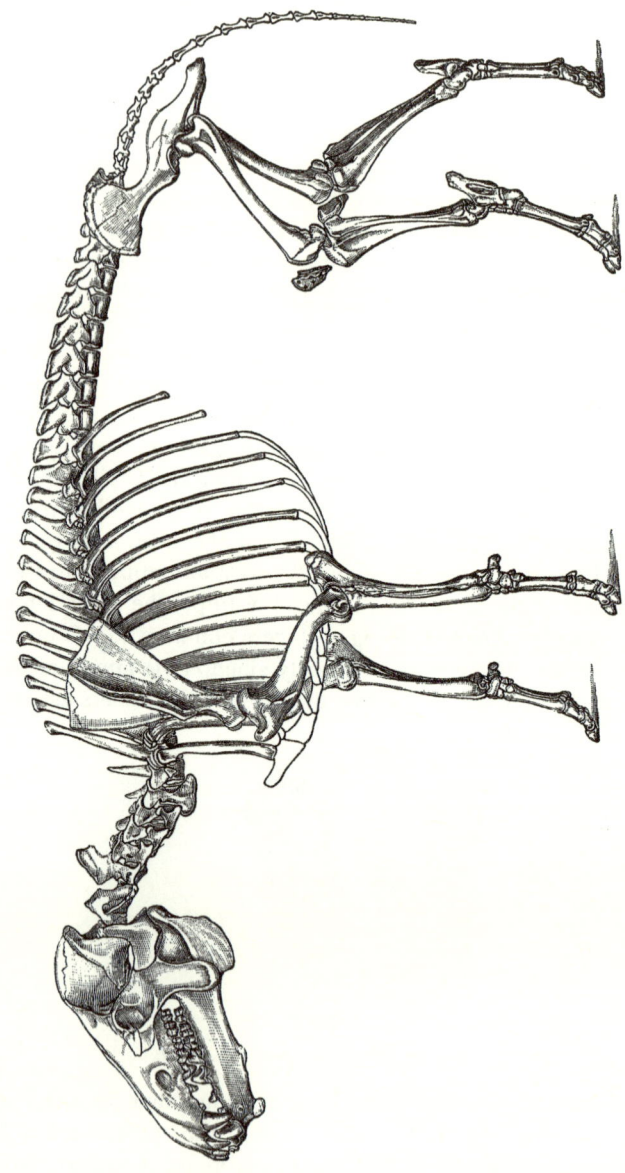

FIG. 35. SKELETON OF *Elotherium crassum*. $\frac{1}{16}$ nat size.

potamus) has the same approximate distribution as *Elotherium*, although it also occurs in the Miocene of northern India, together with *Anthracotherium* and a genus (*Tetraconodon*) closely allied to *Elotherium*.

A group of smaller Eocene and Oligocene mammals, constituting the family *Cænotheriidæ*, differ from the preceding in that their upper molars have two cusps on the front, and three on the hinder half of the crown. Here none of the genera are common to the two sides of the Atlantic, *Cænotherium* and *Dichobunus* being European, while the latter is represented in the Bridger Eocene by the closely allied *Homacodon*. It has been suggested that this group includes the ancestors of the camel tribe.

The latter group (*Camelidæ*), now represented only by the camels (*Camelus*) in the Old World, and the guanacos, vicuñas, and their domesticated allies the llamas in South America, was formerly widely distributed in Arctogæa, which was doubtless its original home, since, as we have seen in the preceding chapter, the llamas and their allies are but comparatively recent immigrants into Neogæa. *Camelus.* itself is represented in a fossil state in the Pliocene of northern India and the Plistocene of Algeria; but no other Old World representatives of the family are known. Very different, however, is the case with North America, where, from the Plistocene downwards, we meet with a host of extinct types, such as *Pliauchenia, Procamelus, Protolabis*, etc., gradually connecting the existing forms with a small animal from the middle Oligocene known as *Poëbrotherium*, which exhibits many very generalised characters. A still earlier representative of the family in *Leptotragulus*, of the lower or Uinta Oligocene, which itself may be sprung from the aforesaid *Homacodon* of the underlying Bridger beds. It is thus perfectly evident that the cameloids were originally a N. American group. One branch crossed the area now occupied by Bering Strait to found the camels of the Old World; while, probably at a later date, a second branch passed over the isthmus of Panama to persist in the guanacos, vicuñas, and llamas of South America. The disappearance of the whole group from the northern half of the New World is a very remarkable fact, but is paralleled by that of the elephants, rhinoceroses, lemurs, and several other groups.

The *Tragulidæ*, or chevrotains, which form a group distinct both from the cameloids and the true ruminants, are now confined to the Oriental and Ethiopian regions, being represented in the former area by the true chevrotains (*Tragulus*), and in the latter by the single existing species of water-chevrotain (*Dorcatherium*). Representatives of the latter genus occur, however, in the Miocene and Pliocene strata of Europe; while in the Oligocene we meet with the more generalised extinct genera *Prodremotherium* and *Bachitherium*. In North America the group is but poorly represented, and apparently confined to the White River Oligocene, where we find two types described under the names of *Leptomeryx* and *Hypertragulus*; the latter differing from the existing forms by having both the third and fourth metacarpals and the corresponding metatarsals separate, instead of being fused together to form a cannon-bone. It is difficult to decide whether the group was originally an Old or a New World one; but, on the whole, it is probable that it originated in the former area.

We now come to the true ruminants, or Pecora, forming the most specialised group of all the artiodactyle section of the ungulates, and characterised by the completely crescentic, or selenodont conformation of the columns of their molar teeth, which are frequently of great height; and likewise by the fusion of the third and fourth metacarpals and metatarsals into a cannon-bone, and by the imperfect development or disappearance of the lateral metacarpals and metatarsals. In the family of the *Cervidæ* the typical deer, or those included in the genus *Cervus*, are almost exclusively Arctogæic[1], being found in all the regions of the realm, except the Sonoran, Ethiopian, and Malagasy. The reindeer (*Rangifer*) and elk (*Alces*) are also solely Arctogæic forms, but have a more restricted range, being confined to the more northern portions of the Holarctic region. A more striking case is afforded by the hollow-horned ruminants, or *Bovidæ*, the whole of the numerous members of which are confined to the realm under consideration, with the sole exception of the anoa (*Bos depressicornis*) of Celebes; and even the latter, as we have seen in an earlier chapter, is very closely related to certain extinct Indian forms. It

[1] The only exception is *Cervus timoriensis*, which may have been introduced into its present habitat.

may be noticed, however, that *Bovidæ* are much more numerously represented in Eastern than in Western Arctogæa ; the latter area having only the genera *Bos*, *Ovibos*, *Ovis*, and *Haploceros*, each with a single species confined to the more northern portions of the continent ; the last genus being peculiar to this area.

We now pass to the perissodactyle, or odd-toed ungulates, which while agreeing with the artiodactyle section in the inter-locking arrangement of the bones of the wrist and ankle joints, and the pulley-like upper surface of the astragalus bone in the latter, differ by the third or middle toe and its supporting meta-carpal or metatarsal bone being symmetrical in itself, and larger than the lateral ones, when such are retained. In this suborder the family of the tapirs (*Tapiridæ*), although now mainly Neogæic, with one outlying Malayan species, was formerly widely spread in northern Arctogæa, fossil remains belonging to the single existing genus *Tapirus* being abundant in the Pliocene of Europe, although none appear to have been recorded from North America. In both Europe and America there occurs, however, the ancestral genus *Protapirus* ; its remains having been obtained in the former area from the upper Oligocene phosphorites of France, and in the latter from the nearly equivalent Uinta beds. Possibly a doubtful form (*Palæotapirus*) from the middle Eocene of France should also be included in the same family. The Uinta and Bridger deposits have also yielded a more or less nearly allied form known as *Isectolophus*, which apparently also occurs in the European Eocene, where it has been described as *Lophiodon*. Indeed, in our view, both this genus, and the still earlier American *Systemodon*, which appears to have been the earliest known representative of the tapiroid stock, may be included in the family *Lophiodontidæ*, where we should also place the ancestral types of the horses. In this family the genus *Lophiodon*, as now restricted, seems to have been confined to the Eocene of Europe, where it died out without giving rise to descendants, the same being the case with the allied Eocene genus *Helaletes*[1]. The well-known *Hyracotherium*, which was an animal of the size of a fox first described from the London Clay but subsequently recorded from the North

[1] Recorded from Europe by Osborn and Wortman, *Bull. Amer. Mus.* Vol. VII. p. 360 (1895).

American Eocene, is, however, the proximate ancestor of the
horse-family (*Equidæ*); and we have thus evidence that the forerunners of both the horses and tapirs were widely spread over the
whole of northern Arctogæa. *Hyracotherium*, it may be observed,
had the typical forty-four teeth characteristic of the earlier eutherians, and four toes to the front, and three to the hind feet; but in
the still earlier *Phenacodus*, which seems to be the ultimate
ancestor of the horses, each foot was furnished with five complete
toes.   As other instances of the community between the early
Tertiary mammalian fauna of the northern halves of the two
hemispheres, we may cite the genus *Pachynolophus* of the middle
and upper Eocene of Europe and the Bridger Eocene of the United
States, which connects *Hyracotherium* with the under-mentioned
horse-like animals; and also *Hyrachyus*, typically from the
Bridger, but probably occurring in the French phosphorites; the
latter being more nearly related to the rhinoceroses.

FIG. 36.   LEFT UPPER CHEEK-TEETH OF *Palæotherium*.

Passing by several less important genera confined to one or
the other hemispheres, we come to the family *Palæotheriidæ*,
which is an ill-defined one including forms connecting the preceding with the undoubted *Equidæ*.   Here the typical Oligocene
genus *Palæotherium*, which includes large tapir-like animals with
three toes to each foot, is exclusively European.   The same is the
case with the contemporary *Anchilophus*, represented by smaller
forms with more decidedly horse-like affinities; but with the
Miocene and upper Oligocene *Anchitherium*, used in its wider
sense, we once more come to a genus common to Western and
Eastern Arctogæa.   In this genus the jaws are provided with the
typical forty-four teeth; but the last lower molar has generally lost
the third lobe found in the preceding forms; the fifth metacarpal
bone being still represented by a splint.   In the small *A*.

(*Mesohippus*) *bairdi* of the White River Oligocene of the United States, the incisor teeth lack the infoldings of the summit of the crown characteristic of the horses, and the lateral digits are relatively large; the whole size of the creature being comparable

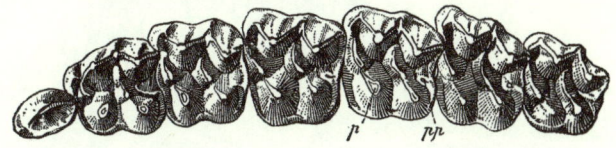

FIG. 37.   LEFT UPPER CHEEK-TEETH OF *Anchitherium*.

to that of a Newfoundland dog.   On the other hand, in the typical *A. aurelianense*, from the European Miocene, the summits of the incisors were infolded, as in the horses.   In spite of this resemblance, Professor Scott, from the structure of the limbs, is of opinion that the latter species was not an ancestor of the modern horses, but that this position was occupied by *A. bairdi*.

Restricting the term *Equidæ* to those members of the suborder in which the crowns of the cheek-teeth are very tall (*hypsodont*), with complicated infoldings of their enamel, and the hollows thus formed completely filled with cement, we have in the lower Pliocene of North America the three-toed genus *Protohippus*,

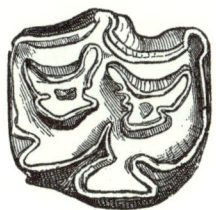

FIG. 38.   RIGHT UPPER MOLAR TOOTH OF A HORSE (*Equus*).

distinguished from the modern horses by the shorter crowns of the cheek-teeth.   The widely-spread genus *Hipparion* differs in having the anterior inner column of the upper molars completely detached[1]; the feet being generally three-toed, although in one Indian species the lateral digits are wanting. These three-toed horses are met with in the Pliocene of Europe, Asia, and North

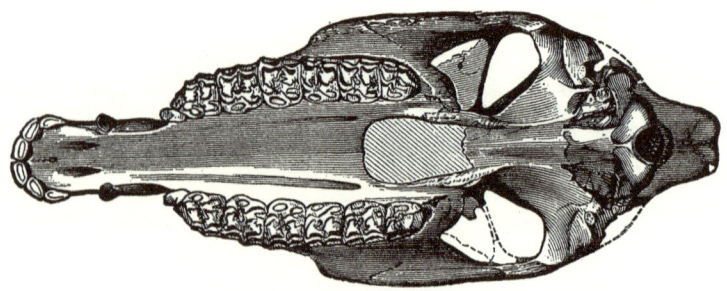

FIG. 39. UNDER SURFACE OF SKULL OF *Hipparion*.

America; and it is suggested by Professor Cope that while in the Old World the intermediate stage between *Anchitherium* and the modern horses was occupied by this genus, in the New World this gap was filled by *Protohippus*. The true horses (*Equus*), characterised by the one-toed feet and the union of the anterior inner column of the upper cheek-teeth with the adjacent middle column, although now confined to the Old World, where they date from the Pliocene, were formerly abundant in North America during the Plistocene, and, as we have seen, were likewise represented during the same epoch in South America. The oldest forms appear to be those from the Siwalik Hills of northern India; and it is thus evident that the group was originally an Arctogæic one[2]. The genus is now represented in all the regions of eastern Arctogæa, with the exception of Madagascar, and its extinction in the New World cannot be satisfactorily explained.

[1] This pillar forms the lowest part of the unshaded area in figure 38.

[2] Professor Scott, to whose views we have already alluded, in a paper published in *Tr. Amer. Phil. Soc.* Vol. XVII. pp. 111—112 (1894), is of opinion that the genus *Equus* was evolved in North America, and that *Anchitherium*, in its restricted sense, was off the direct line. He arranges the direct ancestral forms of the upper Tertiary, from above downwards, in the order *Protohippus, Desmatippus, Miohippus,* and *Mesohippus; Anchitherium* branching off from *Miohippus,* and *Hipparion* from *Protohippus.*

The rhinoceroses (*Rhinocerotidæ*) originally had a distribution very similar to that of the horses, with the exception that they never entered Neogæa; and they also agree with the latter in their present extinction in North America, where they are unknown after the Pliocene.   On both sides of the Atlantic the true rhinoceroses appear to have commenced in the lower Oligocene; and in both areas the earliest forms were hornless.   Early species with a pair of horns placed transversely on the nose are likewise met with both in Europe and North America; but the modern two-horned rhinoceroses appear to be restricted to the Old World, where one extinct species (*Rhinoceros antiquitatis*) ranged as far north as the Arctic circle during the Plistocene period.   On the whole, it seems preferable to include all the living and most of the extinct species in the typical genus *Rhinoceros*; the existing forms

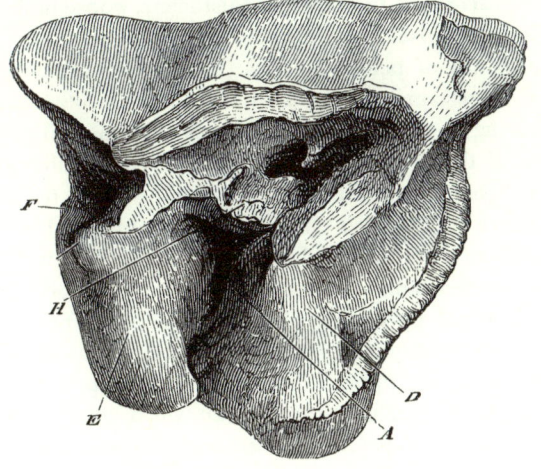

FIG. 40.   SECOND RIGHT UPPER MOLAR OF *Rhinoceros*.
A. median valley; *D*. anterior, and *E*. posterior crests; *F*. posterior valley;
H. crochet.

being confined to the Oriental and Ethiopian regions.   Rhinoceroses, it may be observed, differ from all the preceding families of the suborder by the upper cheek-teeth having a continuous outer wall, instead of being divided by a vertical ridge into two distinct lobes; and while all the living forms have but three toes to each foot, in some of the extinct hornless species the front limb was four-toed. In the typical genus the number of the front teeth is more or less reduced, but in the extinct *Hyracodon* and *Amynodon* of the upper Eocene of North America the full forty-four teeth were developed; and as allied forms also occur in the Oligocene, it would seem highly probable that the group originated in North America, whence it migrated westwards by way of what is now Bering Strait to attain its most specialised development in the Old World. It is, however, noteworthy that the genus *Cadurcotherium*, from the French Oligocene, which should apparently find a place in this family, and is distinguished by the narrowness of its upper molars, makes a curious approximation in the structure of these teeth to the Neogæic *Homalodontotherium*[1]. The most specialised representative of the family is the huge *Elasmotherium*, of the Siberian Plistocene, whose molars shew a resemblance to those of the *Equidæ*.

Another family of perissodactyles (*Titanotheriidæ*) is typically represented by certain huge, somewhat rhinoceros-like, mammals, generally having a pair of transversely-placed tuberosities on the nasal region of the skull, and characterised by a peculiar arrangement of the tubercles on their upper molars. These teeth, which have very short crowns, also differ from those of the rhinoceroses in having the outer wall divided into two lobes by a vertical ridge; while the last lower molar is distinguished from the corresponding tooth of the latter by having a third lobe. The typical genus *Titanotherium* is mainly North American, where it ranges from the lower Oligocene to the upper Eocene, but is also represented in the Tertiaries of the Balkans, although unknown in those of western Europe. An allied genus (*Brachydiastematotherium*) has also been recorded from eastern Europe, but all the other members of the family, such as *Palæosyops*, are North American. This family accordingly appears to have been mainly an American one, but was probably represented in Asia as well as in eastern Europe.

The remarkable genus *Chalicotherium*, whose geological range in the Old World extends from the Oligocene of France to the

---

[1] *Supra*, p. 82.

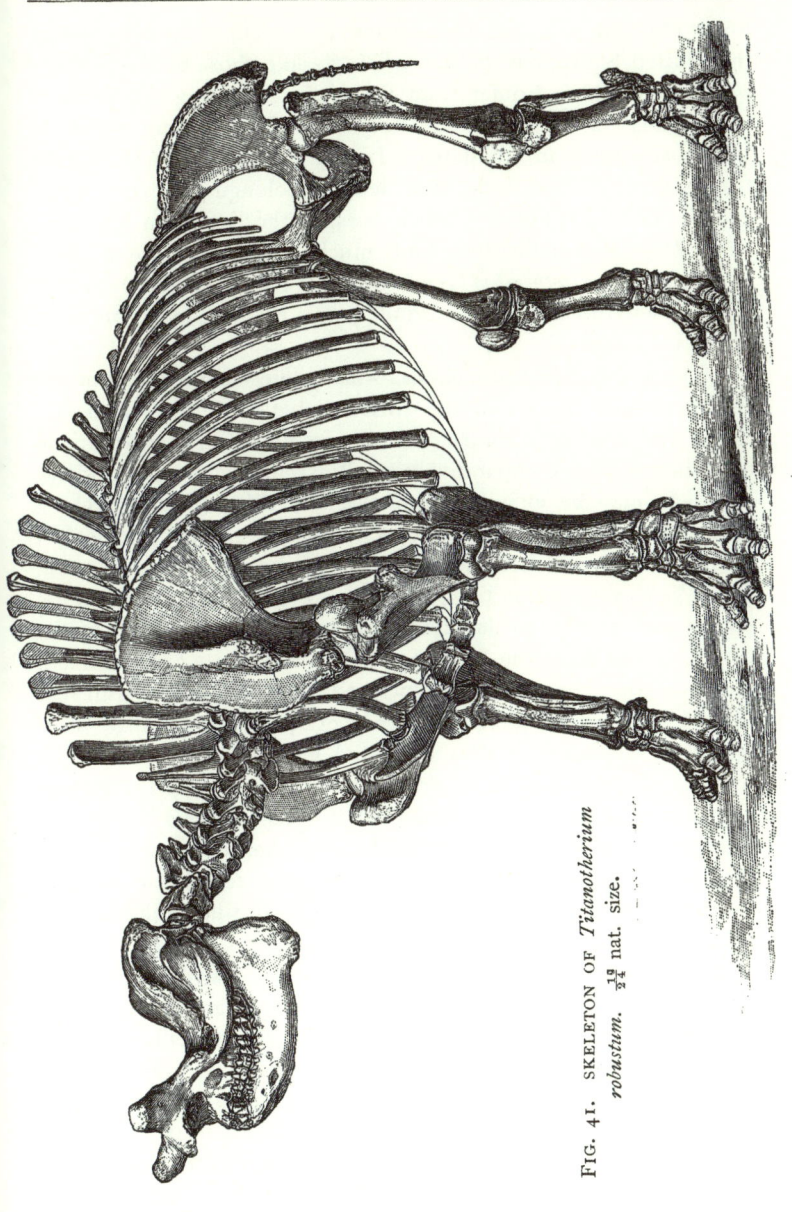

FIG. 41. SKELETON OF *Titanotherium robustum.* $\frac{19}{24}$ nat. size.

lower Pliocene of India, while species also occur in the Miocene of the United States, has molars strikingly like those of the *Titanotheriidæ*, but its feet differ from those of all existing ungulates in terminating in huge curved claws much resembling those of the South American edentates. Indeed, one genus of the family (*Macrotherium*) was long regarded as belonging to the latter group. Of the two genera which occur in the European Miocene, *Macrotherium* has the fore limb much longer than the hinder ; whereas in *Chalicotherium* (*Ancylotherium*) the two are more nearly equal in length. It is to the former genus that the North American forms are assigned.

The proboscidean ungulates, which differ markedly from the two preceding groups in the structure both of their teeth and limbs, and are now represented only by the Indian and African elephant, form a comparatively small assemblage, most of whose members may be included in the family *Elephantidæ*. Among these, the most specialised types constitute the genus *Elephas*, characterised by the complexity of the structure of the cheek-teeth, which generally assume the form of more or less elevated parallel plates, with the intervals filled with cement. In certain of the earlier species from the Pliocene of Asia the plates of these teeth are, however, comparatively low and less numerous, with the intervening valleys almost devoid of cement; so that these stegodont elephants, as they are called, form a complete transition towards the mastodons. Commencing in the Indian Pliocene, elephants ranged over the whole of Europe and Asia during the Plistocene, while we have also evidence of their occurrence during the same epoch in northern Africa; and they were likewise represented by two species in North America, one of which ranged as far south as Texas. One of these American species was identical with the European mammoth (*E. primigenius*), while the second (*E. columbianus*) was nearly allied; both being near relatives of the existing Indian elephant. The extinct stegodont elephants being confined to south-eastern Asia, it is interesting to note that the species of *Mastodon* making the nearest approximation to *Elephas* are met with in this region alone; and from this it may be inferred that the evolution of the latter genus took place in that part of the world. All mastodons, it may be mentioned, have comparatively simple molar teeth, in which the crowns are surmounted by low transverse ridges, frequently separated into more or less distinct tubercles, and with the intervening valleys open, such ridges varying from three to five in number in the majority of the teeth,

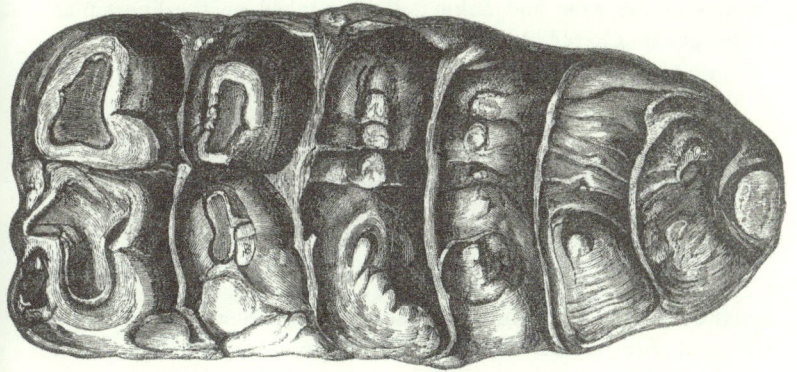

FIG. 42.   LAST UPPER MOLAR TOOTH OF A MASTODON.   ($\frac{1}{2}$ nat. size.)

although more numerous in the last of the series. Both in Europe and North America mastodons make their appearance in the Miocene, although in the latter area they are unknown before the Deep River beds forming the upper portion of that stage; and whereas they disappeared in the Old World with the Pliocene, they persisted in the New till the succeeding Plistocene age. During the Pliocene, as we have seen in the last chapter, they obtained an entrance into South America, so that they cannot be regarded as absolutely characteristic of Arctogæa.   On the whole, it is probable that the group originated in the Old World, although we are still in the dark as to its relationship to other ungulates.

The only other Arctogæic genus of proboscideans is *Dinotherium*, which constitutes a family by itself, and is known from the Miocene and Pliocene of Europe and India, but is unrepresented in America.   In these animals only one of the true molars has three ridges, the others having but two; and tusks were present in the lower jaw alone.

A fourth subordinal group of the ungulates, which is more primitive than any of the foregoing, and has been designated by Professor Cope the Amblypoda, or short-footed group, has one family (*Coryphodontidæ*) common to the lower Eocene of both hemispheres, while a second family (*Uintatheriidæ*), of later age is strictly North American. All these animals had limbs of an elephantine type, each foot being furnished with five toes, and the bones of the wrist and ankle joints arranged on the linear plan. The genus *Coryphodon*, which includes species varying in size from a tapir to a rhinoceros, had forty-four teeth, with the canines

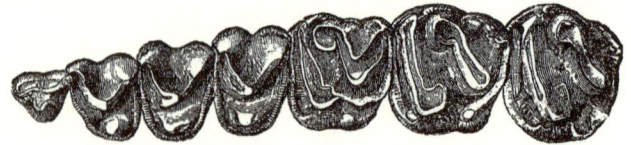

FIG. 43.   LEFT UPPER CHEEK-TEETH OF *Coryphodon*; reduced.

well developed, and the molars bearing prominent oblique ridges. First discovered in the London Clay, and subsequently in the lower Eocene of France, this genus has since been recognised in the lower Eocene of the United States, where nearly perfect skeletons have been obtained.

In a still more primitive group known as the Condylarthra, which apparently contains the ancestral stock of both the artio-dactyles and perissodactyles, it is believed that the genus *Phena-codus* (the ultimate parent of the horses), typically occurring in the Wahsatch Eocene of the United States, is represented in the Swiss Eocene; and the same has been stated to be the case with *Protogonia*.

Summarising the results of the foregoing survey, it may be stated that as regards its mammalian fauna Arctogæa as a whole is characterised by the following features. Absence of monotremes and diprotodont marsu-pials. No existing polyprotodont marsupials except opossums, and these only in its western half. No Tertiary marsupials known except opossums, although other types probably existed in South-eastern Asia. No existing edentates[1], and fossil ones only in North American Plistocene and Pliocene. No marmosets (*Hapalidæ*), and no monkeys of the family *Cebidæ*. Bats of the family *Phyllostomatidæ* wanting, save for a few on the

Summary of the character-istics of the mammalian fauna of Arcto-gæa.

[1] The aard-varks and pangolins are here excluded from the Edentata.

Pacific side of North America. Insectivores abundant. In the following list of more or less widely distributed families the majority are for the most part confined to this realm, while the others contain exclusively Arctogæic genera. Such groups as are practically confined to Arctogæa are printed in italics, those which are extinct having an * prefixed; while the letter H. denotes such of the existing ones as are restricted to the Holarctic region :—

INSECTIVORA.

> *Talpidæ.* H.
> *Soricidæ.* Enters Austro-Malayan region and Mexican sub-region.

CARNIVORA.

> * *Hyænodontidæ.*
> * *Proviverridæ.*
> * *Arctocyonidæ.*

RODENTIA.

> Sciuridæ. Mainly Arctogæic, although species of Sciurus extend in Neogæa as far south as Paraguay.
> *Dipodidæ.*
> *Castoridæ.* H.
> * *Theridomyidæ.* H.
> *Lagomyidæ.* H.

UNGULATA.

> * *Chœropotamidæ.*
> * *Anthracotheriidæ.*
> * *Cænotheriidæ.*
> *Tragulidæ.* Extinct in N. America.
> Cervidæ. The genus *Cervus* exclusively Arctogæic.
> *Bovidæ.* Represented by Bos depressicornis in Celebes.
> * *Lophiodontidæ.*
> * *Palæotheriidæ.*
> *Rhinocerotidæ.* Extinct in W. Arctogæa.
> * *Chalicotheriidæ.*
> * *Titanotheriidæ.* Mainly W. Arctogæic but occurring in the Balkans.
> Elephantidæ. *Elephas.*
> * *Coryphodontidæ.*

† *Multituberculata.*

    * *Plagiaulacidæ.*

    * *Bolodontidæ.*   Only Secondary in Eastern Arctogæa.

The following table exhibits some of the better known Tertiary mammalian genera common to both halves of Arctogæa, together with allied types restricted to the western and eastern hemispheres; such as are still existing being indicated by a † :—

W. Hemisphere.	Both Hemispheres.	E. Hemisphere.
Lemuroidea.		
		Adapis.
Anaptomorphus.	Hyopsodus.	Microchœrus.
Carnivora.		
	Machærodus	
	(Also Neogæa in Plistocene).	
Nimravus.		Ælurictis.
Archælurus.	Eusmilus.	
Temnocyon.		
Hyænocyon.		
Daphænus.		Amphicyon.
		Dinocyon.
		Hyænarctus.
	Hyænodon.	
	Pterodon.	
	Oxyæna.	
	Proviverra.	
Clænodon.		Arctocyon.
Rodentia.		
	Chalicomys.	
	Plesiarctomys.	
Ungulata.		
Achænodon.	Elotherium.	Chœropotamus.
	Ancodus.	Anthracotherium.
Homacodon.		Dichobunus.
		Cænotherium.
Procamelus, etc.		† Camelus.
		† Dorcatherium.

UNGULATA (*continued*).

Leptomeryx.		Prodremotherium.
Hypertragulus.		
	Protapirus.	
Colodon.	Helaletes.	Lophiodon.
Systemodon.	Hyracotherium.	
	Pachynolophus.	
	Hyrachyus.	
		Palæotherium.
	Anchitherium.	Anchilophus.
Protohippus.		
	Hipparion.	
	† Equus (also Neogæic).	
	† Rhinoceros.	
Hyracodon.		
Amynodon.		Cadurcotherium.
		Elasmotherium.
	Titanotherium.	
Palæosyops.		Brachydiastematotherium.
	Chalicotherium.	Macrotherium.
	† Elephas.	
	Mastodon (also Neogæic).	
		Dinotherium.
	Coryphodon.	

Phenacodus ⎰ Typically N. American,
Protogonia ⎱ but stated also to occur in Europe.

In the foregoing table, only some of the better known types have been selected, but these suffice to show that throughout the Tertiary epoch a certain number of genera were common to western and eastern Arctogæa. It is true that, with the exception of those still existing, we have no evidence that any of these ever reached the Ethiopian region; and it is quite probable that many or the whole of them never did so. By its present fauna that region is, however, so closely connected with the Pliocene and modern mammals of Asia and Europe, that there can be no question of its right to inclusion in the same realm; and this being so,

Madagascar cannot be excluded. Still, it is quite probable that during the later Tertiary epoch the Ethiopian and Malagasy regions were almost as distinct from the rest of Arctogæa as are Neogæa and Notogæa at the present day, and if such conditions had continued, the former areas would have been entitled to constitute a realm by themselves.

In the sequel we shall discuss the whole number of existing genera of non-volant terrestrial mammals common to the eastern and western halves of the Holarctic region, and likewise such living and extinct types as are respectively peculiar to Eastern and Western Arctogæa. These, taken in conjunction with the foregoing tables, will show that, in spite of the forms common to the two latter areas, there has always been a large number of types restricted to one side or the other of the Atlantic basin. And this leads to the conclusion that, although during a considerable portion, or the whole of the Tertiary period, there was a free land-communication between North America and Eastern Asia by way of Bering Strait, yet that this connecting land must have been comparatively narrow, so that the faunas of the more southern portions of both areas developed to a great extent independently of one another.

Not the least curious feature in connection with the community of types on the two sides of the Atlantic is the precise parallelism in the development of many of the groups in both areas. In both, for instance, the *phyla* of the horses and rhinoceroses were practically similar, although it is thought that the stage occupied in the one area by *Hipparion* was held in the other by *Protohippus*. If this particular suggestion should prove to be well founded, it will be self-evident that the true horses have been independently evolved in the two areas; and it almost seems as if the same had been the case with the rhinoceroses and certain other groups. Had the culminating forms been devolved in only one of the two areas, we should not expect to find the whole of the ancestral links present in both.

CHAPTER V.

EASTERN ARCTOGÆA.

Mammalian groups peculiar to Eastern Arctogæa—Tertiary Mammalian Faunas of Eastern Arctogæa—Oligocene Fauna—Miocene Fauna—Older Pliocene Fauna—Pikermi and allied Faunas—Siwalik Fauna—Higher Pliocene Faunas.

ALTHOUGH northern Europe and Asia forms but one zoological region with the corresponding part of North America, yet there are numerous groups of mammals confined respectively to its eastern and western portions, which clearly show that the communication between the two areas was always more or less limited. In this chapter attention will be first directed to some of the most striking peculiarities of the mammalian fauna of Eastern Arctogæa, after which the whole fossil fauna may be taken into consideration.

Mammalian groups peculiar to Eastern Arctogæa.

In addition to the total absence of existing opossums (*Didelphyidæ*), and the presence in its warmer portions of fruit-bats (*Pteropodidæ*), which, however, are common to Notogæa, Eastern Arctogæa is especially characterised as being the home of all the higher Primates; namely the family *Simiidæ*, which includes the man-like apes and gibbons, and the *Cercopithecidæ*, embracing all the other Old World monkeys. From the South American monkeys (*Cebidæ*) both these families are broadly distinguished by having two pairs of premolar and three of molar teeth, whereas in the former group there are three pairs of both premolar and molar teeth. Not only are the two families in question confined at the present day to the Eastern hemisphere, but the same appears to have been the case at all epochs, since no trace of a fossil monkey has ever been recorded from North America. This remarkable isolation of the distributional areas of the *Simiidæ* and *Cercopithecidæ* on the one hand, and of the *Cebidæ* (and *Hapalidæ*) on the other, points unmistakeably, in spite of their external similarity, to the dual origin of the monkeys of the Old and New Worlds. Both may, however, have originated from different groups of lemuroids, which, as indicated in an earlier chapter, ranged over the whole of Northern Arctogæa during the earlier part of the Tertiary epoch.

At the present day the man-like apes, which are few in number, have an extremely limited distribution. The chimpanzees (*Anthropopithecus*) are restricted to Equatorial Africa, the gorilla (*Gorilla*) is found only in the hottest regions of Western Africa, the orangs (*Simia*) are confined to the islands of Borneo and Sumatra, and the smaller gibbons (*Hylobates*) are inhabitants of South-eastern Asia, from Assam and Burma to Hainan. Extinct species of chimpanzees and orangs occur, however, in the Pliocene of Northern India; while gibbons are met with in the Miocene of France and Baden, although there is some difference of opinion whether these are generically identical with the Asiatic forms, or should be assigned to a genus apart (*Pliopithecus*). The former deposits have also yielded remains of a large extinct ape (*Dryopithecus*), apparently of a somewhat more generalised type than all the existing forms.

The ordinary monkeys and apes (*Cercopithecidæ*) have a wider distribution, ranging over most of the warmer parts of Eastern Arctogæa, and being represented by a single species at Gibraltar, and by two others in Moupin, in Eastern Tibet, and a fourth in Japan. This family, by the way, is not exclusively Arctogæic, since, as we have seen, one species of a peculiar genus (*Cynopithecus*) inhabits Celebes. Apart from the latter, the family is represented by eight living genera, among which the Ethiopian *Papio* has extinct representatives in the Pliocene and Plistocene of India, as have also the Asiatic *Macacus* and *Semnopithecus*. The two latter genera also occur in the Pliocene of France and Italy; and a tooth of a species of *Macacus* has been obtained from the Plistocene brick-earths of Essex. In addition to these, there are certain extinct genera from the European Tertiaries; among which *Mesopithecus* from the lower Pliocene of Greece agrees with *Macacus* in its short and stout limbs, but approximates to *Semnopithecus* in the character of its skull and dentition. *Dolicho-*

pithecus, from the French Pliocene, has a longer muzzle; while *Oreopithecus*, from the Italian Miocene, seems to connect the *Cercopithecidæ* with the *Simiidæ*.

It is thus evident that during the latter portion of the Tertiary epoch monkeys and apes were spread over the greater part of Eastern Arctogæa; and their extensive diffusion is a proof that this half of the realm could only have been connected with North America by land situated so far north that it formed an impassable barrier to these animals. Although the smaller extinct European monkeys do not necessarily indicate a very high temperature in the regions they inhabited, there can be little doubt that at the era when *Dryopithecus* flourished southern Europe at least enjoyed a moist tropical climate.

Not less characteristic of Eastern Arctogæa are the existing lemuroids, of which there are three families; the largest (*Lemuridæ*) ranging over the Oriental, Ethiopian, and Malagasy regions, the *Tarsiidæ*, with a single genus, being exclusively Oriental, while the sole representative of the *Chiromyidæ* is Malagasy. The numerous existing members of the *Lemuridæ* are all characterised by the first of the three lower premolar teeth assuming the form and functions of a canine; and as this feature is unknown in any of the Tertiary representatives of the sub-order, this family appears to be exclusively confined to the area under consideration. For the most part, the Oligocene lemuroids of Europe seem likewise to have been markedly distinct from those of North America. *Microchœrus*[1], for instance, which is represented both in France and England, indicates a family characterised, among other features, by the general presence of only three pairs of premolar teeth in each jaw; and *Adapis*, which is likewise common to the same two countries, has four such teeth.

Although there are several families of Insectivora peculiar to the eastern hemisphere, the only one of these with a wide distribution is that of the hedgehogs and their allies (*Erinaceidæ*), which has representatives in the eastern Holarctic, Oriental, and Ethiopian regions. Although none are known from America, extinct representatives of this family are common in the European Oligocene.

---

[1] Teeth figured on p. 156.

Of these *Palæoerinaceus* appears to be allied to the true hedgehogs (*Erinaceus*); whereas other genera, such as *Necrogymnurus*, connect the former with the existing long-tailed and spineless *Gymnura* of the Malayan islands. This group is thus essentially characteristic of Eastern Arctogæa as a whole.

Turning to the Carnivora, we have, in addition to the *Proteleidæ*, of which the sole representative is the African aard-wolf, two important families practically confined to this half of the realm. The first of these is the extensive group of the civets, mungooses, and their allies (*Viverridæ*), which has no representatives in the New World, although a single species in two genera ranges into the Austro-Malayan region. Out of a total of twenty-three genera included in this family only one mungoose (*Herpestes*) and the common genet (*Genetta*) range into Europe, most of the other forms being confined to the Oriental, Ethiopian, and Malagasy regions. Civets (*Viverra*) and ichneumons, together with several remarkable extinct genera, such as *Stenoplesictis*, were, however, common in the European Tertiaries, from the lower Oligocene upwards. And from the circumstance that the last-named genus presents characters connecting the *Viverridæ* with the weasel tribe (*Mustelidæ*), it would seem probable that the latter family was originally evolved in Eastern Arctogæa, although it has now a considerable number of American representatives.

By means of certain extinct forms from the lower Pliocene of Southern Europe and Northern India known as *Ictitherium*, the civets are very closely connected indeed with the hyænas (*Hyænidæ*); the three living representatives of which may be included in the single genus *Hyæna*. Although the striped hyæna (*H. striata*) is now confined to Southern Asia and Northern Africa, it occurred in the Pliocene epoch in France and England; while the larger spotted hyæna (*H. crocuta*) of Southern Africa ranged during the Plistocene era over the greater part of temperate Europe, and likewise extended eastwards as far as India. Numerous extinct species of the same genus are found in the Pliocene of Europe and India; and two extinct genera from the same deposits—*Hyænictis* and *Palhyæna*—connect the living forms with the aforesaid *Ictitherium*. Although one extinct Tertiary North American genus has been tentatively assigned to it, the family is thus essentially an Eastern Arctogæic one; and it may be assumed that, as its living representatives are inhabitants of hot climates, the extinct forms were unable to exist sufficiently far north to permit them to cross by the bridge of land *viâ* Bering Strait.

Among the rodents, in addition to the two widely spread families—*Myoxidæ* and *Spalacidæ*—confined to Eastern Arctogæa, we find in the *Muridæ* the sub-family of the gerbils (*Gerbillinæ*) similarly restricted, their range including the whole of Eastern Arctogæa with the exception of the Malagasy region.

The subfamily *Murinæ*, which includes the true rats and mice (*Mus*) is likewise restricted to the Old World. These rodents differ from the hamsters (*Cricetus*) and the New World white-footed mice (*Sitomys*), which, with other forms, constitute the sub-family *Cricetinæ*, by the molar teeth in the upper jaw having their tubercles arranged in three longitudinal rows; whereas in the latter they form only a double series. Distributed over all the regions of Eastern Arctogæa, with the exception of Madagascar, this group is also represented in the Australian region. Of the two extensively distributed families restricted to the area under consideration, the first is that of the dormice (*Myoxidæ*) whose range includes the Eastern Holarctic and Ethiopian regions. Distinguished from all other rodents by the absence of a cæcum or blind appendage to the intestine, and further characterised by the complicated infoldings of the enamel on the crowns of their molar teeth, these beautiful little creatures now attain their maximum development in Africa. In a fossil state they are first known from the lower Oligocene of Europe, and are likewise common in the Miocene. Another family which does not range beyond the limits of Eastern Arctogæa is that of the *Spalacidæ*, typically represented by the great mole-rat (*Spalax typhlus*)—a blind creature, ranging over south-eastern Europe, Persia, Mesopotamia, Syria, and Egypt. The allied genus (*Rhizomys*), in which the eyes, although minute, are not covered with skin, includes several species, whose distributional area embraces the north of India, Tibet, China, Burma, Malaysia, and Abyssinia; but the three remaining genera are restricted to the Ethiopian region. The whole of the foregoing belong to the mouse-like, or Myomorphous section of the order, but among the Hystricomorphous rodents we have the typical porcupines (*Hystricinæ*), in which the tail is never prehensile, practically confined to Eastern Arctogæa, where they range over south-eastern Europe, and the Ethiopian and Oriental regions. The Javan species of the typical genus (*Hystrix javanica*) is, however, found in the island of Timor, in the Austro-Malayan region, although it is doubtless of late introduction there, and may not improbably have been transported by human agency. In a fossil state porcupines of this sub-family are common in the European Tertiary as far down as the lower Oligocene.

Turning to the ungulates, we have in the artiodactyle section two closely allied families, which—if we except certain pigs from the Austro-Malayan region and Papua, which may have been originally introduced by man—are restricted to the area under considera- tion. Both these families, moreover, have representatives, either living or extinct, in all the regions of Eastern Arctogæa, inclusive even of Madagascar, so that they may be reckoned among its most characteristic mammals. The *Hippopotamidæ*—now restricted to the Ethiopian region, where they are represented by the widely- spread common *Hippopotamus amphibius*, and the much smaller terrestrial *H. liberiensis* of the West Coast—ranged during the Plistocene and upper Pliocene epochs over the greater part of Europe as far north as England ; one species from these deposits being apparently indistinguishable from the common African form. Extinct species are met with in the Plistocene of Algeria, the Plistocene and Pliocene of India, the Pliocene of Burma, and the superficial deposits of Madagascar; some of these differing from the common hippopotamus by the presence of three, in place of two, pairs of incisor teeth in each jaw. Whatever may have been the case with the swine, it is evident that the hippopotami were never able to exist sufficiently far north to cross by way of Bering Strait into the New World. In the pigs (*Suidæ*)—which among other features differ from the *Hippopotamidæ* by the nostrils being perforated in a fleshy disc at the extremity of the muzzle, and like- wise by the structure of the teeth—the typical group of the genus *Sus* ranges over most of the Eastern Holarctic and the whole of the Oriental region, being replaced in the Ethiopian and

Malagasy regions by the potamochœrine group, frequently reckoned as a distinct genus. The Ethiopian region is, however, the sole habitat of the wart-hogs (*Phacochœrus*).

Among extinct artiodactyles we have two well-marked families, distinguished from the foregoing by the crescentic columns of their short-crowned cheek teeth—the *Anoplotheriidæ* and *Dichodontidæ*—likewise confined to Eastern Arctogæa, although their remains have hitherto been obtained only from the eastern section of the Holarctic region. The first of these includes several genera from Oligocene strata, characterised by having three columns on the front half, and two on the hinder half of their upper molars. In the typical *Anoplotherium* there were forty-four teeth, arranged

FIG. 44. LAST FIVE RIGHT UPPER CHEEK-TEETH OF AN *Anoplotherium*.

in a continuous even series, and the feet were provided with either three or two toes. *Dacrytherium* differs by the molars being more like those of *Ancodus*, and also by the deep cavity for the reception of a gland on each side of the face in front of the eye; while the small and elegantly formed animals described as *Xiphodon* have the crowns of the first three premolar teeth elongated and trenchant, the feet being two-toed. In the *Dichodontidæ*, of which there are likewise several genera, the cheek-teeth are more completely selenodont, with only four columns on the crowns of the upper molars; and it is not improbable that in this family we have the ancestral types of both the chevrotains and the deer.

In the *Camelidæ*, as we have already seen, the typical genus *Camelus*, which is found living (although not in a wild state) in the Eastern Holarctic, Oriental, and Ethiopian regions, and fossil in the Plistocene of Algeria and the Pliocene of India, is likewise confined to Eastern Arctogæa. And the same is true, both in the recent and fossil state, with the genera *Tragulus* and *Dorcatherium*, which at the present day alone represent the *Tragulidæ*.

The family of giraffes (*Giraffidæ*), of which the Ethiopian *Giraffa camelopardalis* is the sole existing survivor, was formerly extensively distributed over the area under consideration, to which it appears to have been always restricted, albeit represented by a considerable number of generic types. True giraffes (*Giraffa*) ranged during the Pliocene epoch over Greece, Persia, India, and China, and allied types are to be found in *Visnutherium* of the Pliocene of India and Burma, and *Helladotherium* from the corresponding formation of Greece. Still more gigantic than the latter were the huge *Hydaspitherium*, *Bramatherium*, and *Sivatherium*, of the Indian Pliocene, in all of which the simple horns of the giraffes were replaced by large antler-like appendages, differing considerably in their arrangement in the different genera. Other members of the family are *Samotherium*, of the Pliocene of the Isle of Samos, and *Palæotragus* from the equivalent deposits of Attica, in both of which the females appear to have been hornless, although the males had a pair of simple, compressed, and nearly upright horn-cores. The former is represented by a species rivalling the giraffe in the size of its skull, but the latter was a much smaller animal. This group likewise extended to Northern Africa, where a large species from the Algerian Pliocene has been described under the name of *Libytherium*.

Although the extensive family of the *Bovidæ*, including the oxen, sheep, goats, antelopes, etc., is now represented in the northern part of the western Holarctic region by the American bison (*Bos americanus*), the bighorn (*Ovis canadensis*), the musk-ox (*Ovibos moschatus*), and the so-called Rocky Mountain goat (*Haploceros montanus*), together with a few extinct forms from the superficial deposits, while the anoa (*Bos depressicornis*) is peculiar to Celebes, the greater number of its representatives belong to Eastern Arctogæa. The whole of the numerous genera of antelopes, together with the true goats, as well as the great majority of the sheep, are, for instance, restricted to this area. Moreover, whereas the musk-ox is now solely North American, it was common in Europe during the Plistocene; while the bighorn is closely allied to the Kamschatkan wild sheep (*O. nivicola*), and the

American bison not far removed from its Caucasian cousin (*Bos bison*), so that all these forms are probably descended from ancestors inhabiting Eastern Arctogæa.

In the perissodactyle section of the ungulates, if we take fossil forms into account, there are no families peculiar to this area; but among extinct forms we have the large Oligocene genus *Palæotherium*[1], and the Eocene *Lophiodon* absolutely restricted to it.

Although occurring only in Syria and the Ethiopian region, and at present unknown in a fossil state, the peculiar subordinal group of ungulates represented solely by the hyraces (*Procaviidæ*) perhaps deserves mention among the types characteristic of Eastern Arctogæa. A nearly similar observation applies to the extinct proboscidean family *Dinotheriidæ*, in which the single known genus (*Dinotherium*) ranges from the Miocene and Pliocene of Europe to the Pliocene of Northern India.

Of the edentates, with the exception of certain very doubtful forms from the French phosphorites, which may prove to be reptilian, we have no evidence of the existence of any representatives in the Old World. There are, however, in Eastern Arctogæa two very peculiar families commonly assigned to the same order as the latter, although it seems preferable to regard them as indicating an ordinal group (Effodientia) by themselves. Of these the pangolins (*Manidæ*), which are distinguished from all other mammals by their covering of overlapping horny scales, are now confined to the Oriental and Ethiopian regions, to which the one living genus *Manis* is common; but they appear to have been represented in the Oligocene phosphorites of France by smaller extinct forms, to which the names *Necromanis* and *Leptomanis*[2] have been given. The second family, *Orycteropodidæ*, of which the only living members are the Ethiopian aard-varks (*Orycteropus*), differ very widely from the last, the body being nearly naked, and the molar teeth characterised by a peculiarly complex structure which is unique in the whole mammalian class. A fossil species of the existing genus has been discovered in the lower Pliocene of Samos and Maraga, in Persia; while the extinct genus *Palæorycte-*

---

[1] Teeth figured on p. 166.

[2] The so-called *Palæomanis*, from the Pliocene of Samos, turns out to have been founded on remains of an ungulate.

ropus, from the French phosphorites, is believed to have belonged to the same family, so that both groups appear formerly to have been widely distributed.

Summarising the results of the foregoing survey, we may put in a tabular form the leading features of the mammalian fauna of Eastern as distinct from that of Western Arctogæa. In the sub-joined table the letters E., M., O., H. respectively indicate the Ethiopian, Malagasy, Oriental, and Eastern Holarctic regions; and when a family is represented in any of such regions only in a fossil state, a † is added to the denoting letter. The names of such families or groups as are practically peculiar to the area under consideration are printed in italic type ; while extinct groups have an * prefixed.

PRIMATES.

> *Simiidæ.* O. E. H.†
> *Cercopithecidæ.* O. E. H. ; also extending into the Austro-Malayan region.
> *Lemuridæ.* O. E. M.
> * *Microchœridæ.* H.
> * *Adapidæ.* H.

CHIROPTERA.

> Pteropodidæ. O. E. ; also common to Notogæa.

INSECTIVORA.

> *Erinaceidæ.* O. E. H.

CARNIVORA.

> *Viverridæ.* O. E. H. M.; two species extending their range into the Austro-Malayan region.
> *Hyænidæ.* O. E. H.†

RODENTIA.

> Muridæ ; the sub-family *Gerbillinæ*, O. E. H., is restricted to Eastern Arctogæa, while the Murinæ are exclusively confined to the Old World, but range into Notogæa.
> *Myoxidæ.* H. E.
> *Spalacidæ.* H. O. E.

Hystricidæ ; in this family the *Hystricinæ* are practically restricted to Eastern Arctogæa, although a Javan species is found in Timor.

UNGULATA.

Hippopotamidæ.   E. O.† H.† M.†

Suidæ.   H. O. E. M., also extending into the Austro-Malayan region.

* Anoplotheriidæ.   H.

* Dichodontidæ.   H.

Camelidæ ; in this group the genus *Camelus*, H. E. O., is peculiar to Eastern Arctogæa.

Tragulidæ ; the existing genera *Tragulus*, O., and *Dorcatherium*, E. H.†, as well as several extinct ones, are restricted to this area.

Giraffidæ.   E. O.† H.†

Bovidæ ; the whole of the true antelopes and goats, as well as most of the sheep and oxen, are restricted to Eastern Arctogæa, North America now preserving only one species of Bos, Ovis, Ovibos, and Haploceros.

* Palæotheriidæ ; *Palæotherium.*

* Lophiodontidæ ; *Lophiodon.*

Procaviidæ.   E. H.

* Dinotheriidæ.   H. O.

EFFODIENTIA.

Manidæ.   E. O. H.†

Orycteropodidæ.   E. H.†

To these may be added the absence of living opossums (*Didelphyidæ*). It will be observed that in this list such existing families as are confined to one of the regions of the area considered are for the most part omitted. Tertiary families which are at present unknown beyond the Eastern Holarctic region have, however, been included, since the limitation is probably in great part due to our want of knowledge of the early Tertiary faunas of the other regions.

Although the differences indicated between the faunas of

Eastern and Western Arctogæa, which will be still more apparent after the consideration of the mammals of North America, seem at first sight to indicate that these two areas should form distinct divisions of the realm, yet the community of the fauna of the northern portion of the two hemispheres forbids this view. This question may, however, be more fully discussed in the chapter devoted to the Holarctic region.

Before entering upon the consideration of the different zoological regions into which the realm is divided, it is essential to take a brief survey of the Tertiary mammalian faunas of Eastern Arctogæa. By this alone it is possible to understand the true relations of the existing faunas to one another ; while such a survey also serves to demonstrate that the regions in question are but features of the present epoch of the earth's history ; and that even as late as the Pliocene portion of the Tertiary epoch the distinctions now obtaining between the Holarctic, Oriental, and Ethiopian regions had no existence. In our survey we may omit the Eocene period, and commence with the lower Oligocene ; and it will simplify matters to give lists of some of the more important and better known generic types characterising the faunas of the different horizons. The leading affinities of many of the genera mentioned have been already alluded to in the present or preceding chapters ; but it would vastly exceed the limits of our space to attempt to point out the distinctive features of the others. Accordingly, the reader must either take them on trust, and treat them practically as abstract terms, or he must refer to some palæontological treatise in order to find the real nature of the animals indicated by such generic names.

Although it is essential to our purpose to notice the Oligocene faunas of Eastern Arctogæa, it is important to observe that our knowledge of these is practically limited to Western Europe. We are consequently quite unable to say how far the geographical range of such faunas extended, although it is probable that this embraced a large portion of the Eastern Holarctic region. Whether, however, at this epoch Ethiopian Africa had received a large mammalian fauna must be left for future discoveries to determine.

Under the term of lower Oligocene (the upper Eocene of many writers) are included a large series of strata, such as the freshwater beds of Bembridge and Hordwell in the south of England, the gypsum of Montmartre near Paris, and the corresponding black lignite beds of Débruge in Vaucluse[1]. A considerable part of the fauna of the Quercy phosphorites of Central France likewise comes under the same category, only we have here a mixture of Middle and Upper Oligocene forms. And in the case of the siderolites, or bone-earths of Switzerland, this admixture is carried to a still greater degree, undoubted Eocene types occurring with those properly characteristic of the Oligocene. In the following list such genera as are found only in the phosphorites have the letter P. after them; while after those peculiar to the siderolites the letter S. is added; both letters being given when the genera are common to the two formations. As already said, only some of the better-known forms are selected.

Among the lemuroid Primates, we have the genera *Adapis* and *Microchœrus*, both of which occur in the English beds as well as in the phosphorites; these being the last European representatives of the group. The Insectivora include *Necrogymnurus* (P.S.), allied to the Malayan *Gymnura*, *Amphidozotherium*, together with the existing genera *Sorex* and *Talpa*. More remarkable is the occurrence in the phosphorites of an insectivore described as *Pseudorhynchocyon*, which is believed to be a member of the family of jumping shrews (*Macroscelididæ*), now confined to the Ethiopian region.

The true Carnivora are represented by *Eusmilus* (P.), a highly specialised ally of the sabre-toothed tigers, as well as by the more cat-like *Ælurictis*, and the generalised *Pseudælurus*. In addition to species of true civets, referred to the living genus *Viverra*, the *Viverridæ* include *Amphictis* (P.), *Stenoplesictis* (P.), and *Palæoprionodon* (P.); the two latter being generalised forms closely connecting the family with the *Mustelidæ*, which is represented by *Plesictis* (P.). To the *Canidæ* may be assigned the genera *Cynodon* (P.S.), *Cephalogale* (P.), and *Cynodictis*, together with a species which may be included in *Amphicyon*

---

[1] See table on p. 117.

(P.); while the creodont division of the order is represented by *Hyænodon*, *Pterodon*, *Oxyæna* (P.), and *Proviverra*. Among the rodents we may note the squirrel-like *Sciuroides* (P.S.) and *Pseudosciurus* (S.), the existing genus *Sciurus*, and the extinct *Plesiarctomys* and *Plesiospermophilus*, which likewise belong to the same family. Among the *Muridæ*, *Cricetus* includes ancestral types of the hamsters; while the dormice are represented by the existing genus *Myoxus*. As noticed previously, *Theridomys*, *Nesocerodon*, and *Protechinomys* seem to be ancestral forms allied to some of the existing South American rodents.

The ungulates are very strongly represented; the pig-like group including *Cebochœrus*, *Chœropotamus* and *Elotherium* (P.) among the *Chœropotamidæ*, *Acotherulum*, and *Anthracotherium* (P.) and *Ancodus* in the *Anthracotheriidæ*. The anoplotheroids comprise *Anoplotherium*, *Dacrytherium*, and *Xiphodon*; *Dichobunus* and *Cænotherium* (P.) are the characteristic forms among the *Cænotheriidæ*; and *Dichodon*, *Gelocus*, and *Lophiomeryx* among the *Dichodontidæ*; while the chevrotains are represented by *Prodremotherium* (P.) and *Bachitherium* (P.). In the perissodactyle section of the same order, we have *Pachynolophus* (P.) to represent the *Lophiodontidæ*, *Palæotherium* and *Anchilophus* in the *Palæotheriidæ*, *Protapirus* (P.) as an ancestral form of the tapirs, and *Rhinoceros* (P.), *Cadurcotherium* (P.), and *Hyrachyus* (P.) as representatives of the rhinoceroses. The aberrant *Chalicotherium* has also one species from the phosphorites. The Effodientia include *Leptomanis* (P.), *Necromanis* (P.), and *Palæorycteropus* (P.); while the existing genus *Didelphys* alone represents the marsupials.

It will be seen that in this fauna the existing generic types are very few, and if the whole of the extinct ones had been given, their relative proportion would have been still less. The ungulates were abundant, and among these the perissodactyles proportionately more numerous than at the present day; while the anoplotheres are in some respect transitional between the latter and the typical artiodactyles. All the ungulates had brachydont teeth, and annectant types between the modern pigs and ruminants were abundant; the traguloids being the highest development among the artiodactyle section of the order. Creodont carnivores still persisted, although more modern types had already made their appearance on the scene; and opossums flourished.

The middle stage of the Oligocene is represented in Europe by the freshwater marls and clays of Hempsted in the Isle of Wight and the corresponding beds of Ronzon, near Puy-en-Velay, the lignitiferous strata of Cadibona in Liguria, the deposits of Fontainebleau and Ferte-Alais in France, and likewise by certain beds in Hungary and at Monte Promina in Dalmatia. Among the small fauna of this stage we may notice the following. In the Insectivora, *Tetracus*, an ally of the hedgehogs; *Cynodon*, *Amphicynodon*, *Plesictis*, and *Hyænodon* among the carnivores; *Anthracotherium*, *Ancodus*, *Elotherium*, *Cænotherium*, *Gelocus*, and *Rhinoceros* in the ungulates; and opossums (*Didelphys*). While this fauna is closely related to the preceding, it has lost a number of early ungulate types, such as *Anoplotherium* and *Xiphodon* among the artiodactyles, and *Palæotherium* and *Anchilophus* in the perissodactyles. On the other hand, the pig-like forms, such as *Ancodus*, *Anthracotherium*, and *Elotherium*, attained an extraordinary degree of development. Among the creodont carnivores, we may note the final disappearance of the genera *Pterodon* and *Proviverra*, although *Hyænodon* still survived.

The upper Oligocene (lower Miocene) fauna is a large and characteristic one, well represented in the freshwater beds of St Gérand le Puy, in the Allier, as well as in those of Weisenau and other localities in the neighbourhood of Mayence. Among the mammals may be mentioned the following; existing genera being denoted by the prefix of a †.

INSECTIVORA.	† Talpa.	† Sorex.
	Geotrypus.	Dimylus.
	† Myogale.	Palæoerinaceus.
	Plesiosorex.	† Erinaceus.
CARNIVORA.	Cephalogale.	† Viverra.
	Amphicyon.	† Herpestes.
	Plesictis.	Proælurus.
	Potamotherium.	Hyænodon.
	Amphictis.	

RODENTIA.	Theridomys.	† Spermophilus.
	Archæomys.	† Sciurus.
	Issiodoromys.	Chalicomys.
	† Myoxus.	Titanomys.
	† Cricetus.	
UNGULATA.	Anthracotherium.	Cænotherium.
	Hyotherium.	† Tapirus.
	Amphitragulus.	† Rhinoceros.
	Dremotherium.	
MARSUPIALIA.	† Didelphys.	

Of this fauna, Professor von Zittel writes that it seems at first sight closely akin to those of the middle and lower Oligocene; the same ordinal and subordinal groups, and in many instances the same genera characterising the whole three horizons. In the lack of lemuroids, the reduced number or final disappearance of opossums, creodonts, and anoplotherioids, in the greater abundance of forms like *Anthracotherium*, *Hyotherium*, and *Dremotherium*, which were but poorly represented in the lower Oligocene, and in the number of new types, such as *Tapirus*, *Amphitragulus* (an ancestral chevrotain), *Chalicomys* (an early beaver), *Titanomys* (an ally of the picas), *Erinaceus*, *Dimylus* (a form connecting the shrews with the hedgehogs), *Potamotherium* (a generalised otter), *Herpestes* (mungooses), *Prœlurus* (a primitive type of cat), we notice, however, the marked difference of this fauna from its forerunners. Among the incoming genera it is noteworthy that there is none for which an ancestral type cannot be found in the lower Oligocene; the main difference occurring in the more specialised characters of the members of the later fauna. With the exception of certain bats, insectivores, rodents, and the opossums (such as *Vespertilio*, *Erinaceus*, *Sorex*, *Myogale*, *Talpa*, *Sciurus*, *Spermophilus*, *Cricetus*, *Myoxus*, and *Didelphys*), the majority of the genera are, however, still extinct.

It is probable that the beds in the Balkans which have yielded remains of the North American Tertiary genus *Titanotherium* belong to some portion of the Oligocene epoch.

We now come to the Miocene epoch, which, as at present restricted, forms in Europe but a small section of the Tertiary era.

It includes the well-known freshwater strata of Sansan in Gers (the middle Miocene of the older geological classifica-
tions), together with the corresponding beds of Miocene
Fauna. Steinheim in Styria, and likewise the somewhat newer (upper Miocene) deposits of Œningen, in Baden. Grive-St-Alban, in the valley of the Rhone, is likewise another well-known locality where mammaliferous strata of this age are developed; and, among other places, we may also mention Monte Bamboli in Italy, San Isidro in Spain, and Oran in Algeria.

For the first time in Europe we meet with remains of true Primates, of which there are three genera belonging to the *Simiidæ*, two of which, *Dryopithecus* and *Oreopithecus*[1], are extinct, but the third seems scarcely separable from the existing Oriental *Hylobates*. In the Insectivora we meet with the existing European genera *Talpa*, *Myogale* (desmans), *Erinaceus*, *Sorex*, and *Crocidura*; while the extinct *Lanthanotherium* seems to be allied to the tree-shrews (*Tupaia*) of the Oriental region, and *Galerix* intermediate between the latter and the jumping-shrews (*Macroscelididæ*) of Ethiopian Africa. Among the Carnivora, where the creodonts have disappeared, the cats are represented by the sabre-toothed tigers (*Machærodus*) and *Pseudælurus*. In addition to the existing genera *Viverra* and *Herpestes*, we have among the civet tribe the extinct *Progenetta*. The dogs include the existing *Canis*, together with the extinct *Hemicyon* and *Pseudocyon*; while the larger forms described as *Dinocyon* and *Hyænarctus* connect the former with the bears. The *Mustelidæ* are represented by species of the typical living genus *Mustela*, together with certain more or less closely allied extinct types; and *Enhydriodon* filled the place of the modern otters.

From among the rodents the generalised types allied to those now characteristic of Neogæa have all disappeared, nearly all the recorded forms apparently pertaining to existing genera. In the *Sciuridæ* not only have we true squirrels (*Sciurus*), but the Ethiopian spiny squirrels (*Xerus*) are likewise represented, as are also the more widely distributed flying-squirrels of the genus *Sciuropterus*, which now inhabit both Eastern and Western Arctogæa.

---

[1] Some of the characters of these genera have been already mentioned on pages 180, 181.

Chalicomys, *Cricetus*, and *Myoxus* are survivors from the Oligocene; but porcupines (*Hystrix*) are new comers. Picas of the existing genus *Lagomys* are likewise to the fore; and it is a question whether those distinguished by the name of *Myolagus* might not be included under the same title.

A marked approximation to the modern type is likewise the characteristic feature of the ungulates of the European Miocene; although in this group living genera still remain in the minority. The pigs (*Suidæ*) include, for instance, the genus *Hyotherium*, in which the molar teeth are tuberculated and of the general type of those of some of the living members of the family; and also the more aberrant *Listriodon*, characterised by the presence of a pair of transverse ridges on each of the teeth of the same series. A species of the existing West African genus *Dorcatherium* alone represents the chevrotains (*Tragulidæ*); while we have forerunners of the deer (*Cervidæ*) in the extinct *Palæomeryx* and *Dicroceros*, both characterised by the simple structure of their antlers; and *Protragoceros*—a generalised type of antelope—marks the first appearance of the hollow-horned ruminants (*Bovidæ*), which now form such a numerous and characteristic group in the fauna of Eastern Arctogæa. Perissodactyle ungulates are less numerous. *Anchitherium*, to some of whose distinctive characters allusion has been made above, constitutes the representative of the equine line at this stage; and tapirs and rhinoceroses belonging to the existing genera were likewise common. Some of the latter were, however, still hornless, and in none was more than a single horn developed. The aberrant *Chalicotheriidæ*, forming the last family of this section of the order, and characterised by the extraordinary resemblance presented by their claws and toes to those of edentates, are here represented by the gigantic *Macrotherium*. Finally the Miocene is notable as being the stage at which proboscideans first made their appearance on the scene in Europe. In this group we have species of *Mastodon*, which, as already explained, includes the ancestors of the modern elephants; and likewise one of the more aberrant *Dinotherium*.

Compared with the Oligocene, the loss of so many antiquated types, coupled with the appearance of proboscideans and man-like apes, and the general modern facies of all the mammals of the

Miocene, indicates the lapse of a considerable interval of time between the deposition of the two series of strata. And that this is really the case, is demonstrated by the fact that there occurs between the two a considerable thickness of marine deposits which have not hitherto yielded remains of land mammals. It may be noticed that while many of the insectivores and rodents from this horizon belong to genera now inhabiting the Eastern Holarctic region, among other forms we have marked instances of Oriental (*Lanthanotherium, Hylobates*) or Ethiopian (*Galerix, Dorcatherium,* and *Xerus*) affinities in this assemblage; and it is thus evident that at the epoch in question there was no trace of the differentiation of Eastern Arctogæa into regions.

Still more markedly are the same features displayed by the older Pliocene fauna of Europe and Southern Asia. This fauna, which was formerly regarded as of upper Miocene age until shewn by Dr Blanford to be unquestionably referable to the succeeding era of geological history, had a very wide distribution ; and it is represented at certain localities, mostly at long distances from one another, by an extraordinary profusion of remains. One of these charnel-holes occurs at the village of Pikermi, near Athens, a second in the Isle of Samos, in the Turkish Archipelago, and a third at Mont Léberon, in Provence. This fauna is also met with locally in the valley of the Rhone, at the foot of the Pyrenees, in Spain, Asia Minor, and at Maraga in Persia. It is likewise represented in the regions lying to the north of the Alps, only here the number of forms is less, and the antelopes and giraffe-like ruminants, fitted for roaming over the open plains of the south, are conspicuous for their absence; their place being taken by forest-haunting deer. The sand-beds of Eppelsheim in Hessen-Darmstadt, together with strata in the neighbourhood of Vienna, and others in Hungary and Rumania, may be cited as localities where the northern section of this fauna is preserved.

Taking first the European and Western Asiatic portion of this fauna, and leaving its Oriental members for subsequent consideration, we find the Primates represented solely by an extinct genus of monkey, taking its name of *Mesopithecus* on account of presenting certain features intermediate between the existing *Semnopithecus* and *Macacus*. The insectivores are likewise known only by a solitary form, a shrew (*Sorex*); but this is probably due to the nature of the strata being unfitted for the preservation of the remains of such small creatures. The Carnivora, on the other hand, were abundant, the *Felidæ* being represented not only by the sabre-tooths (*Machærodus*), but true cats (*Felis*) likewise making their appearance on the scene. Hyænas display a great variety of development, there being one species of the typical genus *Hyæna*, with certain resemblances to the existing Cape form, while the more generalised types known as *Lycyæna* and *Hyænictis* were likewise present, as were also species of *Ictitherium*, and the allied *Palhyæna*, which, as already mentioned, formed a connecting link between the hyænas and the civets. True dogs seem to have been absent from this assemblage; but *Amphicyon* still survived from the Miocene, and an aberrant form known as *Simocyon* made its appearance. *Hyænarctus* was likewise another survivor from the Miocene, and may be regarded as a forerunner of the true bears. Finally, in the weasel tribe (*Mustelidæ*) we have representatives of the existing genus *Mustela*, as well as the extinct *Palæomephitis* and *Promeles*, the latter being an ancestral type of the badgers.

Rodents make but a poor show, as we have only the extinct beaver-like *Chalicomys*, a species of porcupine (*Hystrix*), and a representative of the curious little spiny mice (*Acomys*), now characteristic of Syria, Palestine, and north-eastern Africa.

A remarkable advance over their Miocene forerunners is displayed by the ungulates, especially those from Pikermi and Maraga. Here, in the artiodactyle section, we meet for the first time with true pigs of the genus *Sus*, which at this period ranged over the greater portion of Europe, and some of which attained very large dimensions. Water-chevrotains (*Dorcatherium*) serve to connect the Miocene representatives of their genus with the existing West African form; while muntjacs (*Cervulus*), now confined to the Oriental region, filled the place of the stags. Giraffe-like creatures were numerous, for not only have we true giraffes belonging to the existing Ethiopian genus *Giraffa*, but the gigantic hornless *Helladotherium* stalked over the plains of Greece, and the allied but horned *Samotherium* inhabited the area now occupied by the Turkish Archipelago, and extended eastwards as far as Persia; *Palæotragus* being a smaller but allied form. The *Bovidæ* are represented by antelopes, most of which present a marked Ethiopian facies, although *Tragoceros* (probably the direct descendant of the Miocene *Protragoceros*) is an aberrant form, with compressed horn-cores like those of the goats. And it may be remarked that most of the Pikermi antelopes have short-crowned molar teeth, in which respect they resemble the existing eland, kudu, and their allies. Of the Pliocene forms, *Palæorias*, which is common to Southern Europe and Algeria, seems to be inter-

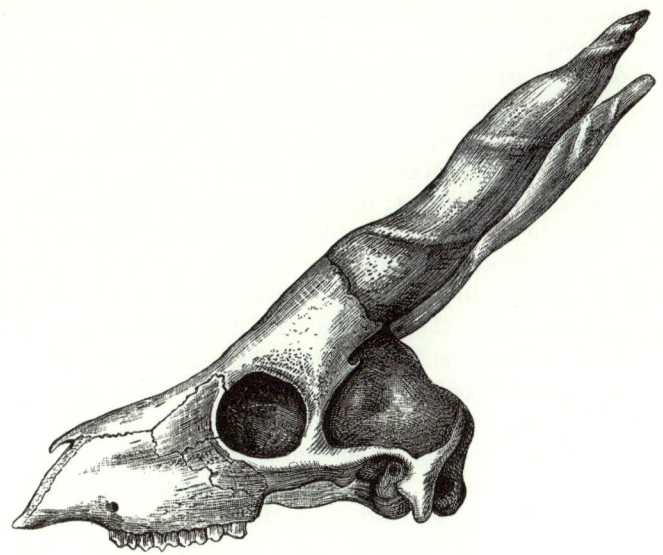

FIG. 45. SKULL OF *Palæorias*.

mediate between the kudus (*Strepsiceros*) and elands (*Orias*); while the so-called *Protragelaphus* is so closely allied to the existing Ethiopian harnessed antelopes (*Tragelaphus*) as to be included by some in the same genus. On the other hand, *Palæoryx* is nearly related to the gemsbok and its allies (*Oryx*), although with certain resemblances to the sable antelope group (*Hippotragus*). Gazelles (*Gazella*), which are essentially inhabi-

tants of open plains, were likewise abundant; one being considered
a near relative of the South African springbok. The genus
Helicophora, on the other hand, closely resembles the water-buck
group (*Cobus*), which is exclusively Ethiopian. In the perissodactyle division, the three-toed horses (*Hipparion*) seem to have
approximated in general structure to the Ethiopian zebras, and,
like those animals, may have been ornamented with dark and light
stripes. While some of the Pliocene rhinoceroses were hornless,
another was a two-horned species closely allied to the common
African *Rhinoceros bicornis*, of which it may be regarded as the
parent form. There is also an extinct genus (*Leptodon*), of somewhat uncertain affinity; while tapirs are found in the Eppelsheim
beds, although not apparently in the southern area. The *Chalicotheriidæ* were represented by the typical genus *Chalicotherium*
(*Ancylotherium*), which, as we have seen, was a near ally of the
Miocene *Macrotherium*, and also occurs in the Oligocene phosphorites. As in the Miocene, the proboscideans include only
Mastodon and *Dinotherium*; the one species of the former ranging
from Greece to Persia, but being different from all the Indian forms
of the same epoch. Finally, the occurrence of an aard-vark
(*Orycteropus*) both in Samos and Persia serves to accentuate the
Ethiopian affinities of the southern section of this fauna.

We have thus evidence that one and the same fauna extended
from Spain and Algeria across Southern Europe to Asia Minor
and Persia; and we may infer from the deposits at Samos, that
what is now the Ægean sea formed a tract of land connecting
Greece with Turkey. It is further evident that there must have
been free communication across the Mediterranean basin (which
in Cretaceous times is known to have been a *mare clausum*, in the
physical, and not the political sense of the term) between Europe
and Africa. This communication may have existed both by way
of Gibraltar, and also between Italy, Sicily, and Malta on the
one hand, and Tunis on the other; since the Plistocene mammals
of the islands in question clearly indicate continental connection.
While the antelopes and hipparions of this fauna prove the existence of open plains during the lower Pliocene epoch, the host of
individuals of *Mesopithecus* as unmistakeably point to the presence
of extensive forest-tracts. In the northern section of the fauna, as

displayed at Eppelsheim, the Ethiopian affinities are much less apparent, aard-varks and the whole of the giraffe-group being absent, while tapirs and deer were abundant. That there was a more or less marked separation between the two areas thus seems evident; and the tapirs and muntjac-like deer, both of which seem wanting in the Siwalik fauna, are indicative, so far as they go, of Malayan affinities.

Nearly related to that of Pikermi, Samos, and Persia, the celebrated Siwalik fauna of India and the adjacent countries presents certain well-marked differences; **Siwalik Fauna.** this being specially shown by the occurrence of several essentially modern types quite unknown in the former. Moreover, there are a considerable number of peculiar genera which do not occur in the western fauna; while we also come across certain Miocene, and even Oligocene types, which are equally strange to the latter. Although in some cases these occur in beds which are not improbably of upper Miocene age, in others they appear mingled with the later forms; but, in any case, they indicate a survival in this area of archaic types which at that time had completely disappeared from Europe.

Originally discovered in the outer ranges of the typical Himalayan area, the Siwalik fauna has been traced towards the north-west into the Punjab, Kach, Sind, and the north-eastern frontier of Baluchistan; the beds from the two latter areas being lower in the series than those from the typical Siwalik hills, and containing an older assemblage of forms, although several are common to all. An outlier of the same fauna occurs in Perim Island in the gulf of Cambay. Eastwards the Siwalik fauna ranged through Sylhet and Assam to Burma, whence it has been traced at intervals, as in Java, Sumatra, and the Philippines, into China and Japan. In China it extended from Yunnan in the south-west northwards through Szechuen to Kansu, and thence eastwards through Shensi to Shansi, its extreme eastern limit being indicated by the discovery of a Siwalik elephant's tooth at Shanghai. Northward of Kansu the fauna ranged into Mongolia, probably by way of the gap formed by the course of the Hwang-ho through the Ala-shan mountains—if such mountains existed at the time. And it is not a little remarkable that of the few Mongolian forms at present known, two (*Hyæna macrostoma* and *Equus sivalensis*) are identical with species from the Siwalik Hills[1].

As regards the fauna itself, we find, in the first place, the Primates much more fully represented than at Pikermi, and all by existing generic types. Of the man-like apes (*Simiidæ*), there is a chimpanzee (*Anthropopithecus*) presenting a more human type of dentition than its living Ethiopian cousins; while a single tusk indicates the former existence of an orang (*Simia*) allied to the living Bornean and Sumatran species. The other three generic types belong to the *Cercopithecidæ*, and include baboons of the Ethiopian genus *Papio* (*Cynocephalus*), together with species of *Semnopithecus* and *Macacus*, the former genus being exclusively, and the latter mainly, Oriental at the present day, although both occur in the later Pliocene of Europe. Doubtless owing to the unsuitability of the strata for the preservation of small specimens, no remains of insectivores have hitherto been obtained. The Carnivora are, on the other hand, well represented; the *Felidæ* including large and small species of the typical genus *Felis*, and apparently one of the allied *Cynælurus* (hunting-leopard), now exclusively Oriental and Ethiopian. *Machærodus* had two species; and another form has been identified with the European Oligocene genus *Ælurictis*. Civets include species of *Viverra* larger than any now existing; this genus being also one now confined to the Ethiopian and Oriental regions, although more abundant in the latter than in the former. The *Canidæ*, in addition to a survivor of the Miocene *Amphicyon*, were represented by wolves and jackals (*Canis*), as well as by a species apparently allied to the long-eared fox (*Otocyon*) of Africa. While in the *Ursidæ* the generalised *Hyænarctus* still survived, true bears (*Ursus*) make their appearance for the first time, the single known Siwalik species presenting, however, a marked approximation in the characters of its skull and dentition to the Indian sloth-bear (*Melursus*). Among the few known representatives of the *Mustelidæ*, we have a large marten (*Mustela*), probably allied to the living yellow-throated Indian species; a ratel, belonging to a genus (*Mellivora*) now restricted to India and Africa; and likewise an otter (*Lutra*) whose nearest affinities are with an existing Sumatran species. The same family also

[1] Lydekker, *Rec. Geol. Surv. India*, Vol. XXIV. pp. 207—211 (1891).

includes a member of the otter-like genus *Enhydriodon*, the other species being from the Italian Miocene. Among the most remarkable features of the Siwalik Carnivora is the survival of a species of *Hyænodon*, of which the remains have been discovered in the Punjab.

The Rodentia are but very imperfectly known. They include a representative of the bamboo-rats (*Rhizomys*), which are now exclusively Oriental, and belong to the family *Spalacidæ*; and, among the *Muridæ*, a species of *Nesocia*, which genus is likewise confined to the Oriental region. The other forms are a porcupine (*Hystrix*) and a hare (*Lepus*).

A very long list is presented by the ungulates, which are numerous not only in generic, but likewise in specific types. The pig-like artiodactyles include, among the family *Suidæ* several representatives of the true pigs (*Sus*), some of which attained gigantic dimensions, while others are remarkable for the complex structure of their molar teeth, which show a marked resemblance to those of the existing Ethiopian wart-hogs (*Phacochærus*). A still more elaborate structure is displayed by the corresponding teeth of the allied genus *Hippohyus*, which is peculiar to this fauna; and the family is also represented by species of the European Miocene genera *Hyotherium* and *Listriodon*, the remains of the two latter being mostly obtained from the Punjab and districts to the west. The same areas are mainly those which have yielded remains of *Anthracotheriidæ*, although some of these have been discovered in Sylhet. In this family we have species of the European genera *Anthracotherium* and *Ancodus*, the former of

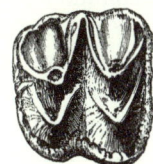

FIG. 46.   RIGHT UPPER MOLAR OF A SMALL SPECIES OF *Merycopotamus*.

which is elsewhere unknown above the Middle Oligocene; and there are also three peculiar types, respectively known as *Merycopotamus*, *Hemimeryx*, and *Chœromeryx*, differing from all the rest in having only four columns on the crowns of the molars, as shown in the annexed figure, and thus presenting a marked approximation to the ruminants. The earlier Tertiary *Chæropotamidæ* likewise had a survivor in the genus *Tetraconodon*, which was represented by a large pig-like creature remarkable for the .enormous size of its simple conical premolar teeth. The pig-like group closes with *Hippopotamus*, which makes its appearance on the scene for the first time in this formation, where it is represented by a generalised species with three pairs of incisor teeth in each jaw. Turning to the groups with fully-developed selenodont molars, we have first to notice the occurrence of fossil camels of the existing genus *Camelus*, which are unknown elsewhere except in the Algerian Plistocene. As we have seen that the *Camelidæ* were originally a New World group, it is interesting to note that these earliest Old World representatives occur in Asia instead of Europe; and it is further noteworthy that in the structure of their molar teeth the Siwalik camels retain evidences of affinity with the South American guanacos and vicuñas which are lost in their living descendants. The *Tragulidæ* contain representatives of the true chevrotains (*Tragulus*) and water-chevrotains (*Dorcatherium*), now respectively characteristic of the Oriental and Ethiopian regions, while among the deer (*Cervidæ*) we have species of the Oligocene European genus *Palæomeryx*, together with others belonging to *Cervus*, the representatives of the latter being all closely allied to existing Oriental types. Not improbably also a musk-deer (*Moschus*) should be included among the Siwalik *Cervidæ*. Among the *Giraffidæ*, in addition to true giraffes (*Giraffa*), which are common to the Pikermi beds, and extended eastwards into China, we have the peculiar gigantic antlered types respectively known as *Vishnutherium, Sivatherium, Hydaspotherium*, and *Bramatherium*, of which the first seems common to the Siwaliks of Burma and the Punjab, while the second is confined to the more easterly Himalaya, the third to the Punjab, and the fourth to Perim Island. They include the most gigantic of all ruminants, *Sivatherium* almost rivalling an elephant in bulk.

Not one of the least curious features in this marvellous fauna is that while, as we have seen, deer of Oriental types were abundant, antelopes closely allied to those now inhabiting the Ethiopian

FIG. 47. RESTORATION OF *Sivatherium*.

region—where deer are totally absent—were likewise extraordinarily numerous. Of the African genera we have a species of *Bubalis* intermediate between the hartebeests and the blesbok, a member of the sable antelope group (*Hippotragus*), a kudu (*Strepsiceros*), an eland (*Orias*), and probably a representative of the water-buck group (*Cobus*). On the other hand, Oriental forms are not wanting, as proved by the occurrence of a nilgai (*Boselaphus*), and probably of a four-horned antelope (*Tetraceros*); while the widely-spread gazelles (*Gazella*) were likewise present. Goats and oxen for the first time made their appearance; the former group being represented not only by species belonging to the typical *Capra*, but likewise to the shorter-horned genus *Hemitragus*, now confined to India and Arabia. The oxen (*Bos*) included members of all the existing groups, that is to say typical oxen, bison, buffalo, and smaller forms with upright triangular horns nearly allied to the anoa of Celebes.

The perissodactyle ungulates, so numerous in the earlier Tertiary formations, have now become proportionately much fewer as compared with the artiodactyles. While typical forms of *Hipparion* were present, one species differs from the rest by the loss of the lateral toes, and thus resembles the modern horses (*Equus*), which here make their appearance for the first time. Rhinoceroses include not only hornless forms, but likewise one species allied to the existing Oriental *Rhinoceros unicornis* and *R. sondaicus*, and a third as closely related to the African Burchell's rhinoceros (*R. simus*). In the same group *Chalicotherium* is a survivor from older formations.

Finally, the proboscideans exhibit a development unparalleled in any other formation or epoch. *Dinotherium* appears for the last time in the Siwaliks of Perim Island, Kach, Sind, and the Punjab; while the mastodons include a large number of species, some of which present such a close approximation to the so-called stegodont elephants (which, as already mentioned[1], are peculiar to this fauna) as to render it impossible to draw any well-defined demarcation between the genera *Mastodon* and *Elephas*. Not only does the Siwalak fauna include the aforesaid stegodont, or

[1] *Supra*, p. 172.

transitional elephants, but likewise one which may well have been the ancestor of the species now inhabiting India. Eastwards these transitional elephants and mastodons have been traced into Java, Borneo, China, and Japan ; and, as stated in an earlier chapter, there can be no doubt that the modern elephants were evolved in this area.

Although, as shown in the foregoing survey, the Siwalik fauna differs in certain respects from that of Pikermi, Samos, Léberon, etc., yet there can be no hesitation in regarding the whole lower Pliocene fauna of Europe, North Africa, Asia Minor, and South and East Asia as essentially one ; and consequently at this epoch there was no possibility of distinguishing between the Palæarctic and Oriental regions. Whence Ethiopian Africa had by this time received the forerunners of its present higher mammalian fauna, we have, unfortunately, no decisive evidence. Writing some years ago, Dr Blanford[1] seems to suggest that the irruption of the modern African fauna was anterior to the Pliocene. After referring to certain peculiarities connected with the existing mammalian fauna of India and the Malayan area, he observes that " these cases of isolation probably indicate that the animals belong to an older fauna, now partly replaced by newer types, and that the older fauna was common to India and Africa. It is very probable that these animals are descended from the ancient tropical fauna of the early Tertiary times. But, so far as it is possible to judge, the process of variation would have caused a greater distinction between forms so widely separated and exposed to such different conditions, if the period of isolation were great ; and it is difficult to suppose that the lands inhabited by the ancestors of the *Simiidæ*, *Lemuridæ*, *Tragulidæ*, and *Manidæ* of the Oriental and Ethiopian regions can have been separated prior to the early part of the Miocene period."

This is perfectly true so far as it goes, but since, as we have seen, genera like *Hippopotamus*, *Bos*, *Capra*, *Equus*, and *Elephas* are unknown previous to the Siwalik epoch, and some of them at least were evolved at or about that time in the Indian area, it seems necessary to assume the existence of a free land communi-

[1] *Manual of Geology of India*, 1st ed. p. lxviii. (1879).

cation between the Ethiopian and Oriental regions at least as late
as the lower Pliocene epoch. With regard to where this connection was situated, we may note, in the first place, that Dr Wallace[1]
was of opinion that even the Pikermi fauna made its way into
Africa chiefly through Syria, although a brief connection of Europe
with Tunis is admitted. When the passage in question was
written, little or nothing was, however, known as to the Pliocene
fauna of Algeria. And although this undoubtedly indicates a
western connection between Europe and Africa, yet even in the
Pliocene the Sahara probably formed, as now, a barrier[2] across
which the fauna of northern Africa could not pass south. Accordingly, even the Pikermi fauna may have come round by way of Egypt.
Be this as it may, it seems clear that the Siwalik fauna entered
Africa by way of Syria or Arabia, or possibly by both. The most
direct line of communication would be *viâ* the Gulfs of Oman and
Aden; and some indication that such a line of connection may
have existed is afforded by the distribution of the goats of the
genus *Hemitragus*. As already stated, fossil species of this genus
occur in the Siwaliks of Perim Island and the Himalaya, while of
the three existing forms, one is Himalayan, a second confined to
the Nilgiri and certain other South Indian ranges, and the third
inhabits Oman. So far as it goes, the evidence of these goats is
strongly suggestive of the former existence of a land-bridge across
the mouth of the Persian Gulf, as otherwise we should expect to
find living species in Persia and other parts of western Asia. If
the existence of such a bridge be admitted, we only require another
across the narrow strait of Bâb-el-Mandeb to give a free line of
communication between India and Africa in this direction.

Whether, however, the migration from India to Africa took
place at the north or south end of the Red Sea, or at both ends,
it is certain that the connecting land must have been of considerable width, and suited to the passage of mammals of all kinds.
In referring to the nature of the connection, Dr Wallace[3] remarks
that "we may now perhaps see the reason of the singular absence

---

[1] *Geographical Distribution of Animals*, Vol. I. p. 288.

[2] The idea that there was a Tertiary sea in the Sahara is incorrect; see
Blanford, *Quart. Journ. Geol. Soc.* Vol. XLVI. p. 90 (1890).

[3] *Op. cit.* p. 291.

from tropical Africa of deer and bears; for these are both groups which live in fertile or well-wooded countries, whereas the line of immigration from Europe to Africa was probably always, as now, to a great extent a dry and desert tract, suited to antelopes and large felines, but almost impassable to deer and bears." The Siwalik chimpanzee, however, indicates most unmistakably that the communication by way of Arabia or Syria between the Ethiopian and Oriental regions must have embraced a forest-area, and accordingly have been of considerable width.

With regard to the question why so many genera which existed in India and southern Europe during the Pliocene should have disappeared from those areas to live on in Africa, all we can say is that it is quite evident that a southern migration of the fauna has certainly taken place, and that this was probably induced by the cold heralding the approach of the glacial period. Although we have few, if any, decisive physical evidences of a cold period in India, yet the existence of a goat (*Hemitragus*) nearly allied to a Himalayan species in the ranges of southern India seems to indicate that such must have occurred, as it would be quite impossible for the ancestral form to have crossed the intervening plains under present conditions of temperature. It is further noteworthy that many of the animals which have disappeared from India, such as chimpanzees, hippopotami, giraffes, water-chevrotains, and ostriches, are precisely those which are now restricted to very hot climates; whereas the lion, tiger, rhinoceroses, elephants, and monkeys, which both now or during the Plistocene are known to be capable of existing in cold climates, have persisted.

Leaving these exceedingly difficult questions, two other points may be noticed in connection with the Siwalik fauna. In the first place, since the Siwalik hills themselves form ranges of considerable height on its southern flank, it is evident that the Himalaya was much lower during the lower Pliocene epoch than it is at present; Dr Blanford[1] stating that the movement which led to its elevation "has been distributed over the Tertiary and post-Tertiary period, and a great portion in post-Plistocene." This will account for the community between the lower Pliocene

---

[1] See *Geol. Mag.* Decade 3, Vol. IX. p. 166, note (1892).

fauna of the Himalayan area and Mongolia, the Himalaya at that epoch not forming, as now, an impassable barrier to the north of the Oriental region. The second point relates to the survival in the Siwalik fauna of archaic forms, which had disappeared at that date from Europe. This fact, especially since old types such as lemurs and gymnuras are even now met with in the Oriental region, lends support to the view advanced in an earlier chapter[1] that marsupials may have lived on in south-eastern Asia long after they had completely disappeared from Europe.

Our knowledge of the later Pliocene faunas of Eastern Arctogæa is mainly confined to Europe, where at this period the general distribution of land and sea was apparently very much the same as at the present day. Spain was, however, connected with Africa, as was probably also Italy by way of Sardinia and Malta. A portion of Italy was, however, submerged, while in Belgium, Holland, and the south-east of England the sea intruded upon what is now land; but, on the other hand, Britain was joined to the Continent. Few mammaliferous deposits of this age have been preserved to us, but among these are the Crags of the east coast of England (which contain numerous fossils derived from earlier formations), the fresh-water beds of the Val d' Arno in Italy, as well as others in the Auvergne, in the Rhone valley, at Roussillon, and in the neighbourhood of Montpellier. The following genera are included in this fauna, those which are extinct having an asterisk prefixed.

PRIMATES.   Semnopithecus.
                      * Dolichopithecus.
                      Macacus.

INSECTIVORA.   Sorex (Shrews).

CARNIVORA.   * Machærodus (Sabre-tooths).
                      Felis (Cats).
                      Viverra (Civets).
                      Hyæna.
                      Canis (Wolves and Foxes).
                  * Hyænarctus.

Later Plio-cene Faunas.

---

[1] _Supra_, p. 57.

CARNIVORA (*cont.*).

　　　　Ursus (Bears).
　　　　Ælurus (Cat-bears).
　　　　Mustela (Martens and Weasels).
　　　　Lutra (Otters).

RODENTIA.

　　　　Arctomys (Marmots).
　　　*Chalicomys.
　　　　Castor (Beaver).
　　　*Trogontherium (Giant Beaver).
　　　　Cricetus (Hamsters).
　　　　Microtus (Voles).
　　　　Mus (Rats and Mice).
　　　　Hystrix (Porcupines).
　　　*Pellegrinia.
　　　*Myolagus ⎱
　　　　Lagomys ⎰ (Picas).
　　　　Lepus (Hares)

UNGULATA.

　　　　Sus (Pigs).
　　　　Hippopotamus.
　　　　Cervus (Deer).
　　　　Alces (Elk).
　　　　Cervulus (Muntjacs).
　　　*Palæoryx.
　　　　Gazella (Gazelles).
　　　　Bos (Oxen).
　　　　Tapirus (Tapirs).
　　　　Rhinoceros.
　　　　Equus (Horses).
　　　*Hipparion—very rare.
　　　*Mastodon ⎱
　　　　Elephas ⎰ (Elephants).

In this list by far the greater number of the genera are living ones, and if we removed from it types like *Hyæna*, *Hippopotamus*, *Rhinoceros* and *Elephas*, which were spread during the Pliocene and Plistocene epochs over the greater part of Eastern Arctogæa, its Ethiopian resemblances are by no means strongly marked. Although the larger forms (as in the succeeding Plistocene epoch) include a considerable number of genera now mainly confined to tropical or subtropical countries, the rodent fauna exhibits a marked Palæarctic facies, thus indicating an approximation to the existing state of things. Among the extinct rodents, *Pellegrinia*, from the Sardinian Pliocene, belongs, however, to the *Octodontidæ*, and is probably allied to the existing African *Ctenodactylus*. *Trogontherium* is a gigantic extinct type of beaver, which also persisted into the Plistocene. The deer include northern types unknown in the lower Pliocene.

One of the most remarkable features of this fauna is the occurrence of a large species of *Ælurus*,—a genus represented elsewhere only by the cat-bear or panda (*Æ. fulgens*) of the eastern Himalaya, which, although formerly regarded as the type of a family by itself, is now included in the American *Procyonidæ* (raccoons). The fossil species has been hitherto detected only in the English Crag; the genus may, however, be expected to occur in the Siwaliks, since it is quite clear that it must have been originally connected with the American representatives of the family by forms inhabiting Eastern Asia.

With the end of the Pliocene epoch this brief survey of the Tertiary mammalian faunas of Eastern Arctogæa may be brought to a close, since the Plistocene mammals can be more conveniently considered under the headings of the different regions of this great province. While throughout the Oligocene, Miocene, and lower Pliocene epochs no trace of the present zoological regions of this half of the Arctogæic realm is shown, when the Upper Pliocene is reached there are faint indications of the demarcation of the Eastern Holarctic. At the time of the Plistocene, as will be shown in a later chapter, the Eastern Holarctic, Oriental and Ethiopian regions appear to have assumed a still more marked distinction, although this is to a great extent obscured by the wide range even at that epoch of genera like *Hippopotamus*, *Rhinoceros*, *Elephas*, *Macacus*, etc. Moreover, several species which are now confined to one of the three regions in question had then a more extensive distribution, so that it is only during the recent epoch that the Holarctic, Oriental, and Ethiopian regions attained the full faunistic peculiarities by which they are now characterised.

CHAPTER VI.

THE MALAGASY REGION.

Limits—Mammalian Fauna—Relations of Madagascar to the Mainland.

INCLUDED by Drs Sclater and Wallace within the Ethiopian region, Madagascar and the adjacent groups of islands were referred to a region apart by Dr Blanford[1]; this separation being justified not only by the mammalian fauna, but likewise by many other groups of animals. To quote Dr Wallace, this region "comprises, besides Madagascar, the islands of Mauritius, Bourbon, and Rodriguez, the Seychelles, and Comoro Islands. Madagascar itself is an island of the first class, being a thousand miles long, and about two hundred and fifty miles in average width. It lies parallel to the coast of Africa, near the southern tropic, and is separated by 230 miles of sea from the nearest part of the continent, although a bank of soundings projecting from its western coast reduces this distance to about 160 miles. Madagascar is a mountainous island, and the greater part of the interior consists of open elevated plateaus; but between these and the coast there intervene broad belts of luxuriant tropical forests." It is this forest-district which forms the home of most of its peculiar fauna. As regards geological structure, it appears from the researches of Messrs Cortese and Baron that, roughly speaking, a line drawn from north to south so as to divide the island into two longitudinal halves, gives an area of granitic and volcanic rocks on the right or eastern side, and on the left or western side one of sedimentary deposits, containing beds belonging to the Jurassic, Cretaceous, Eocene and recent epochs. Blown sand occurs in abundance

[1] Appendix, No. 8, p. 76.

around the coast, and numerous old lake-basins or marshes, some of very large dimensions, form receptacles where remains of the later faunas have been preserved. With the exception of the Comoro group, which contain a few species, the non-volant mammalian fauna is confined to Madagascar, so that the other islands do not properly come within the province of the present work. It is, however, important to observe that the Seychelles differ from almost all oceanic islands in consisting largely of granitic and other crystalline rocks.

In an island lying so close to the African continent as Mada-gascar, the natural assumption would be that, if it possessed a mammalian fauna at all, such fauna would be closely allied to that of the mainland. As a matter of fact, precisely the reverse is the case, and out of a total of fully 28 genera of non-volant mammals now or recently inhabiting the island, only three are common to Africa. This, however, is by no means all, for out of these three genera two (*Hippopotamus* and *Sus*) are such as have probably crossed the intervening channel, although at a time when it was narrower than at present, while it is quite possible that the third (*Crocidura*) may have been introduced by human agency. Even this, however, scarcely gives a true idea of the case. In the first place, not only are the peculiar genera unknown in Africa, but they are equally strange to all the other regions of the world. In the second place, these genera belong to groups which form only a very small por-tion of the existing mammalian fauna of the Ethiopian region. At the present day, as will be more fully indicated in the following chapter, Ethiopian Africa is especially characterised by its nume-rous antelopes, as well by giraffes, zebras, rhinoceroses, elephants, hippopotami, wart-hogs, bush-pigs, lions, leopards and various other large cats, baboons, anthropoid apes, aard-varks, and ostriches. But, with the exception of the aforesaid bush-pig and extinct hippopotamus, not a single representative of any one of these groups is found in Madagascar. In place of such animals, Madagascar is populated by a host of lemurs, so numerous that the number of their species considerably exceeds that of all the other non-volant mammalian inhabitants of the island. Civet- and mungoose-like species, all pertaining to peculiar genera, alone represent the numerous Carnivora of the mainland; the Insectivora, in addition to the aforesaid *Crocidura*, or musk-shrew, include only the peculiar family of the tenrecs (*Centetidæ*), which is confined to the island, and a representative of the Ethiopian family *Potamogalidæ*; while the rodents comprise five genera of the cosmopolitan mouse-family (*Muridæ*), more or less closely allied to one another, but different from any found elsewhere.

The following is a list of the genera of non-volant Malagasy mammals; those which are extinct being indicated by an asterisk, and the names of all the groups peculiar to the island printed in italics.

PRIMATES.—LEMUROIDEA.

    LEMURIDÆ.

        Chirogale (Mouse-lemurs); 4 species.

        Microcebus (Dwarf Lemurs); 5 species.

        Opolemur (Fat-tailed Lemurs); 2 species.

        Lemur (True Lemurs); 8 species.

        Mixocebus (Hattock); 1 species.

        Hapalemur (Gentle Lemurs); 2 species.

        Lepidolemur (Sportive Lemurs); 8 species.

        Avahis (Avahi); 1 species.

        Propithecus (Sifakas); 4 species.

        Indris (Endrina); 1 species.

   * *MEGALADAPIDÆ.*

        * *Megaladapis* (Giant Lemur); 1 species.

    CHIROMYIDÆ.

        Chiromys (Aye-aye); 1 species.

INSECTIVORA.

    SORICIDÆ.

        Crocidura (Musk-shrews); 1 species.

    CENTETIDÆ.

        Centetes (Tenrec); 1 species.

        Hemicentetes; 2 species.

        Ericulus (Hedgehog-tenrec); 1 species.

        Echinops; 1 species.

        Microgale (Long-tailed tenrecs); 3 species.

        Oryzorictes (Rice-tenrecs): 2 species.

INSECTIVORA (*cont.*).

    POTAMOGALIDÆ.

        Geogale; 1 species.

CARNIVORA.

    VIVERRIDÆ.

        Cryptoproctinæ.

            Cryptoprocta; 1 species.

        Viverrinæ.

            Fossa; 1 species.

        Herpestinæ.

            Galidictis (Striped Mungooses); 2 species.

            Galidia (Ring-tailed Mungoose); 1 species.

            Hemigalidia (Brown-tailed Mungoose); 1 species.

        Euplerinæ.

            Eupleres (Small-toothed Mungoose); 1 species.

RODENTIA.

    MURIDÆ.

        Hypogeomys; 1 species.

        Nesomys; 2 species.

        Brachytarsomys; 1 species.

        Hallomys; 1 species.

        Eliurus; 2 species.

UNGULATA.

    SUIDÆ.

        Sus (Potamochœrus); 1 species.

    HIPPOPOTAMIDÆ.

        Hippopotamus; 1 species (extinct).

Considering the fauna in more detail, it may be first mentioned that the lemurs (Lemuroidea) differ from the higher, or Anthropoid Primates by their generally lower grade of organisation, as well as by certain features of the skull and internal anatomy which need not be more fully noticed here. They all have fox-like, expression-less faces; and, with the exception of the aye-aye and the Asiatic tarsiers, they are characterised by the innermost pair of upper incisor teeth being separated from one another in the middle line.

At the present day lemuroids are represented elsewhere only
in the Ethiopian and Oriental regions ; the African forms being
more nearly allied to the Malagasy types than are those of Asia.
As stated in an earlier chapter, the group was, however, well
represented in the lower Oligocene of western Europe, where
certain forms (*Microchœrus*) distinctly approximate some of the
living kinds, although differing in the conformation of the first
lower premolar tooth, which in the existing *Lemuridæ* assumes

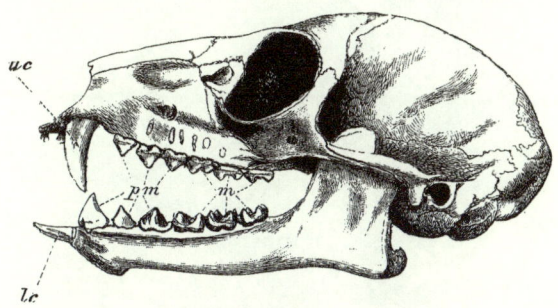

FIG. 48. SKULL OF LEMUR.
uc. upper canine ; *lc.* lower canine ; *pm.* premolars ; *m.* molars.

the form and function of a canine or tusk.   In the latter family (of
which the distribution is coextensive with that of the suborder) the
first three genera in the foregoing list belong to a subfamily
(*Galaginæ*) distinguished by the elongation of the bones of the
tarsus, and represented by an allied genus (*Galago*) on the African
mainland.   The next four genera constitute the typical subfamily
(*Lemurinæ*), which is absolutely confined to Madagascar and some
of the islands of the Comoro group, and of which the ring-tailed
lemur (*Lemur catta*) is one of the most familiar examples in
European menageries.   All these lemurs, which have long,
although non-prehensile tails, differ from the first subfamily by the
normal structure of the bones of the ankle.   The third subfamily
(*Indrisinæ*), which is likewise peculiar to this region, includes the
avahi, sifakas, and the endrina, all of which differ from the two
preceding groups by having only thirty, in place of thirty-six teeth ;
while the endrina is peculiar in having the tail rudimentary.   The
group includes the largest living lemurs ; the sifakas and endrina

differing from other members of the suborder by their diurnal habits. They form a characteristic feature in every wooded Malagasy landscape, there being scarcely a copse in the island which is not tenanted by one or more of these strange creatures ; and when

FIG. 49.   RING-TAILED LEMUR (*Lemur catta*).

walking from covert to covert, they do so in an erect posture, with their hands clasped behind their necks.

Whereas the endrina (the largest living lemur) is only two feet in length exclusive of the rudimentary tail, the extinct *Megaladapis*, whose remains have been obtained from the Ambolisatra marsh, had a skull three times the size of that of the latter, so that the whole animal might be compared in size to a mandrill. The skull of this species is characterised by the great elongation of the face, and in several respects shows resemblances to that of the European Oligocene genus *Adapis*; although the upper molar teeth are peculiar in having tritubercular crowns, whereas those of all modern lemurs are quadrangular. There is considerable reason to believe that the giant lemur was actually living in the middle of the seventeenth century, an otherwise unknown animal being described by De Flacourt in 1658 under the name of *trétrétrétré*, or *tratratratra*, which accords fairly well with the fossil remains. The giant lemur is, however, not the sole extinct member of the group from Madagascar, since the hinder part of a

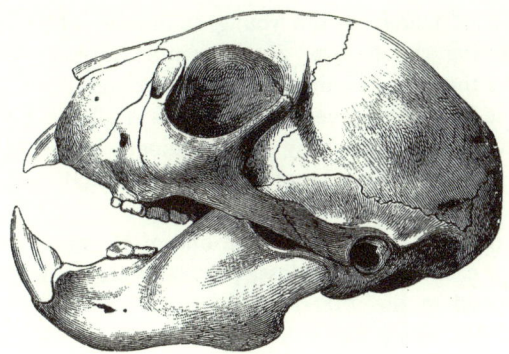

FIG. 50. SKULL OF AYE-AYE.

skull indicates another, but at present unnamed genus, apparently allied to *Hapalemur*. The last of the Malagasy lemurs is the singular aye-aye (*Chiromys*); a creature representing by itself a separate family, broadly distinguished from all other members of the suborder by the curious resemblance of the dentition to that of the rodents, to say nothing of the extreme elongation and slenderness of the middle finger of the hand.

Apart from the single musk-shrew, the Malagasy insectivores all belong to the group with tritubercular upper molar teeth, which, as already mentioned, is now confined to the more southern portions of the world, and is evidently a very primitive one. The small mouse-like creature (*Geogale aurita*) representing the *Pota-*

mogalidæ differs from its Ethiopian cousin, not only in its inferior dimensions, but likewise in having but thirty-four in place of forty teeth; and it is possible that, when more fully known, it will have to be assigned to a family by itself. As stated in a previous chapter[1], the tenrecs (*Centetidæ*) appear to have their nearest allies in the West Indian solenodons, although the relationship is now believed to be somewhat less close than was formerly supposed. The common tenrec (*Centetes*), which is the largest member of its order, measuring from a foot to sixteen inches in length, is a tail-less creature, remarkable for the possession of four pairs of upper molar teeth, as in marsupials. Much smaller are the two species of *Hemicentetes*, which, in addition to differences in the dentition, are distinguished by having rows of spines along the back at all ages, instead of merely in the young condition. The hedgehog-tenrecs, forming the genera *Ericulus* and *Echinops*, are small forms having the whole of the back and tail covered with close-set spines. The two other genera are spineless at all ages; *Microgale* being readily distinguished by the inordinate length of the tail which is equal to twice that of the head and body, while *Oryzorictes* has this appendage relatively short.

The largest, and at the same time one of the most peculiar of the Malagasy carnivores is the fossa (*Cryptoprocta*), which, although usually included in the *Viverridæ*, is so different from all other members of that group that it has been regarded as constituting a family by itself, specially characterised by the feline type of dentition. On the other hand, Daubenton's civet (*Fossa*), although representing a genus by itself, has its nearest relative in the widely distributed Oriental rasse (*Viverra malaccensis*). The latter species, although now found both in Madagascar and the Comoro group, has in all probability been introduced there. Of the four remaining genera, *Galidictis*, *Galidia*, and *Hemigalidia* are more or less closely allied to the mungooses, although presenting certain structural differences from other genera; but the fourth (*Eupleres*) is so markedly distinct as to constitute a subfamily by itself.

The five genera of murine rodents call for but little remark, although it is noteworthy that they are all more or less closely

[1] *Supra*, p. 70.

allied, and belong to the cricetine section, which contains the oldest members of the family. Nothing need be said in regard to the two ungulates, except that they both belong to Ethiopian types. Although bats are not taken much into account in the present volume, it is important to notice a peculiar distribution of the fruit-bats or *Pteropodidæ* ; more especially as this coincides with that of many Malagasy birds. On this point Dr Blanford writes that "the only African genus belonging to the family is *Epomo-*

FIG. 51. THE FOSSA (*Cryptoprocta ferox*).

phorus[1], which is confined to the continent, whilst throughout the Mascarene archipelago, and even in the Comoro islands in the Mozambique channel, the typically Oriental genus *Pteropus* occurs, and is represented in various islands by five species, one or two of them only distinguished by critical characters from the common ' flying-fox ' of the Indian peninsula."

In groups other than mammals, certain common features between the reptiles of Madagascar and South America have been

---

[1] A second genus, *Scotonycteris*, has been described from the Cameruns since this passage was written ; and the author has omitted mention of *Trygenycteris* (*Megaloglossus*).

mentioned in an earlier chapter[1], where it was attempted to show that although these instances of discontinuous distribution might be explained by parallel migration from a common northern centre, yet that the Tertiary mammalian evidence indicated that the American forms had reached their present habitat by way of Madagascar and Africa. It will suffice to add here that giant land-tortoises, which existed in the Mascarenes during the present epoch, are represented by extinct species from the superficial deposits of Madagascar; and that the latter have also yielded remains of gigantic flightless birds (*Æpyornis*) markedly distinct from any other known type. And here it may be mentioned that the chamæleons (*Chamæleontidæ*) present a certain similarity in their distribution to the lemuroids, the Malagasy region including 23 out of the 49 species, while nearly all the others are Ethiopian. As a whole, the Malagasy reptiles, with the exception of the snakes, are stated to be more nearly allied to those of the mainland than are either the mammals or the birds; but the amphibians exhibit more decided traces of Oriental affinities.

Concentrating our attention mainly on the mammals alone, their distinctness from those of all other parts of the world are quite sufficient to indicate the right of Madagascar to form the centre of a separate zoological region. In the survey of the lower Oligocene fauna of Europe it has been shown that both lemuroids and civet-like carnivores were common, one of the latter having been referred to the existing genus *Viverra*. Hence it is probable that to this fauna we must look for the ancestors of the Malagasy mammals. The only lemuroids closely allied to those of Madagascar are the African galagos, and as the civet-family (*Viverridæ*) is better represented in Africa than elsewhere, it may be taken for granted that Madagascar received its mammalian fauna from the mainland. Putting aside the hippopotamus and bush-pig, which doubtless arrived later, the Malagasy fauna can, however, have been derived from Africa only at a time anterior to the introduction of the modern types of ungulates into that continent, when it was chiefly popu·lated by lemuroids and civet-like carnivores[2]. The question then

Relations of Madagascar to the Mainland.

---

[1] *Supra*, p. 131.

[2] It is of course probable that some of the Oligocene primitive ungulates

narrows itself as to the probable date of the connection between the island and the continent. Now, so far as can be determined, none of the European Oligocene lemuroids are referable to the family *Lemuridæ*; and since both the Ethiopian and Malagasy representatives of the subfamily *Galaginæ* resemble one another in the peculiar structure of the ankle, or tarsus, it is pretty evident that not only was the family, but likewise the subfamily differentiated before the separation of Madagascar. Allowing time for the southward migration of the Oligocene lemuroids and civets, and the modification of the former into the *Galaginæ*, it seems impossible to put the separation at an earlier date than the Upper Oligocene, while it might well be Miocene[1]. Confirmation of this comparatively late separation of the island is afforded by some observations of Dr Blanford with regard to the passage of the bush-pig across the intervening strait, for it is evident that both that animal and the hippopotamus must have reached Madagascar by swimming, as otherwise more ungulates would assuredly have

migrated into Africa with the lemuroids and *Viverridæ*; but if so, all have died out. The Tertiary palæontological history of Africa or Madagascar can alone decide this point; but if ancestors of the South American extinct ungulates reached their home by way of Africa, it is certain that primitive members of that order must have first passed into that continent.

[1] It must be remembered that we are here dealing with the mammalian evidence alone. In regard to the molluscan Mr A. H. Cooke (*Conchologist*, 1893, p. 131) states that this region possesses sufficient individuality from that of the mainland to entitle it to separation. The *Helicidæ* are peculiar, not being found in the Mascarenes, Seychelles, or Comoros. They seem to be related to certain Cingalese and Australian types. Upwards of fifty-four species of *Cyclostoma* are known, distributed over Madagascar, the Comoros, Seychelles, Mauritius, and Bourbon. The African *Bulimi* are represented by two species, but *Achatina* (so common there) is scarce; and groups of *Bulimi* are peculiar. A single species of the genus *Caliella* is identified with an Indian form; and unmistakable indications of Oriental affinities are afforded by the freshwater molluscs. There are two species of *Paludomus*, *Bithynia* occurs, and while several of the *Melaniæ* are of a type common in the Indo-Malayan countries, the *Melanatriæ*, which are peculiar to Madagascar, have their nearest allies in Ceylon or India. Although not a single African freshwater bivalve has yet been recorded from Madagascar, yet several Ethiopian genera of gastropods occur there, and, in common with the land-molluscs, indicate a former connection between Madagascar and Africa, and this, in Mr Cooke's opinion, occurred at an immeasurably remote epoch.

been found in the island.   After remarking that bush-pigs are stated to be more aquatic in their habits than ordinary swine, Dr Blanford[1] asks "how far could *Potamochœrus* swim?   Surely it is not likely that it could cross the Straits of Dover.   I think we are justified in assuming about ten miles as a probable limit of its power of crossing the sea, but, to be safe, let us suppose double as much.   Then, in Pliocene or Plistocene times, quite as probably the latter as the former, when *Potamochœrus* reached South Africa, Madagascar was separated by a channel not more than twenty miles broad.   The conclusion is inevitable, that nearly the whole depression of upwards of a thousand fathoms is of Pliocene or Post-pliocene date.   Of course it must not be assumed that this date is proved.   What we may consider, however, as beyond any doubt is that the depression cannot be older than the Middle Tertiary."   This view may be taken as practically identical with the one here advanced, namely that Africa and Madagascar were united till the period of the upper Oligocene or Miocene.

With the exception of the fruit-bats and Daubenton's civet, which, as already mentioned, is more nearly allied to the Oriental rasse than to the Ethiopian *Viverridæ*, the Malagasy mammals do not exhibit any well-marked alliance with those of India.   But the case is different with the birds, molluscs, and certain other groups; while we have no evidence that giant land-tortoises ever inhabited the African mainland, although an extinct species is known from the Indian Pliocene.

Basing his conclusion on evidence drawn from several sources, Dr Blanford, in the communication last cited, is of opinion that there was formerly a direct land-connection between India and South Africa, and that this connection "included the Archæan masses of the Seychelles and Madagascar, that it continued throughout upper Cretaceous times, and was broken up into islands at an early Tertiary date.   Great depression must have taken place, and the last remnants of the islands are now doubtless marked by the coral atolls of the Laccadives, Maldives, and Chagos, and by the Saya de Malha bank.   It is immaterial whether Bourbon, Mauritius, and Rodriguez ever formed part of

[1] Appendix, No. 8, p. 88.

the Mascarene land or not." It is added that if future soundings should indicate the absence of a bank extending the whole way from India to Africa, it may be a question whether the whole of the ocean-bed between those two countries has not sunk to its present depth since the Cretaceous era[1].

This presumed connection satisfactorily explains much in regard to the distribution of the molluscs. It is, however, certain that fruit-bats did not exist in the early Tertiary, and the *Pteropus* must accordingly have made the journey across the sea from India, aided by what remained of the chain of islands, which may have been more extensive during the Pliocene. The same explanation also holds good with regard to most of the Oriental types of birds. The case of the land-tortoises is, however, more difficult. Nearly allied forms have been found in Mauritius, Rodriguez, Madagascar, and Aldabra; and since this group is unknown, even in Europe, before the Oligocene, it is evident that they could not have travelled from India by means of the con-necting land-bridge, which is considered to have been broken up at the commencement of the Tertiary epoch. This being so, the probability is that they originally came from Africa; but whether they entered Madagascar with the ancestral lemurs, or whether they, or their eggs, were transported across the channel when narrower than at present, there is no evidence to show. Be this as it may, it is probable that they reached Rodriguez and Mauritius across the intervening sea, since even if these islands ever joined Madagascar, such union must apparently have been at a date anterior to the existence of true tortoises. That none of these tortoises could have been transported by sea from India is proved by an observation of Dr Blanford to the effect that on this line the currents invariably set from the Seychelles to India. It may be added that some writers have considered it probable that the giant tortoises of the Malagasy region, like those of the Galapagos Islands, attained their large dimensions after they had reached the islands they respectively inhabit. The existence of gigantic

---

[1] Neumayr (*Erdgeschichte*, 2nd ed. vol. II. p. 262, 1895) considers that when India was connected with Madagascar during the Jurassic era, only the southern extremity of that island was joined to South Africa.

species on nearly all the great continents during the Tertiary epoch seems, however, an insuperable objection to this view.

In the absence of any evidence as to the Tertiary vertebrate palæontology of eastern Africa, I have no suggestion to offer as to the origin of the gigantic birds of the genus *Æpyornis*, which during the late Plistocene or recent epoch formed such a marked feature in the Malagasy avifauna.

NOTE.—The Author has reason to believe that several new Malagasy mammals have been discovered by Dr C. I. Forsyth-Major; but as no description of these had appeared when this chapter was passed for press, they could not be noticed.

CHAPTER VII.

THE ETHIOPIAN REGION.

Extent—Characteristics of the Mammalian Fauna—Birds—Past History of
Ethiopia—Subregions.

IT would be difficult to find a much greater contrast to the
mammalian fauna of Madagascar than is presented by that of
Africa south of the tropic of Cancer ; the one area, as shown in the
last chapter, being characterised by the number of lemurs, together
with its peculiar *Viverridæ* and insectivores, while the other is
distinguished from all other parts of the world by the extraordinary
number (both as regards genera, species, and individuals) of large
ungulates which roamed through its plains and forests until deci-
mated or exterminated by the hand of man.   As regards the
number of individuals of large animals inhabiting equal areas, it is
quite probable that at the date when the bison flourished in its
millions on the North American prairies, the balance in this
respect may have been in favour of the New World ; but whereas
the prairies had but a single species, Ethiopian Africa was populated (for it is unfortunately necessary to write in the past tense)
with a host of species of antelopes, together with buffaloes, giraffes,
hippopotami, zebras, rhinoceroses, and elephants.   Such a fauna
has existed during the recent epoch in no other part of the world,
and in past times has only been paralleled by the lower Pliocene
fauna of southern Europe and Asia, although even this, as regards
the number of generic and specific types of antelopes, is by no
means its equal.

Separating Madagascar and the associated islands as a distinct
division, the Ethiopian region may be taken to include such
portions of Africa and Arabia as lie to the south of the tropic of

Cancer; northern Africa, as it did in the Pliocene, clearly forming
a part of the Holarctic region. The greater part of
the Sahara, as well as the northern portion of
the Nubian desert, although included in the Holarctic, will form
a kind of transition zone towards that region, as is also the case
with Syria, where a considerable number of Ethiopian types of
mammals are met with, while western Arabia shows a decided
approximation to the Oriental region, as is well exemplified by
the occurrence there of a species of the short-horned goats constituting the genus *Hemitragus*. As has been stated in an earlier
chapter, the Sahara and Nubian deserts, although they have
apparently never been submerged since the Cretaceous epoch, seem
always to have formed a more or less complete barrier to the passage
of the mammals of Algeria and the adjacent countries into the
Ethiopian region; and the main migration from the north and east
has thus taken place along the north-eastern side of the continent.

With the exception of its southern extremity, the whole of this
vast area lies within the tropics. As regards its physical features,
Dr Heilprin writes that "it presents several well-marked physical
peculiarities. In the first place, we have the vast expanse of
desert, which in the north occupies a transverse band varying in
width from about four to nearly ten degrees of latitude. This is
succeeded by what may not improperly be termed the open
pasture-lands, which as a narrow belt bounds the Sahara on the
south, curves southwards at about the position of Kordofan, and
occupies the greater portion of the continent lying east of the
thirtieth parallel of east longitude and south of the fifth parallel of
south latitude. A very considerable portion of this pasture-tract
forms a plateau of from four thousand to five thousand feet elevation. Included within it, and bounded on the west by the Atlantic
Ocean, is the region of the great equatorial forests, to the present
day a *terra incognita* in great part both to geographers and
naturalists. That portion of the African continent lying south of the
tropic of Capricorn differs in many respects, both as to its physical
configuration and its vegetable products, from the region to the
northward, and is characterised by a vegetation which is one of
the richest and most remarkable on the globe. With this marked
peculiarity in its vegetable development there is of necessity a

certain amount of faunal peculiarity superadded as well, but this is not sufficiently pronounced to permit of the separation of this tract from the tract lying immediately to the north.   We have thus on the continent three strictly defined faunal sub-regions : (1) the pasture-lands already described, constituting the *East Central African* sub-region, through whose vast expanse there is manifest a strong identity in the character of the animal products, the same or very closely related forms being in many instances found at the extreme points of this sub-region ; (2) the forest-tract, constituting the *West African* sub-region, whose animal products naturally differ very essentially from those of the last ; and (3) the desert or *Saharan* sub-region, containing a comparatively limited fauna, which, with almost insensible gradations, merges into the fauna of the Mediterranean tract.   To the same division belong in great measure the desert tracts of Arabia, or that portion of the peninsula lying to the south of the tropic of Cancer."

Although Dr Wallace had previously divided continental Ethiopia into an East African, West African, and South African sub-region, the foregoing arrangement seems, on the whole, preferable.   There are, however, considerable reasons for regarding Somaliland as a sub-region by itself, and South Arabia should perhaps constitute another.   Although the precise determination of such areas does not come within the province of this work, it is most important to notice that the West African or Equatorial forest tract continues right across the continent as far eastwards as the Congo-Nile watershed, that is to say, close up to Wadelai, where all traces of the West African fauna are suddenly lost.   On this point Mr O. Thomas[1] writes that "the abruptness with which the change of fauna occurs on the watershed is, considering the insignificant nature of the physical barriers, very remarkable, and almost unequalled in the distribution of the mammals of any part of the world.   The reason of the change is, however, clear enough, being not the occurrence of such barriers to migration as mountains or rivers, but the abrupt ending of the great West African forest, which, as we know from the travels of Schweinfurth and others, extends quite into this region, but abruptly ceases before the slopes of the upper Nile basin are reached."

[1] *Proc. Zool. Soc.* 1888, p. 17.

Before considering the leading characteristics of the Ethiopian mammal fauna and its relations to that of other regions, both in the present and the past, it is desirable to make reference to certain deficiencies, which are very difficult, if not impossible to explain adequately with our present knowledge.

Characteristics of Mammalian Fauna.

Although deer (*Cervus*), typical pigs (the genus *Sus* in its restricted sense), and bears are met with in northern Africa, no member of any one of these genera with the single exception of a pig (*Sus sennaarensis*) from the Sennaar district of Upper Nubia, inhabits Ethiopia. Even the entire family of the *Cervidæ* is unrepresented. These deficiencies form a most marked contrast between the Ethiopian region on the one hand, and both the Oriental and the Holarctic on the other. Almost equally conspicuous is the absence of goats and sheep ; the only exceptions being the occurrence of a species of *Capra* in the highlands of Abyssinia, and one of *Hemitragus* in Oman, in south-eastern Arabia. The absence of sheep and goats is, however, by no means so remarkable as that of the other groups above mentioned, since the former are exclusively mountain animals, and probably need some general lowering of the temperature to enable them to pass from one chain to another, and of the existence of such cold period there seems no evidence in Ethiopian Africa. A somewhat similar explanation will probably apply to the total absence of marmots (*Arctomys*), susliks (*Spermophilus*), chipmunks (*Tamias*), beavers (*Castoridæ*), voles (*Microtinæ*), and picas (*Lagomys*), since all these are inhabitants of elevated or northern areas. More difficult to explain is the absence of all shrews (*Soricidæ*), with the exception of one genus peculiar to the region ; but the deficiency of moles (*Talpidæ*) may perhaps be accounted for by the slow travelling powers of these animals, which did not allow them time to pass into Ethiopia during the (probably short) period when its connection with other regions was of such a nature as to permit their living in the intermediate lands. Possibly also the absence of moles from peninsular India has something to do with this deficiency. In this connection it is worth remark that the place held in the Holarctic region by moles is by no means unoccupied in the Ethiopian, both the golden moles (*Chrysochloris*), and the so-called Cape mole (*Bathyergus*), with its allies, having similar subterranean habits.

Together with the Oriental, the Ethiopian region shews a marked distinction from all others as the sole habitat of the man-like apes (*Simiidæ*). The Ethiopian forms comprise the chimpanzees (*Anthropopithecus*) and gorilla (*Gorilla*), both of which are restricted to the equatorial forest-region, where the former ranges as far east as Uganda, although the latter has a more circumscribed distribution. The occurrence of a fossil chimpanzee in the Indian Pliocene affords the most convincing evidence of the derivation of a large part of the Ethiopian fauna from what is now the Oriental region. Among the ordinary monkeys and baboons (*Cercopithecidæ*) there are five genera confined to this region. Of these, *Colobus* differs from the Oriental langurs (*Semnopithecus*) by the absence or rudimentary condition of the thumb, which frequently has lost all trace of a nail. On the other hand, the large genus *Cercopithecus*, which is most fully represented in the forest region, is as nearly related to the Oriental *Macacus*, from which it differs in the less prominent muzzle, and the absence of a projecting heel, or hinder lobe to the last lower molar tooth. This heel is, however, present in the mangabeys, or white-eyelid monkeys (*Cercocebus*), all of which are exclusively confined to the forest tract. Although the dog-faced baboons (*Papio*[1]) have a wider distribution, ranging from the Cape to Arabia, some of the largest and most peculiar forms, such as the mandrill and drill, are confined to West Africa. This genus is one of those common to the Ethiopian region and the Indian Pliocene. The nearly-allied gelada baboons (*Theropithecus*), of which there are two representatives, are, on the other hand, exclusively north-eastern types, one being confined to Abyssinia. Among the lemuroids, the galagos (*Galago*[2])—which, as stated in the last chapter, belong to a sub-family which attains its maximum development in Madagascar—extend right across the equatorial portion of the continent, descending somewhat on the east coast, where they are very numerously represented. The pottos (*Perodicticus*), which are nearly related to the lorises of the Oriental region, are, how-

---

[1] Syn. *Cynocephalus*.   [2] Including *Otogale*.

ever, exclusively confined to West Africa, where they are known by two species, regarded by some as representing as many genera.

It will be unnecessary to say more with regard to the Chiroptera than that the fruit-bats (*Pteropodidæ*) are represented solely by three peculiar genera in the Ethiopian region, of which *Epomophorus* has the greater number and the most peculiar of its species confined to the western forest tract, while the single species of *Scotonycteris* is solely found in this part of this continent, as is also the case with that of *Trygenycteris*.

Of great importance from a distributional point of view are the Ethiopian Insectivora, since of the five families found within the

FIG. 52. AFRICAN JUMPING-SHREW (*Macroscelides tetradactylus*).

area under consideration, two are almost or exclusively confined to it, while the third has only one aberrant representative in Madagascar, and even this may prove entitled to constitute a family by itself when its structure is more fully known than is at present the case. It is further noticeable that two of the families belong to the primitive group characterised by their tritubercular upper molar teeth, and were accordingly in all probability very early immigrants into the country. The first family almost peculiar to the region is that of the jumping-shrews (*Macroscelididæ*), the members of which are easily recognised by their elongated hind limbs, long snout, and leaping habits. While the typical genus *Macroscelides* has representatives throughout the region and also extends into Holarctic Africa, the four species of *Rhynchocyon*, in which the legs are shorter and the snout longer, are restricted to East Africa. To the latter is closely allied the European Oligocene genus *Pseudorhynchocyon*, and it would thus seem that the family, although unrepresented in Madagascar, arrived at a relatively early date in Ethiopia. On the other hand, since the hedgehogs are represented only by the widely-spread typical genus *Erinaceus*, together with a West African species which has been separated as

FIG. 53. WEST AFRICAN POTAMOGALE (*Potamogale velox*).

Proechinus, it is probable that they did not make their appearance on the scene till a later epoch. In addition to the widely spread *Crocidura*, the *Soricidæ*, or shrews, are represented in Ethiopia only by three species belonging to the peculiar genus *Myosorex*, differing from all the other genera in which the teeth are white by the absence of long hairs on the tail[1]. With the West African genus *Potamogale*, we come to the first of the two families with tritubercular upper molars; the present one (*Potamogalidæ*) being represented elsewhere only by the Malagasy *Microgale*, of which, as already said, the systematic position is doubtful. The potamogales, which attain a couple of feet in length, are thoroughly aquatic in their habits, swimming by the aid of the highly compressed tail. It has generally been considered that there is only a single species, but it has recently been suggested that there may be two. The family is probably an ancient one, although we have no fossil evidence to this effect, *Microgale*, even if it belong to another family, indicating that the group was among the earlier mammalian colonists of Ethiopia. The golden moles (*Chrysochloridæ*), which take their name from the brilliant metallic lustre of the fur in the majority of the species, are blind, earless, fossorial insectivores, having the middle toe of the fore foot furnished with an enormously powerful claw. As the moles (*Talpidæ*) form a group nearly related to the shrews, so the golden moles are equally nearly related to the tenrecs (*Centetidæ*) of Madagascar, from which they may be regarded as a highly specialised offshoot. Accordingly, it may be taken for granted that their ancestors obtained an entry into Ethiopia with the ancestral lemurs. It is not improbable that the prevalence of higher types of mammalian life has been the cause of the assumption of mole-like habits in the *Chrysochloridæ*; the tenrecs, which live in an island where the competition is much less severe, having retained the original primitive type. The golden moles, which may all be included in the single genus *Chrysochloris*, are mainly confined to South Africa, although one species extends on the east coast as far north as Ugogo.

Turning to the Carnivora, it is unnecessary to say anything with regard to the *Felidæ*, except that three species of *Felis* are common to Ethiopia and India; while the single species of hunting-leopard (*Cynælurus*) is likewise found in both countries, the genus being apparently also represented in the Indian Pliocene. In the civet family (*Viverridæ*) the true civets (*Viverra*) and

---

[1] See Dobson, *Proc. Zool. Soc.* 1887, p. 575.

mungooses (*Herpestes*) are common to the Ethiopian and Oriental regions; but the whole group attains a far greater development within the former area than elsewhere. Although the common genet is an inhabitant of the southern portion of the eastern Holarctic region, all the other species of *Genetta* are Ethiopian. The West African linsang (*Poiana*) is the Ethiopian representative of the beautiful linsangs (*Linsanga*) of the eastern portion of the Oriental region; the distribution of this group being a well-marked instance of the close alliance of the fauna of the Malayan countries to that of West Africa, to which reference will again be made. The Oriental palm-civets (*Paradoxurus*) are represented by the nearly-allied Ethiopian genus *Nandinia*, of which one species is West African, while the other comes from Nyasaland. The small-toothed mungoose (*Helogale*) is common to West and East Africa; and the allied genus *Bdeogale* has representatives on both sides of the continent. Two other peculiar Ethiopian genera, *Cynictis* and *Rhynchogale*, have each but a single species : the former being South African and the latter East African. The cusimanses (*Crossarchus*), although mainly characteristic of the forest tract, have one representative in Abyssinia. Lastly, the meerkat, the sole representative of the genus *Suricata*, is an exclusively South African form. A peculiar family (*Proteleidæ*) is constituted by the aard-wolf (*Proteles*), ranging from the Cape to Somaliland, and a near ally of the hyænas, from which it is distinguished by the extremely feeble development of the dentition. Both the spotted hyæna (*Hyæna crocuta*) and the brown hyæna (*H. fusca*) are now confined to Ethiopia, but the former ranged over a large portion of Europe as well as southern India during the Plistocene epoch; and as all the three living species are included in the same genus, there is no generic type in this family restricted to the Ethiopian region.

In the *Canidæ* wolves are absent, but the jackals are represented by species allied to *Canis aureus*, which occurs in North Africa; wild dogs (sub-genus *Cyon*) are, however, wanting. Although the long-eared foxes or fennecs, such as *Canis chama*, are common in Ethiopia, they are by no means characteristic, since they range into North Africa, Syria, Persia, and Afghanistan; being, in fact, like the gazelles, desert-haunting forms. There are, however, two genera of the family, each with a single species, now confined to this region; the first being represented by the Cape hunting-dog (*Lycaon pictus*), which is a large, somewhat hyæna-like animal, easily recognised by its spotted coloration and long bushy tail, and distinguished from the other genera by having only four toes on both the fore and hind feet. That the genus is of northern

FIG. 54. CAPE HUNTING-DOG (*Lycaon pictus*).

origin is proved by the occurrence of remains of an extinct species in the Glamorganshire caves. The small Lalande's fox (*Otocyon megalotis*) of South Africa, in addition to the enormous size of its ears, is peculiar in having four pairs of molar teeth in the lower jaw, and either three or four in the upper. Possibly a fossil species from the Indian Pliocene may be an allied type.

Passing by the *Ursidæ* and *Procyonidæ* as unrepresented in this region, we find the *Mustelidæ* very poorly developed, martens (*Mustela*) being absent, and true weasels very scarce. The striped Cape weasel (*Pœcilogale*) constitutes, however, a genus by itself; while the similarly coloured Cape polecat (*Ictonyx*), is one of two representatives of a small genus, the second of which ranges from

Sennaar to Egypt, and is also stated to occur in Asia Minor. A fossil species from the European Miocene may perhaps belong to this genus. The ratels (*Mellivora*) which are now represented by one Ethiopian and a second Indian species, are proved to be comparatively late immigrants from the north into this region by the occurrence of a fossil species in the lower Pliocene of Northern India. As the animal described under the name of *Galeriscus* is at present known only by a skin, it is not even certain that it belongs to the *Mustelidæ* at all.

Among the squirrel-like rodents, the most striking feature is the absence of the true flying-squirrels and their replacement by a

FIG. 55.   FULGENT AFRICAN FLYING-SQUIRREL (*Anomalurus fulgens*).

distinct family (*Anomaluridæ*), characterised, among other features, by the presence of scales on the under surface of the root of the tail. Mainly characteristic of the forest area, the typical genus *Anomalurus* has, nevertheless, representatives on the eastern side of the continent; although the mouse-like long-tailed flying-squirrel (*Idiurus*) is exclusively West African. While true squirrels (*Sciurus*) are common to this and the other regions of Arctogæa, the pigmy squirrels constituting the genus *Nannosciurus* are represented by one species (*N. minutus*) in West Africa, while all the others are Malayan. The spiny squirrels of the genus *Xerus* are now, on the other hand, exclusively Ethiopian, although their northern origin is proclaimed by the occurrence of an extinct species in the French Miocene. Dormice (*Myoxidæ*) are exceedingly abundant in Ethiopia, and if it is considered desirable to split up the family into more than two genera, the genus *Graphiurus*, characterised by the short, cylindrical, and tufted tail, and the simple structure of the molar teeth, will be peculiar to this region, as will also be the single West African form described as *Claviglis*. Dormice, as mentioned earlier, date in Europe from the lower Oligocene, and therefore they might well have entered Ethiopia with the ancestral lemuroids, although, so far as it goes, their absence from Madagascar is against this view. In the mouse-family (*Muridæ*) five sub-families are met with in Ethiopia, two of which are peculiar to the region. Curiously enough, the cricetine sub-family, which is the oldest of all, and the only one met with in Madagascar, is represented only by a single highly specialised genus (*Trilophomys*). The inference from this would seem to be that the ancestral cricetines and lemuroids entered Ethiopia together, whence some migrated to Madagascar; the former group, with the exception of the one peculiar genus, having completely died out on the continent, where the remaining murines are more recent immigrants from the north. In this family the Eastern Arctogæic group of the gerbils (*Gerbillinæ*) includes six genera, out of which no less than five, namely *Pachyuromys*, *Mystromys*, *Otomys*, *Dasymys*, and *Malacomys*, are exclusively Ethiopian, the last being West African. The elongated hind limbs and the transverse laminæ of the molars readily serve to distinguish the gerbils from the exclusively Ethiopian sub-family *Dendromyinæ*, in which the molars are rooted and tuberculated and the ears remarkably hairy. The typical genus *Dendromys* includes two dormouse-like forms, one from South, and the other from East Africa; the other genera being *Lima-*

comys, *Steatomys*, and *Lophuromys*, the last having the fur replaced by fine flattened bristles. Although there is no palæontological history of either of the preceding sub-families, the *Cricetinæ* have already been shown to date from the lower Oligocene of Europe. Their sole Ethiopian representative is the single species of the genus *Trilophomys*[1], from Upper Nubia and the Red Sea littoral in the neighbourhood of Suakin, and perhaps ranging into southern Arabia. The crested rat, as the creature is called, takes its name from the prominent crest of stiff hair running down the back; while it is specially characterised by the roofing over of the whole upper surface of the skull with bone, on which account it has (quite unnecessarily) been made the type of a family by itself. Perhaps, however, the most interesting member of the family inhabiting Ethiopia is a species of mouse known as *Deomys*, representing both a genus and sub-family by itself, and characterised by having its upper molars intermediate between those of the cricetines and murines. Doubtless, therefore, this rodent is a somewhat modified descendant of the true cricetines which entered Africa while it was still united to Madagascar; its habitat being that refuge for ancient types, the lower Congo valley. The two other generic representatives of the family, *Cricetomys* and *Saccostomus*, although resembling the hamsters in the presence of cheek-pouches, have molars like other *Murinæ*, and are accordingly referred to that sub-family. While there are two species of the second genus, there is but one of the first. By some zoologists the striped mice, as typically represented by the Barbary mouse, are separated from *Mus* to form a genus *Arvicanthis*[2], which is mainly Ethiopian.

The small family of the *Spalacidæ*, or burrowing-rats, has four genera out of six exclusively Ethiopian, while a fifth (*Rhizomys*), which is mainly Oriental, enters Abyssinia. Of the four Ethiopian genera, which constitute a sub-family by themselves, *Bathyergus* includes only the great Cape mole-rat of South Africa; and *Myoscalops* has also but a single species, although there are several of *Georhychus*. Closely allied to the latter are two tiny little burrowing and nearly naked creatures (*Heterocephalus*) from

---

[1] Syn. *Lophiomys*.   [2] Syn. *Isomys*.

Somaliland, which may be regarded as degraded descendants from that type. The family is unknown either in Madagascar or in a fossil state in Europe, although an extinct species of *Rhizomys* occurs in the Indian Pliocene, and it is, therefore, in all probability a comparatively late immigrant into Ethiopia. In addition to one genus ranging from Nubia to Siberia, the jerboa-family (*Dipodidæ*) has one peculiar Ethiopian genus, represented solely by the Cape jumping-hare (*Pedetes*);—a form so different from all the others that it must constitute a sub-family apart. With this genus we leave the mouse-like, and come to the porcupine-like group of the rodent order, in which the family *Octodontidæ* has nearly all of its representatives which are not Neogæic confined to this region, although the gundi (*Ctenodactylus*) of North Africa, in the neighbourhood of Tripoli, is south Holarctic. This genus and an allied type (*Pectinator*) from Somaliland, together with the next form, alone represent a sub-family typically characterised by the presence of a horny comb-like appendage and stiff bristles on each of the hind feet; *Pectinator* thus being Ethiopian. The South African rock-rat (*Petromys*), although now included in the same group, approximates to a sub-family of which all the members are South American; the resemblance between one of the latter and the Ethiopian species being curiously close. The two species of cane-rat (*Triaulacodus*[1]) constitute the sole African representatives of a sub-family containing a large number of Neogæic genera. Probably, as there has already been occasion to remark, both the Neogæic and Ethiopian representatives of this family trace their origin to the extinct *Theridomyidæ* of the Oligocene of the Holarctic region or some nearly allied forms; and as certain forms occur in the Santa Cruz beds of Patagonia, it is probable that the migration into Ethiopia was at least as early as that of the early lemuroids, the extinct *Pellegrinia* of the Sicilian Pliocene being the last survivor of the group in the northern half of the Old World[2]. Although there is no generic type of porcupine (*Hystri-*

[1] This name is proposed to replace *Aulacodus*, Temm. (1827), which was preoccupied in 1822 by Eschscholtz for a genus of Coleoptera.

[2] There is difficulty in this respect on account of the absence of hystricomorphous rodents from Madagascar; but perhaps the early forms entered the west side of the continent, while the lemurs travelled in along the east.

cidæ) peculiar to Ethiopia, it is not improbable from the wide distribution and antiquity of the group, that these rodents also entered tropical Africa at a comparatively early epoch. The deductions drawn from these rodents as to a connection between Africa and South America have been mentioned in Chapter III.

Among all the striking features of the mammalian fauna of Ethiopian Africa, none is more remarkable than the enormous preponderance of ungulates, many of which are of great corporeal bulk. Of these a large number of genera and two families are absolutely peculiar to this region. As may be gathered both from their absence at the present day in Madagascar, and the late epoch at which their remains are found in the Tertiaries of the Holarctic and Oriental regions, all these creatures have reached the Ethiopian region but recently. The *Hippopotamidæ* is one of the two families now practically peculiar to the region, the common species (*Hippopotamus amphibius*) having ranged over a considerable portion of Europe during the Plistocene and upper Pliocene ages, while even in the beginning of this century it

FIG. 56.   HEAD OF WART HOG (*Phacochœrus æthiopicus*).

frequented lower Egypt. The pigmy hippopotamus (*H. liberiensis*) of western Africa, which is referred by many writers to a genus apart, and more resembles the pigs in its mode of life,

L.   16

appears to be more nearly allied to a small species from the Sicilian and Maltese Pliocene. As already mentioned, with the exception of a single species (*Sus sennaariensis*), true pigs are unknown in Ethiopia, their place being taken by the two species of bush-pigs, forming the potamochœrine group of the same genus, and distinguished by the simpler structure of the molar teeth, as well as by the tendency of the front premolars to fall out in the adult. The reason for the occurrence of a third species of the group in Madagascar has been already sufficiently discussed[1]. Still more distinctive of the region are the hideous wart-hogs (*Phacochærus*), specially characterised by the facial warts from which they take their name, the huge tusks, and the great complexity of the last molar tooth in each jaw; the tusks and these molars being frequently the only teeth remaining in aged animals. It is, however, very noteworthy that certain extinct species of *Sus* from the Pliocene of India and Algeria have their last molars of a type which could easily be developed into those of the wart-hogs; and it would accordingly seem that the latter are comparatively recent descendants of ordinary pigs. Although wild camels are unknown in the region, the *Tragulidæ* are represented in West Africa by the water-chevrotain, which is now the only existing species of the genus *Dorcatherium*, although fossil forms occur in the Pliocene of India and the European Miocene; the ancestral forms having probably entered the region from India. The second ungulate family now confined to Ethiopia is the *Giraffidæ*, of which there appears to be only a single living species, although the North African form shows a decided difference in coloration from its southern brother. The occurrence of species belonging to the existing genus *Giraffa* in the lower Pliocene of Greece, Persia, India, and China, shows that giraffes came into Africa with the other ruminants; the African species being very probably the direct descendant of the extinct Indian one.

Abounding as it does in ungulates in general, Ethiopian Africa is the especial home of the antelope group, which here takes the place of the sheep and goats so characteristic of the elevated districts of the eastern half of the Holarctic region. Regarding

[1] *Supra*, p. 223.

their distribution in the Ethiopian region, Dr Sclater writes[1] that
although "antelopes are to be met with in every part of Africa,
they are most numerous where the country is comparatively open,
and where there are grassy plains interspersed with sheltering
bushes.   South of the tropic of Capricorn this condition generally
prevails, and throughout the Cape Colony and its adjoining territory they are—or, at all events, before the advent of a European

FIG. 57.   WATER-CHEVROTAIN (*Dorcatherium aquaticum*).

population were,—everywhere abundant.   The early settlers at the
Cape describe antelopes as to be met with in herds of thousands
on the *veldt*, and in parts of Africa where the white man and his
destructive firearms have not yet penetrated a similar condition
prevails even at the present day.   When we advance further north
and meet with the dense forests of the Niger and Congo basins,
we find the mass of antelopes holding rather to the more open
lands on the eastern coast, throughout which they are to be met

[1] *Natural Science*, vol. I. p. 255 (1892).

with in great abundance up to Cape Guardafui. The vast plains traversed by the Upper Nile and its tributaries are likewise well stocked with antelope life ; but in the great Sahara only some of the more desert-loving forms are to be found. In Senegambia again, and in the more open districts on the West Coast, many forms of antelopes occur, but they cannot rival the numbers and varieties of those of Eastern and Southern Africa." Although most of the genera of Ethiopian antelopes are peculiar to the continent, a few desert-haunting types range into southern Arabia, and hence northwards into Syria, where they enter the Holarctic region. Many of the groups have extinct representatives in the lower Pliocene of southern Europe and India, and since existing Ethiopian genera are more common in the latter than in the former area, it seems probable that the great migration into Ethiopian Africa has taken place from the east, by way of Syria or Arabia. As such extinct genera or species have been already noticed[1], it will suffice to take a very brief survey of the genera now mainly or exclusively confined to Ethiopian Africa.

The first section includes the hartebeests and their allies the bontebok and blesbok, all of which may well be included in the genus *Bubalis*, although the two latter are often separated as *Damaliscus*. One species of the typical group ranges into Syria, while a second is an inhabitant of Tunis. To the same section also belong the wildebeests, or gnus (*Connochætes*). On the other hand the numerous species of duikerboks (*Cephalophus*) constitute, so far as Africa is concerned, a section to themselves. They are, however, allied to the Indian four-horned antelope (*Tetraceros*), and it is not improbable that they are represented in the Pliocene of the Siwalik Hills. While many species of the genus are found in East and South Africa, the largest kinds are confined to the forest-districts of the West Coast. The small African antelopes classed by Sir V. Brooke in the *Cervicaprinæ* and included in the genera *Neotragus* and *Nanotragus* are now referred by Messrs Sclater and Thomas to a section apart, under the name of *Nanotraginæ*, and are classed in six genera. Of these, *Madoqua*[2], with six species, includes Salt's antelope (*M. saltiana*); *Nanotragus*

---

[1] *Supra*, pp. 197—206.   [2] Syn. *Neotragus*.

is represented only by the minute royal antelope (*N. pygmæus*) of Guinea; *Nesotragus* is typified by the Zanzibar steinbok (*N. moschatus*); the true steinbok (*Raphiceros campestris*) forms the fourth genus; the oribi of South Africa (*Oribia scoparia*) is still more distinct; while the well-known klipspringer (*Oreotragus saltator*), which ranges from the Cape to Abyssinia, differs from all the rest by its coarse brittle fur.

The *Cervicaprinæ* include larger forms. Foremost among these is the South African rheebok (*Pelea*); while the water-buck and its allies (*Cobus*) are some of the largest of all antelopes. The last representative of this section (*Cervicapra*) is typified by the South African rietbok, but there are also other species in West and East Africa. The fine South African antelope known as the pala (*Æpyceros*), together with an allied species from the West Coast, form the first representatives of another section. Here belong the true gazelles (*Gazella*), more characteristic of the desert tracts of the eastern half of the Holarctic region, although represented in South Africa by the somewhat aberrant springbok, as well as by several more typical species on the East Coast. Clarke's gazelle (*Ammodorcas*) of Somaliland is, however, the sole species of an exclusively Ethiopian genus; and the same is the case with Waller's gazelle (*Lithocranius*), of which the range extends on the East Coast from Somaliland to Kilimanjaro. The single species of the East African genus *Dorcatragus* seems to be an aberrant gazelle, with the trunk-like muzzle of *Madoqua*, but retaining the small gland-pits of the type. Yet another section of antelopes is typified by the beautiful sable antelope and its allies (*Hippotragus*), in which the horns sweep backwards in a graceful curve, and are ringed nearly to their tips. This genus is exclusively Ethiopian, but in the one typified by the gemsbok (*Oryx*), and characterised by the long and slender horns being either straight or but slightly curved, and ringed only at the base, the range includes all the desert regions of Africa, Arabia, and Syria, although the majority of the species are Ethiopian. To the same section belongs the addax antelope (*Addax*), but although this animal occurs as far south as latitude 18° N., it is mainly an inhabitant of the deserts of North Africa, Arabia, and Syria. The last section includes some of the largest and at the same time the most beautiful of all antelopes, the three Ethiopian genera being all characterised by the more or less strongly-marked spiral twisting of their horns, and the short crowns of their molar teeth ; the last feature distinguishing them sharply from the sable antelope and gemsbok group. The first of the genera in question includes the

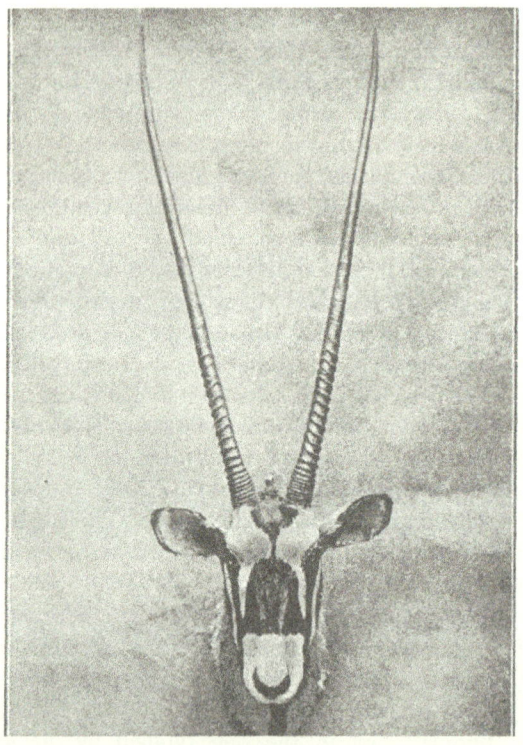

FIG. 58.   HEAD OF GEMSBOK (*Oryx gazella*).

harnessed antelopes (*Tragelaphus*), in which the spiral twisting of the horns is less marked than in the other two ; these antelopes frequenting forest or jungle, and being most numerous in western Africa. The kudus (*Strepsiceros*) agree with the last in that the females are hornless, but the horns of the males form a more corkscrew-like spiral than is generally the case with the harnessed antelopes; while in the elands (*Orias*)[1] the horns are present in both sexes and form a close spiral.

Turning to the perissodactyle section of the order, we find Ethiopian Africa the home of several species differing markedly from any now living in other parts of the world, although, according to the system here adopted, these do not constitute generic groups by themselves. There are two, or possibly three species of *Rhinoceros*, both of which are two-horned, and differ from those of the Oriental region in the absence of canine and incisor teeth. Of these the common African rhinoceros (*R. bicornis*) is closely allied to an extinct species from the Pikermi beds of Attica; while Burchell's rhinoceros (*R. simus*), which is now nearly exterminated, has its nearest allies in the extinct *R. platyrhinus* of the Siwalik Hills, and the woolly rhinoceros (*R. antiquitatis*) of the Plistocene of Europe and northern Asia. In addition to being the habitat of the parent form of the domestic ass, which is confined to Somaliland and the adjacent regions, Ethiopian Africa is also characterised by containing all the striped horses, or zebras, separated by some authorities as a distinct genus, under the name of *Hippotigris*. The best known representatives of this group are the true or mountain zebra (*Equus zebra*), Burchell's zebra (*E. burchelli*), Grevy's zebra (*E. grevyi*) of the Galla country, and the quagga (*E. quagga*). The latter, although formerly abundant in the southern extremity of the continent, is now fast verging on extinction, and serves to connect the more typical members of the group with the true asses. In spite of the difference in their coloration, the zebras are indeed indistinguishable in their osteology and dentition from the latter, and it is quite possible that they are represented in a fossil state in the later Tertiaries of Europe, while they are not improbably the direct descendants of the ancestral genus *Hipparion*. The absence of tapirs from the Ethiopian region is as remarkable as the want of deer; but it is noteworthy that the former are unknown among both the Pikermi and Siwalik faunas.

Although the African elephant (*Elephas africanus*) is markedly distinct from the Oriental species, yet the two are so closely connected by intermediate extinct forms that it is impossible to regard

[1] The name is usually spelt *Oreas*, but as it is derived from ὀρειάς, the proper orthography is *Orias*.

them as the representatives of separate genera. The existing species is now confined to the Ethiopian region, but since its fossilised remains occur in deposits of Plistocene age in Algeria, Spain, and Sicily, it is evident that, like the spotted hyæna and lion, it formerly enjoyed a much more extensive range.

With the exception that a single species is found in Syria, the small rodent-like ungulates, known as hyraces, which constitute not only a family (*Procaviidæ*) but likewise a sub-order (Hyracoidea) by themselves, are especially characteristic of the Ethiopian region, where they are represented by a large number of species, more

Fig. 59. Cape hyrax (*Procavia capensis*).

particularly in the southern portion of the continent. Although these animals closely resemble the rhinoceroses in the structure of their molar teeth, they differ markedly from all the perissodactyles in having the carpus constructed on the linear type[1], and from all other living forms of the order in that their single pair of upper incisor teeth grow continuously throughout life, as in the rodents. They were formerly divided into at least two generic groups, but both the terrestrial and arboreal forms are now included in the single genus *Procavia*. Nothing definite is

[1] See figure 12 on p. 78.

known as to the past history of the group[1]; but it has been suggested (p. 85) that they may be allied to certain extinct South American ungulates.

The list of peculiar Ethiopian mammals is brought to a close by the aard-varks (*Orycteropodidæ*), which although generally included in the Edentata have nothing to do with the typical South American representatives of that order, and are here, together with the pangolins, regarded as forming an ordinal group—Effodientia— by themselves. Only a single genus (*Orycteropus*) now exists, of which there are two living Ethiopian species, and there are extinct species in the Pliocene of Persia and Samos. A skull from the Plistocene of Madagascar has been described as *Plesiorycteropus*, and another genus occurs in the French Oligocene. Not improbably some members of the family entered Africa and Madagascar with the ancestral lemuroids and civets, but the discovery of the Pliocene forms renders it probable that the existing genus is a later immigrant.

Finally, it may be mentioned that among the more widely-spread genera a few species of mammals are either now common to the Ethiopian region and India, or were so during the Plistocene age. In the *Felidæ* the lion (*Felis leo*), leopard (*F. pardus*), jungle-cat (*F. chaus*), caracal (*F. caracal*), and hunting-leopard (*Cynælurus jubatus*) still range over the two areas; fossilised remains of the first three of these also occurring in the European Plistocene deposits. On the other hand, the spotted hyæna (*Hyæna crocuta*), which lived in Southern India (as well as in Europe) during the Plistocene era, is now restricted to Ethiopian Africa; and the same is the case with the giant pangolin (*Manis gigantea*) of West Africa, fossilised remains of which have been discovered, in company with those of the spotted hyæna, in a cavern in Madras.

The following table shows the genera and family of mammals now more or less exclusively restricted to the Ethiopian region; the names of such as are practically peculiar to this area being printed in italic type.

---

[1] I am informed that a skull belonging to an extinct member of this group has been discovered in the Pliocene of Samos, but no description has been published.

I. **Primates.**

    1. ANTHROPOIDEA.

        SIMIIDÆ.

            Anthropopithecus. Equatorial Africa. Fossil in Indian Pliocene.

            Gorilla. W. African.

        CERCOPITHECIDÆ.

            Colobus.

            Cercopithecus. Largely W. African.

            Cercocebus. W. African.

            Theropithecus. N. E. African.

            Papio. Fossil in Indian Plistocene and Pliocene.

    2. LEMUROIDEA.

        LEMURIDÆ.

            Galago. Equatorial and E. African.

            Perodicticus. W. African.

II. **Insectivora.**

        MACROSCELIDIDÆ. Fossil in European Oligocene.

            Macroscelides (including *Petrodromus*). One N. African species.

            Rhynchocyon. E. African.

        ERINACEIDÆ.

            Proechinus. W. African.

        SORICIDÆ.

            Myosorex.

        POTAMOGALIDÆ. Elsewhere only in Madagascar.

            Potamogale. W. African.

        CHRYSOCHLORIDÆ.

            Chrysochloris (including *Chalcochloris*). S. and E. African.

III. **Carnivora.**

    FELIDÆ.

        Cynælurus. Elsewhere only in India, the same species being common to the two regions.

III. **Carnivora** (*cont.*).

VIVERRIDÆ.

Genetta.　One species in South Holarctic region.

Poiana.　W. African.

Nandinia.　W. and E. African.

Helogale.　W. and E. African.

Bdeogale.　W. and E. African.

Cynictis.　S. African.

Rhynchogale.　E. African.

Crossarchus.

Suricata.　S. Africa.

PROTELEIDÆ.

Proteles.　S. and E. African.

CANIDÆ.

Lycaon.　S. and E. African.　Fossil in European Plistocene.

Otocyon.　S. African.

MUSTELIDÆ.

Mellivora.　One African and one Indian species; and another from the Indian Pliocene.

Ictonyx.　Ranges into Egypt, and perhaps Asia Minor.

Pœcilogale.　S. African.

Galeriscus.　E. African.

IV. **Rodentia.**

ANOMALURIDÆ.

Anomalurus.　W. and E. African.

Idiurus.　W. African.

SCIURIDÆ.

Nannosciurus.　Elsewhere in Malaysia.

Xerus.　Fossil in European Miocene.

MYOXIDÆ.

Graphiurus.

Claviglis.　W. African.　The right of the one species to generic distinction is doubtful.

IV. **Rodentia** (*cont.*).

MURIDÆ.

  Pachyuromys.

  Mystromys. S. African.

  Otomys. S. E. and W. African.

  Dasymys. S. African.

  Malacomys. W. African.

  Dendromys. ⎫
  Limacomys. ⎪ These represent a peculiar Ethiopian
  Steatomys. ⎬   sub-family—the *Dendromyinæ.*
  Lophuromys. ⎭

  Trilophomys. N. E. African, and (?) S. Arabian.

  Deomys. W. African. Alone represents a sub-family.

  Cricetomys.

  Saccostomus. W. African[1].

SPALACIDÆ.

  Bathyergus. ⎫ Constitute a sub-family—the *Ba-*
  Georychus. ⎪ *thyerginæ;* the last of the four
  Myoscalops. ⎬ being confined to Somaliland,
  Heterocephalus. ⎭ while the first is S. African.

DIPODIDÆ.

  Pedetes. The sole representative of a sub-family.

OCTODONTIDÆ. Elsewhere at the present time only in
  the Neogæic realm and N. Africa.

  Pectinator. N. E. African.

  Petromys. S. African.

  Triaulacodus. W., S. and E. African.

V. **Ungulata.**

1. ARTIODACTYLA. ⎧ Fossil in European and Asiatic
  HIPPOPOTAMIDÆ. ⎪ Plistocene and Pliocene, and
           ⎨ also in Madagascar. Formerly
   Hippopotamus. ⎪ in lower Egypt.
           ⎩

SUIDÆ.

  Sus ; the *Potamochœrine* group, frequently regarded as
   a distinct genus, is peculiar to the Ethiopian and
   Malagasy regions.

  Phacochœrus.

---

[1] If *Arvicanthis* be accepted as a genus it should come here.

V. **Ungulata** (*cont.*).

TRAGULIDÆ.

    Dorcatherium. W. African; fossil in European Miocene, and also in the Pliocene of India.

GIRAFFIDÆ. ⎰ Fossil in lower Pliocene of Europe and
    Giraffa. ⎱   Asia.

BOVIDÆ.

    Bubalis (including *Damaliscus*). Ranges into Syria and Tunis; fossil in Indian Pliocene.

    Connochætes.

    Cephalophus.

    Madoqua. N. E. and E. African.

    Nanotragus. W. African.

    Nesotragus.

    Rhapiceros.

    Oribia. S. African.

    Oreotragus. S. and E. African.

    Dorcatragus. Somaliland.

    Pelea. S. African.

    Cobus. Fossil in Indian Pliocene.

    Cervicapra. S. W. and E. African.

    Æpyceros. S. and W. African.

    Ammodorcas. Somaliland.

    Lithocranius. E. African.

    Hippotragus. Fossil in Indian Pliocene.

    Oryx. Ranges into Syria.

    Tragelaphus. Perhaps fossil in European Pliocene.

    Strepsiceros. ⎱ Fossil in Indian Pliocene.
    Orias. ⎰

2. PERISSODACTYLA.

RHINOCEROTIDÆ.

    Rhinoceros; the species without front teeth are now peculiar to the Ethiopian region, although allied forms occurred in the European and Asiatic Pliocene and Plistocene.

EQUIDÆ.

    Equus; all the striped species confined to this region.

V. **Ungulata** (*cont.*).

3. *HYRACOIDEA.* ⎫
       PROCAVIIDÆ.   ⎬ Range into Syria.
       Procavia. ⎭

VI. **Effodientia.**   Ethiopian and Oriental.
      ORYCTEROPODIDÆ.   Fossil in French Oligocene.
           Orycteropus.   Fossil in Pliocene of Samos and Persia.

Among groups other than mammals, attention may be directed

Birds. to the remarkable difference between the birds of Ethiopia and those of Madagascar.   On this point Dr Blanford[1] writes that "the most characteristic African families, such as plantain-eaters (*Musophagidæ*), colies (*Coliidæ*), and wood-hoopoes (*Irrisoridæ*), barbets, hornbills, secretary-birds (*Serpentarius*), and a number of genera, such as *Lamprotornis* (glossy starlings), *Buphaga* (ox-peckers), *Laniarius*, and *Telephonus*, that are the common and familiar birds of every part of Africa south of the Sahara, are entirely wanting in the Mascarene Islands, including the Seychelles, Mauritius, etc., while no fewer than four peculiar families and a number of genera confined to the archipelago replace them.   Amongst the Mascarene birds, too, are found several representatives of Oriental genera, or genera closely allied to Oriental types, and without any near Ethiopian relations. Foremost among these are certain bulbuls, forming the genera *Ixocincla* and *Tylas*, the former composed of species which have been usually referred to the typically Oriental genus *Hypsipetes*, and the latter nearly affined.   In fact, as was shown by Geoffroy St Hilaire, and as Hartlaub has since pointed out, there is in the Mascarene avifauna a more marked connexion with Indian than with Ethiopian types.   In the Seychelles, especially, out of the seven Passerine genera represented by peculiar species, three, *Nectarinia*, *Zosterops*, and *Tchitrea*, are Indian and African, one, *Foudia*, is Ethiopian, but not Indian, and two, *Copsychus* and *Hypsipetes*, or *Ixocincla*, are Indian but not African."

All this is confirmatory, not only of the right of Madagascar and the Mascarenes to form a region by themselves, but likewise

---

[1] Appendix, No. 8, p. 89.   In this quotation the English or Latin names have in some cases been added to the original.

of the distinction between the Ethiopian and Oriental regions, which some have proposed to unite. It is, however, somewhat remarkable that secretary-birds (*Serpentarius*) are unknown in Madagascar, seeing that they are represented in the upper Oligocene of France, and may therefore be presumed to have entered Ethiopia with the ancestral lemuroids and civets. Finally, the ostriches (*Struthio*), which are now mainly confined to Africa and Syria, are evidently recent immigrants into the region, the genus being represented in a fossil state in the Indian Pliocene.

With the exception of the occurrence of remains of certain existing species, such as *Rhinoceros simus, Phaco-choerus*, etc., in the superficial deposits of southern Africa, nothing is known of the mammalian Tertiary palæontology of the Ethiopian region. Fortunately, however, the clue given by the existing fauna of Madagascar and the Tertiary faunas of Europe and southern Asia enables a considerable portion of the past history of the population of the region to be given with a fair degree of completeness. And here, with one important exception, Dr Wallace's explanation, as given in the *Geographical Distribution of Animals*[1], may be accepted almost in its entirety;—the one exception being, as mentioned in an earlier chapter, that the Sahara was never a sea during Tertiary times; although it appears always to have formed a barrier between northern 'Africa and Ethiopia. As already mentioned, the ancestral types of the existing mammalian fauna of Madagascar obtained an entrance into Ethiopia some time during the Oligocene period, and soon after ranged over the whole of what are now the Ethiopian and Malagasy regions, which were then united and possessed a common fauna. During the Pliocene age, when Madagascar had become isolated, came the great irruption into Ethiopia, of the higher and larger mammals, such as apes, monkeys, ungulates, etc., which were then flourishing all along southern Europe and Asia. Finding the country unoccupied and eminently suited to their existence, these rapidly attained a development now unequalled in any other part of the world ; many new genera being apparently evolved within the Ethiopian area, although a large

Past History of Ethiopia.

[1] Vol. I. p. 285—292.

number were already in existence at the time of the southern migration. Several of these existing genera are met with in the Pikermi deposits of Greece, but more were confined, at this epoch, to the Pliocene of Persia, Samos, and India; and it may therefore be assumed that the great migration was by way of Syria or Arabia. Dr Wallace has indeed expressed the opinion that a certain number of types—among them the elephants and rhinoceroses—obtained an entrance to the westward of Tunis; but there are no true elephants in the Pikermi deposits, and apparently none in those of Persia, whereas their remains abound in the Siwalik Hills. As to the rhinoceroses, although the Pikermi species is closely allied to the African *Rhinoceros bicornis*, the Siwalik *R. platyrhinus* is equally close to *R. simus*; and in the Siwaliks we meet with chimpanzees (*Anthropopithecus*), baboons (*Papio*), ratels (*Mellivora*), hippopotami, water-chevrotains (*Dorcatherium*), and several genera of Ethiopian antelopes, all of which are totally unknown in the Pikermi beds. Ostriches, too, are first known in the Siwaliks; while aard-varks occur in the Persian and Samos beds. All the evidence accordingly points to the great immigration having taken place along the eastern side of the continent; and the existence of certain species of mammals which are either still common to India and Africa, or which were so during the Plistocene epoch, lends support to this view. Further testimony in this direction is afforded by the occurrence of closely-allied generic types in the Ethiopian and Oriental regions. Among the lemuroids, for instance, the Oriental lorises (*Nycticebus* and *Loris*) are replaced in Western Africa by the potto and awantibo (*Perodicticus*); while in the *Viverridæ* the true linsangs (*Linsanga*) of the eastern half of the Oriental region are represented in Fernando Po by the allied *Poiana*, and the Oriental palm-civets (*Paradoxurus*) have very close allies in the two species of the Ethiopian genus *Nandinia*. A less marked instance is afforded by the occurrence of the water-chevrotain (*Dorcatherium*) in West Africa and of the true chevrotains in southern India and the eastern half of the Oriental region.

And here it may be remarked that especial stress has been laid upon the much greater resemblance that exists between the fauna of the eastern, or Malayan, division of the Oriental region and

Western Africa, than between that of peninsular India and Eastern and South Africa ; large man-like apes and linsangs being confined in the Oriental region to its eastern half, while palm-civets, lorises, and chevrotains are more abundant there than in other parts of the same region.   This, however, appears to be mainly or entirely due to similarity of climatic conditions, and not to original distributional distinctions.   And, it may be re-marked, increased acquaintance with the fauna of Ethiopia tends to show that types formerly thought to be confined to the western half of Africa really extend far to the eastward ; chimpanzees, for instance, being now known to range as far east as Ugogo, while the genus *Nandinia*, which was originally known solely by a West African species, is now proved to have an eastern representative in Nyasaland.

The similarity between the fauna of the Malayan sub-region and that of Western Africa naturally leads on to the consideration of the former land-connection between India and Africa.   Writing on this subject, Dr Wallace[1] observes that " we may now perhaps see the reason of the singular absence from tropical Africa of deer and bears ; for these are both groups which live in fertile and well-wooded countries, whereas the line of immigration from Europe to Africa was probably always, as now, to a great extent a dry and desert tract, suited to antelopes and large felines, but almost impassable to deer and bears.   We find, too, that whereas remains of antelopes and giraffes abound in the Miocene[2] deposits of Greece, there were no deer (which are perhaps a somewhat later development), neither were there any bears, but numerous forms of *Felidæ*, *Viverridæ*, *Mustelidæ*, and ancestral forms of *Hyæna*, exactly suited to be the progenitors of the most prevalent types of modern African zoology."

As mentioned in an earlier chapter, since this passage was written the discovery of the remains of a species of chimpanzee in the Indian Siwaliks has shown quite clearly that the line of communication between India and Africa must have included a wooded tract comparable to the existing equatorial African forest region ; and this would be true even if the migration

[1] *Op. cit.* p. 291.
[2] The Pikermi beds were formerly universally held to be of Miocene age.

had taken place from Africa to India, which was not the case. Evidence of such a tract is, I believe, afforded by the occurrence of fossilised tree-stems in many districts which are now desert. And along this tract there can be little doubt that the ancestors of the mammalian types now common to the West African and Malayan sub-regions originally wandered from their common Indian home. Subsequently the whole of the countries lying between eastern Africa and India have become deforested, while in Africa itself the forest-area has shrunk away from the eastern side of the continent.

This leaves the question of the absence of bears and deer from Africa without any adequate explanation. Bears are, however, in the main, mountain animals, some of which, like the isabelline bear of the Himalaya, inhabit districts where there is but little forest; and it is noteworthy that, with the exception of the sloth-bear, which forms a genus apart (*Melursus*), there are no bears in India proper, although a fossil species allied to the sloth-bear occurs in the Siwaliks. This being so, when we take into account the absence of ursine remains from the Pikermi and Persian beds, there is nothing very wonderful in the fact that none of these animals entered Ethiopia during the great Pliocene migration. The absence of all *Cervidæ* is more difficult to explain, seeing that deer of an Oriental type are abundant in the Siwaliks, while they are also sparingly represented in the Pikermi beds. Typical deer of the red-deer group are, however, totally wanting in the Siwaliks, as they are in the Oriental region at the present day; and we are, therefore, perfectly able to account for their absence from the Ethiopian region, although they occur in Africa north of the Sahara. With regard to the absence of Oriental types of deer in Ethiopia, it can only be said that it is as difficult to see any reason why these should have continued to flourish since the Pliocene in the Oriental region without ever having entered Africa, as it is to explain why giraffes, hippopotami, and ostriches should have disappeared from the former area to survive in the latter.

Another difficulty is presented by the case of the pigs, but here it may be suggested that the absence of the typical group of the genus *Sus* from the whole of Ethiopia, with the exception of Sennaar, may perhaps be accounted for by all the other species of that genus which originally entered the country having been developed into the more specialised *Phacochœrus*. Attention has already been called to the fact that the molars of some of the Indian Tertiary species of *Sus* show a distinct approximation to those of *Phacochœrus*, and further evolution might easily lead to the development of the latter. Since this genus is unknown in a fossil state from other regions, is it improbable that it has originated from *Sus* within the limits of its present habitat?

Summarising the results of the foregoing survey of the mammalian fauna of Ethiopia, it would appear that the Sahara has for a very long period formed a barrier between Ethiopian Africa and the northern part of the continent. When first populated by Tertiary mammals, Ethiopia and Madagascar were in union, and formed but a single zoological province, which would seem to have been to a great extent isolated from the rest of the Old World during the Miocene epoch. Had such conditions persisted this province would have been entitled to form a primary zoological realm by itself. During the Pliocene epoch, however, Madagascar became separated, while a more complete union of the continent with Asia by way of Syria or Arabia permitted the influx of larger and higher mammals from the eastward. Hence there is a most intimate relationship between the Pliocene fauna of India and the one now inhabiting Ethiopia; but the distinction between the two areas at the present day is fully sufficient to justify the separation of the Ethiopian from the Oriental region. Of all the zoological regions of the world, the Ethiopian may be regarded as the one which has been most recently evolved; and it may be shortly characterised by the sole possession of the gorilla and chimpanzees; the absence of bears and deer; and the presence of the African elephant, hyraces, rhinoceroses devoid of front teeth, zebras, hippopotami, wart-hogs, giraffes, numerous genera of antelopes, and aard-varks, as well as by the great development of its large ungulates in general. It shares with the Oriental region the distinction of being the sole habitat at the present day of man-like apes, true civets (*Viverra*), linsangs, palm-civets[1], ratels (*Mellivora*), elephants, rhinoceroses, and pangolins

---

[1] Also in the Austro-Malayan region.

(*Manis*), together with the rodent genera *Nannosciurus*, *Golunda* and *Atherura*. The probable relationship between Ethiopia and Neogæa has been sufficiently discussed in the third chapter.

Although, from the mammalian point of view, Ethiopia is a very modern region, as a continent it is one of the oldest, the greater portion of its area having been land since the Palæozoic epoch. As has been shown in an earlier chapter, in Palæozoic times southern Africa formed a portion of the great southern or equatorial continent distinguished from more northern lands by the peculiar characteristics of its flora; and it is probable that it remained connected with India until late in the Secondary epoch[1]; as is proved by the identity of the flora and reptiles of the two areas. Early, however, in this epoch there must also have been free communication with Europe, as shewn by the close alliance of the anomodont reptiles of the two continents, and likewise by the occurrence in both of the mammalian genus *Tritylodon*. In the Cretaceous, so far as vertebrates are concerned, our knowledge of the Ethiopian region is practically a blank.

In conformity with the plan adopted in other cases, the subregions into which Ethiopian Africa may be divided will be treated very briefly.

Sub-regions.

Writing of the desert-tracts of the Saharan sub-region, where the necessary conditions for existence are largely wanting, Dr Heilprin observes that "there is a marked impoverishment of the fauna. The more formidable carnivores, such as the lion and the leopard, are absent from most districts, leaving their places to be filled by some minor cats, the hyæna, jackal, fox, and fennec. The hoofed animals are represented (in some parts) by the buffalo, and a limited number of antelopes (*Gazella*, *Oryx*, and *Addax*). Among rodents the families of rats (*Muridæ*) and jumping-mice (*Dipodidæ*) are fairly represented, in addition to which we have the porcupine and hare (*Lepus mediterraneus*). The ostrich is sufficiently abundant throughout most of the sub-region."

In marked contrast to the poverty of the Saharan districts, is the remarkable richness of the great equatorial forest-tract,

---

[1] *Vide supra*, p. 224.

which is the exclusive home of all the man-like apes and of the lemurs of the genus *Perodicticus*. To this sub-region also belong *Potamogale*, the African linsang (*Poiana*), as well as several other genera of *Viverridæ*, such as *Nandinia* and *Helogale*. Among the rodents the flying-squirrels of the family *Anomaluridæ* are very characteristic of the forest tract, and the peculiar murine genus *Deomys*, as also *Saccostomus*, is restricted to it. The water-chevro- tain (*Dorcatherium*) is solely West African, as is the small Liberian hippopotamus; while certain genera of antelopes, such as the duikers (*Cephalophus*), harnessed antelopes (*Tragelaphus*), and elands (*Orias*), have larger and finer representatives here than elsewhere. The giant pangolin (*Manis gigantea*), the largest member of its genus, is likewise a West African form.

Although South Africa has a certain number of mammalian genera, such as *Cynictis, Suricata, Otocyon, Pæcilogale, Bathyergus, Pedetes, Petromys*, and *Pelea*, peculiar to it, others, such as the golden moles (*Chrysochloris*) and aard-wolf (*Proteles*), have been proved to extend far up the east coast, the latter occurring in Somaliland. Hence it seems advisable to unite Dr Wallace's South African sub-region with that portion of his East Central sub-region which is not included in the equatorial forest-tract. Of this East Central sub-region, as it may be collectively called, it will suffice to say that it is the home of the greater number of the characteristic Ethiopian mammals exclusive of those restricted to the forest-tract. Here antelopes attain their greatest numerical development, both as regards genera and species; and here is also the true home of the lion and the spotted hyæna; while to this sub-region are confined the Cape hunting-dog (*Lycaon*) and the aard-wolf (*Proteles*). The distribution of other genera is sufficiently indicated in the table already given.

The light which has recently been thrown upon the mammals of northern Somaliland has shown that, as regards antelopes at least, these are so peculiar that it may be questioned whether this tract is not entitled to be separated as a sub-region by itself. According to the lists given by Dr Sclater[1] and Capt. Swayne[2],

---

[1] *Natural Science*, vol. i. p. 264 (1892).
[2] *Seventeen Trips to Somaliland*, London (1895).

there are no less than sixteen species of antelopes found in northern Somaliland, of which the names are as follows, viz.:

1. Bubalis swaynei.	9. Gazella spekei.
2. Madoqua swaynei.	10. ,, sœmmerringi.
3. ,, phillipsi.	11. Ammodorcas clarkei.
4. ,, guentheri.	12. Lithocranius walleri.
5. Oreotragus saltator.	13. Oryx beisa.
6. Dorcatragus megalotis.	14. Tragelaphus decula.
7. Cobus ellipsiprymnus.	15. Strepsiceros kudu.
8. Gazella pelzelni.	16. ,, imbubis.

Among these Nos. 1, 2, 3, 4, 6, 9, 11 are quite peculiar to this district, while No. 12, which, like Nos. 6 and 11, is the sole representative of its genus, is only found elsewhere along the east coast as far south as the Tana river. Another generic form peculiar to Somaliland is *Heterocephalus*, including two small naked rodents with burrowing habits; while in the same order the single species of *Pectinator* is restricted to this district. Among species may be mentioned two musk-shrews (*Crocidura smithi* and *C. somalica*), a hedgehog (*Erinaceus sclateri*), as well as a banded mungoose (*Crossarchus somalicus*). The Somali ostrich seems likewise to represent a species by itself. On the other hand, in addition to those mentioned in the foregoing list of antelopes, there are several East or South African mammals, such as the aard-wolf, which range into Somaliland, and further evidence is perhaps desirable before the right of that country to form a separate sub-region can be admitted.

The case is more clear with regard to south-eastern Arabia, whence Mr O. Thomas[1] records the following fifteen species of land-mammals, viz.: *Xantharpyia amplexicaudata, Taphozous nudiventris, Rhinopoma microphyllum, Erinaceus niger, Crocidura murina, Herpestes albicauda, Canis pallipes, C. leucopus, Gerbillus dasyurus, Mus rattus, Lepus omanensis, Gazella muscatensis, Oryx beatrix, Hemitragus jayakari,* and *Procavia syriaca.* Of these Mr Thomas remarks that their geographical relationships " are, as might be expected, about equally with Africa and India, three of

---

[1] *Proc. Zool. Soc.* 1894, p. 449. In one case the generic title has been altered, in order to bring it into harmony with the system followed in this work.

the species being distinctly African in affinities, three Indian, and the remainder either peculiar or widely-spread, and of no special significance." The association of a goat belonging to the Oriental genus *Hemitragus* with such an essentially Ethiopian animal as a hyrax (*Procavia*), shows that we are here truly on the border-land between the two regions in question.

In conclusion, from whatever aspect it be regarded, Ethiopia is one of the most interesting of the regions of Arctogæa; and if, as is suggested in an earlier chapter, it has been the feeder by means of which South America received its earliest mammalian Tertiary fauna, it is entitled to an importance above all the other zoological regions of the realm to which it pertains.

CHAPTER VIII.

THE ORIENTAL REGION.

Sub-regions—Characteristics of the Mammalian Fauna—Past History of the Region—Malayan sub-region—Nicobars, Mentawi, and Christmas Islands—Philippine sub-region.

FAR inferior in extent to the Ethiopian, the Oriental region is taken to include those portions of the continent of Asia lying south of the Holarctic region (with the exception of southern Arabia, which is Ethiopian), together with the islands of Ceylon, Formosa, the Philippines, Sumatra, Java, Borneo, and numerous smaller ones. In India the northern limits of the region are formed by the higher ranges of the Himalaya, while "Wallace's line" constitutes the eastern boundary defining it from the Austro-Malayan region of the Notogæic realm. In a region so diversified as the Oriental, it would not be natural to expect a homogeneous fauna; and, as a matter of fact, there are in this respect great diversities between the different portions of the region, many of the peculiar genera having a very restricted distribution. Nevertheless, the positive and negative features of the mammalian fauna of the region as a whole are sufficient to indicate its zoological unity, and also to differentiate it from the Ethiopian region, to which it is now most closely allied. In the Himalaya there is a gradual transition towards the Holarctic fauna; and it is probable that in this portion of the area the differentiation between the Oriental and Holarctic faunas has been largely due to the elevation of the Himalaya itself, which has taken place entirely since the early part of the Tertiary period, and is to a considerable extent of Post-tertiary date. It has already been shown that the older Pliocene fauna of northern India and Burma contained a remarkable admixture of mammalian genera now respectively confined to the Ethiopian and Oriental regions, together with some of a Holarctic facies; and the completion of the elevation of the Himalayan chain was probably an important factor in the dispersal and differentiation of this common fauna. Holarctic types are again met with in force in the open desert regions on the north-western frontier of India. On the other hand, the fauna of the Philippines exhibits an approximation to that of the Austro-Malayan region, and thus shows a blending between the Arctogæic and Notogæic realms. In physical features the Oriental region displays great variation, a large portion of peninsular India consisting of open dry grassy plains, whereas the slopes of the eastern Himalaya, together with the greater part of Assam, Burma, and the Malayan countries are clad with luxuriant forests; these tropical or sub-tropical forest-regions being those where the fauna attains its fullest and richest development.

The poorest part of the region is, as Dr Wallace[1] observes, the great triangular plateau forming the Indian peninsula; this area differing remarkably from the Himalaya in its geological features, and having been land since an extremely ancient date, whereas the Himalayan area consists very largely of marine formations. Since it is stated in the passage cited that peninsular India during the Tertiary period existed as an island entirely disconnected from the Himalaya and Burma, it may be well to quote the later and more authentic views of the authors of the *Manual of the Geology of India*[2] on this subject. After reference to the extent of Eocene rocks in northern India, it is there stated that "the Peninsula of India in Eocene times was part of a tract of land, perhaps of a great continent united to Africa; that there was a sea to the eastward, extending far to the north-east, in the region now occupied by the Assam hills, and another sea to the north-west, covering great part, if not the whole, of Persia, Baluchistan, the Indus plain, and a portion of the upper Ganges plain. An arm of this sea extended from the north-west up the Indus valley in Ladak. The Himalaya, and perhaps Tibet, wholly or in part, were raised above the sea; but formed in all probability

[1] *Geographical Distribution of Animals*, vol. I. p. 314.
[2] First edition, pt. I. p. liii.

land of moderate elevation. Whether the Himalayan land was united to the Peninsula is, of course, uncertain—but very probably it was; for there is no evidence of marine conditions having existed in the Ganges plain to the east of the Dehra Dun; and if the ferruginous bands of the Subathu group be laterite, as they appear to be, the trappean detritus composing them must have been derived, in all probability, from the peninsular area; and the latter must consequently have extended northward to the base of the Himalayas, in the neighbourhood of Umballa.... In Miocene times, although marine conditions prevailed throughout western Sind, the area of the sea was very much smaller than in the Eocene period; for all the marine beds of the Punjab and Sub-Himalayas are destitute of marine fossils, and are probably fluviatile deposits."

From this it would appear probable that during the whole of later Tertiary times, at least, there has been no isolation of peninsular India from the eastern Himalaya and the Burmese countries; and consequently that the differences between the faunas of these areas are mainly or solely due to their differences of physical features and climate.

By Dr Wallace the Oriental region is divided into four sub-regions; namely (1) the Indian, comprising the whole of upper India, (2) the Ceylonese, including southern India and Ceylon, (3) the Indo-Chinese, embracing Assam, Burma, and such portion of China as lies within the limits of the region, and (4) the Indo-Malayan, which includes not only the Malayan archipelago and islands, but likewise the Philippines.

Sub-regions.

So far as India and its dependencies are concerned, the following amended scheme has been proposed by Dr Blanford [1], viz.:

i. *Himalayan.* The southern slopes of the Himalaya, from the base to about the limit of trees.

ii. *Indian.* India from the base of the Himalaya to Cape Comorin, with the exception of the Malabar coast, but with the addition of northern Ceylon.

iii. *Malabar or Ceylonese.* The Malabar coast and the neighbouring hills as far north as the Tapti river, together with southern Ceylon.

[1] *Fauna of British India*—Mammalia, p. v.

iv. *Burmese*. All Burma except south Tenasserim, and with the
       addition of Assam and the intervening countries.

v. *Malayan*. South Tenasserim, the Malay Peninsula, and the
       Malayan Islands as far as Wallace's line.

       Whether the Philippine Islands should be included in this
       sub-region, or should form one by themselves, may be
       doubtful.

vi. *Indo-Chinese*. Although not free from objection, this term
       may be employed for the sub-region indicated by that
       portion of China coming within the limits of the Oriental
       region.

Regarding these sub-regions in general, Dr Blanford observes
that some "may require further subdivision. Thus the fauna of
the North-west Provinces and Punjab differs considerably from
that of southern India, and both areas exhibit zoological distinctions from the forest-clad tracts of south-western Bengal.
There is also much difference between the animals of Pegu and
Arakan, on the one hand, and those of the drier regions of upper
Burma on the other; and even greater distinctions may be traced
between those found in the sub-tropical and those inhabiting the
temperate regions of the Himalaya. On the other hand, the sub-
tropical Himalayas were united with the Burmese sub-region by
Wallace, and the two are, perhaps, zoologically more allied to
each other than to any other sub-region."

Recent discoveries clearly indicate that the Philippine Islands,
exclusive of Palawan and the Calamianes, should form a sub-region
by themselves.

Taking the Oriental region as a whole, it may be stated that
the number of peculiar generic types of mammals is
less than in the case of the Ethiopian; and that **Characteris-**
there are but two families absolutely confined to it, **tics of the**
Mammalian
although a third is very nearly so. While sharing **Fauna.**
with the Ethiopian region the want of several groups of insectivores
and rodents, such as the typical shrews (*Sorex*), marmots (*Arctomys*), and voles (*Microtus*), it lacks some of the other deficiencies
of that region, true pigs (*Sus*) and deer being abundant, although
the latter belong to groups distinct from those of the Holarctic
region, while bears, belonging to two genera, are likewise met

with. Wart-hogs (*Phacochœrus*), aard-varks (*Orycteropus*), hyraces (*Procaviidæ*), and jumping-shrews (*Macroscelididæ*) are among some of the more characteristic Ethiopian animals which are wanting in the Oriental region at the present day, and have not hitherto been obtained there in a fossil state. Giraffes, a number of genera of antelopes, and hippopotami, are equally conspicuous by their absence, although this, as we have seen, is but a comparatively recent feature of the region, since most of these forms are represented in the Pliocene of India and Burma. A notable feature of the Oriental as distinct from the Ethiopian region is the circumstance that the great majority of its fruit-bats (like those of Madagascar) belong to the typical genus *Pteropus*, which is wanting in Ethiopia, while the three genera of the family found in the latter area are absent from the present region. Indeed there is a very curious dissimilarity between the flying-mammals of the two areas, Ethiopia possessing the flying-squirrels of the family *Anomaluridæ*, while the Oriental flying-squirrels all belong to the *Sciuridæ*, the genus *Pteromys* being peculiar to the region. Among the Insectivora, the so-called flying-lemur (*Galeopithecus*) is a peculiar Oriental type which has no Ethiopian representative. A somewhat similar instance to that of the flying-squirrels is afforded by the tree-shrews (*Tupaiidæ*) of the Oriental region, which are represented in Ethiopia by the jumping-shrews (*Macroscelididæ*). From the Holarctic region, the Oriental is distinctly differentiated by the presence of apes and lemurs and the abundance of monkeys, together with the presence of the groups mentioned above as being now restricted to this and the Ethiopian region, and likewise by the absence of the typical elaphine deer, marmots, susliks, voles, etc., and the scarcity of sheep and true goats, which, indeed, enter the region only in the north-western frontier of India.

Commencing with the man-like apes of the family *Simiidæ*, we find the Oriental region destitute of chimpanzees and gorillas, whose place is taken by the orangs (*Simia*) of Borneo and Sumatra, characterised by the reddish, instead of blackish, coloration of their hair, and their more wide departure from the human type. Orangs appear, however, to have inhabited northern India during the Pliocene, and as chimpanzees were then also in existence there, it would seem that the region must be regarded as the original home of the larger man-like apes. The smaller long-armed apes known as gibbons (*Hylobates*) are likewise characteristic of the Oriental region, where they range from Assam through the Burmese and Malayan countries to Hainan and Cambodia. In the upper Miocene these apes occurred in Central Europe, but—perhaps on account of their small size—their remains have not hitherto been obtained from the Indian Pliocene. Among the *Cercopithecidæ* the Oriental genera of monkeys are now entirely distinct from those of Ethiopia, although, as we have seen, the

FIG. 60.   SLOW LORIS (*Nycticebus tardigradus*).

African genus *Papio* occurs in the Indian Pliocene and in Madras survived till the Plistocene. And here it may be noticed that this is the only one of the Ethiopian genera of monkeys that is found in Arabia. Of the other genera, *Macacus* is mainly Oriental, although with representatives in northern Africa, Kashmir, Tibet, and Japan ; while the langurs (*Semnopithecus*) are likewise almost wholly confined to this region, although they range into Kashmir and eastern Tibet. As both these genera are found in the European and Indian Pliocene, they are evidently ancient types which were formerly widely spread in the eastern half of the Old

World. Each represents a sub-family by itself; and as both these sub-families have Ethiopian genera, which are unknown in a fossil state, it may be suggested that the latter (like the wart-hogs) have been evolved within the limits of the Ethiopian region from the Oriental types. To the same sub-family as the langurs belongs the singular proboscis monkey (*Nasalis*) of Borneo. Among the lemuroid Primates there are Oriental representatives of two families. Of these the *Lemuridæ* include the lorises of the genera *Nycticebus* and *Loris*, the former ranging over the Burmese, Malayan and Indo-Chinese sub-regions, while the latter is confined to southern India and Ceylon. Although these animals are nearly allied to the pottos (*Perodicticus*) of western Africa, nothing is known as to their past history. The tarsiers (*Tarsiidæ*), of which there is but a single genus (*Tarsius*), although several specific forms have been recognised, are almost confined to the Malayan sub-region, but are represented in Celebes, as they are in the Philippines.

Of the Insectivora, the most aberrant and remarkable forms are the flying-lemurs (*Galeopithecus*), constituting a sub-order by themselves, and ranging from south Tenasserim through the Malay peninsula and islands to the Philippines. As with the tarsiers, we have no palæontological history of these creatures, which are probably comparatively modern types. The most characteristic Oriental family of this group is that of the tree-shrews (*Tupaiidæ*), whose members have the form and habits of squirrels, with the structure of shrews. The typical genus *Tupaia*[1] ranges from India throughout the Burmese and Malayan regions, but is unknown in Ceylon; while the single representative of the pen-tailed tree-shrews (*Ptilocercus*), characterised by the pen-like extremity of the exceedingly long tail, is confined to Borneo and some of the adjacent islands. As mentioned in an earlier chapter, the European Miocene genera *Lanthanotherium* and *Galerix* appear to be ancestral types of this family; and this distribution of the family is a well-marked instance of the curious affinity existing between certain mammalian genera of the middle Tertiaries of Europe and

---

[1] Two species (*T. murina* and *T. frenata*) are often separated as *Dendrogale*, but as there is an annectent form (see Thomas, *Proc. Zool. Soc.* 1892, p. 225), this appears unnecessary.

those of the Malayan sub-region, the absence of *Tupaia* from
Ceylon probably indicating that the genus is essentially a Malayan
one which has immigrated but recently into India.   The hedgehog
family (*Erinaceidæ*), which, as already shown is an ancient one,
has a very remarkable distribution in the Oriental region ; true
hedgehogs (*Erinaceus*) ranging into India, but apparently not occurring in Ceylon, and being unknown to the west of the Bay of
Bengal.   In the latter districts their place is taken by the spineless
and more rat-like animals forming the genus *Gymnura* and the

FIG. 61.   TREE-SHREW (*Tupaia tana*).

allied *Hylomys ;* and here, again, we have to note the occurrence
of an allied type in the European Oligocene, which has been
described under the name of *Necrogymnurus*.   In passing, it may
be remarked that the survival of these early Tertiary insectivorous
types in the Malayan sub-region serves to lend support to the
suggestion made in a previous chapter[1] that opossums may also
have survived in the same area long after they had ceased to exist
in western Europe.

Passing over the moles (*Talpidæ*), of which the typical genus
Talpa only just impinges on the region in the frontiers of India,

[1] *Supra*, pp. 51, 57.

we come to the *Soricidæ* or shrews. Here the typical shrews (*Sorex*) are wanting, while the section of the family to which that genus belongs (characterised by the reddish-brown tips to the teeth) is represented only by the genus *Soriculus*, ranging from the southern slopes of the Himalaya to China. Of the widely-spread musk-shrews (*Crocidura*) it is unnecessary to speak ; but it may be mentioned that of the two almost tailless and scaly-footed species forming the genus *Anurosorex* one is from Assam and the other from eastern Tibet and Pekin. *Chimarrogale* includes two aquatic shrews, one of which is found in the eastern Himalaya, the hills north of Burma, and Mᵗ Kina Balu in Borneo, while the second is Japanese.

Among the Carnivora the region is especially rich in *Felidæ*, containing more species than any other part of the world. The tiger (*Felis tigris*) is usually regarded as one of the most character-istic mammals of India, but as its range extends northwards to Siberia, while its fossilised remains have been found within the Arctic circle, and it is unknown in Ceylon, there is a great proba-bility that this feline is a comparatively recent immigrant into India from the north-east. The range of the lion (*F. leo*) in this region is limited to India, not extending to the eastward of the Bay of Bengal, and as this animal was widely distributed during the Plistocene in Europe, while it ranges all over Africa, it may be regarded as essentially a western type, or exactly the opposite of the tiger. Possibly certain remains from the Indian Plistocene which have been assigned to the latter animal may really belong to the former. As noticed on p. 234, there are other species of *Felis*, as well as the hunting-leopard (*Cynælurus*), which are com-mon to India and Africa, some of these occurring in the European Plistocene, while only the jungle-cat (*F. chaus*) is found to the eastward of the Bay of Bengal, and there not further east than Burma. On the other hand, there are certain species, like the clouded leopard (*F. nebulosa*) and the marbled cat (*F. marmorata*), which are essentially eastern forms, their range including the Malayan sub-region and India, but not Ceylon. The rusty-spotted cat (*F. rubiginosa*) and the Indian desert-cat (*F. ornata*) are species whose range is limited in one case to India and Ceylon, and in the other to India alone.

In the civets and their allies (*Viverridæ*) the Oriental region approaches the Ethiopian in richness, and thereby stands in marked contrast to the Holarctic, which contains only a single species of *Genetta* and another of *Herpestes* in southern Europe, although the latter genus ranges into Kashmir. Of the true civets (*Viverra*) the whole of the species, with the exception of one from the Ethiopian region, are Oriental, and some are confined to the countries to the east of the Bay of Bengal; one small species being separated by many zoologists as *Viverricula*. The beautifully-coloured linsangs (*Linsanga*) are exclusively Oriental, and are confined to the eastern Himalayan and Malayan sub-regions, although represented in West Africa by the nearly allied *Poiana*. Equally characteristic of the region are the two species of *Hemigale*, which are, however, exclusively Malayan, *H. hosei* being limited to the mountains of North Borneo. The palm-civets of the genus *Paradoxurus* range throughout the region, and have also representatives in Celebes : their place being taken in the Ethiopian region by the allied genus *Nandinia*. On the other hand, the two species of small-toothed palm-civets constituting the genus *Arctogale* are restricted to the Burmese and Malayan sub-regions ; the same being also the case with the binturong, which is the sole representative of the genus *Arctitis*, distinguished by its pencilled ears and prehensile tail. Still more circumscribed is the range of the peculiar genus *Cynogale*, of which the single species is confined to the Malayan sub-region. All the foregoing forms belong to the sub-family *Viverrinæ*; the *Herpestinæ*, which are so numerous in the Ethiopian region, being represented only by species of the large and widely-spread genus *Herpestes*. Both this genus and *Viverra* date from the European Oligocene, the latter also occurring in the Pliocene of France and India ; but none of the others are known in a fossil state. It is, however, probable that most of the other genera are comparatively modern derivatives from the original stock; and the high development in the Malayan sub-region of a group first known from the Oligocene of Europe is another instance of the relationship of the faunas of these countries.

Although the striped hyæna (*Hyæna striata*) is by no means confined to India, its range extending through south-western Asia

to northern Africa, it is unknown in Ceylon or in the countries on
the eastern side of the Bay of Bengal, which forms, indeed, the
present limits of the range of the genus in this direction.   As
remains of the existing African spotted hyæna (*H. crocuta*) have
been met with in a cave in Madras, while they are common in the
Plistocene of southern and central Europe, it is manifest that both
these animals are as essentially western types as is the lion.   And
it is a curious circumstance that nearly all these western types of
mammals ranging into India (of which a list is given in the sequel)
belong to genera which date only from the Miocene or Pliocene,
whereas very many of the Malayan or eastern types date from the
Oligocene.   During the Pliocene a single species of hyæna ranged
as far eastwards as China, and species were exceedingly abundant
in India at the same epoch.

As regards its *Canidæ*, the Oriental region is inferior to
the Ethiopian in lacking any peculiar generic type, although it
possesses a true wolf (*Canis pallipes*), and three species of wild
dog (*C. rutilans*, etc.), the latter, on account of the absence of the
last tooth in the lower jaw and other differences, being frequently
referred to a distinct genus, under the title of *Cyon*.   Whereas,
however, the Indian wolf, which ranges into southern Arabia, is
unknown either in Ceylon or in the countries to the east of the
Bay of Bengal, the wild dogs are found throughout the region, and
have also a representative beyond it in the mountains of Central
Asia, and they are likewise known by fossil species from the
European Plistocene.   The wolf, which is very closely allied to
the European species, may be the descendant of a fossil form
found in the Siwaliks, but its absence from Ceylon would seem to
indicate that it has only reached southern India at a comparatively
modern date.   No foxes are known to the east of the Bay of
Bengal, and the jackal does not range east of Burma.

The Oriental region is the home of three well-marked species
of bears, and thereby presents a decided contrast to the Ethiopian.
Of these the Himalayan black bear (*Ursus torquatus*) ranges from
the forest districts of the Himalaya to Burma, and thence to the
Indo-Chinese sub-region.   The small Malayan bear (*U. malayanus*) is restricted to the Burmese and Malayan sub-regions ; and
the great Indian sloth bear (*Melursus ursinus*), which is the sole

representative of a separate genus, is confined to India and Ceylon, and is known to have been an inhabitant of Madras since the Plistocene era.   Not improbably it may be the descendant of a Siwalik species (*U. theobaldi*), which is the earliest known representative of the true bears; and the Malayan species may be

FIG. 62.   INDIAN SLOTH-BEAR (*Melursus ursinus*).

derived from a small extinct bear whose remains occur in the Plistocene of the Narbada valley.

One of the most remarkable of Oriental carnivores is the panda, or cat-bear (*Ælurus fulgens*), which ranges from the Eastern Himalaya to Yunnan, and is the single existing Old World representative of the *Procyonidæ*.   Curiously enough, the remains of a much larger species of the same genus have been discovered in the English Pliocene; and it is thus evident that *Ælurus* formerly enjoyed a wide range.   From the restriction of all the other known members of the family to the New World, fossil types may be looked for in eastern Asia, as it is quite clear that the distributional area of the group must once have been continuous.

In the *Mustelidæ* there are four generic types very character-

istic of the region, although two of them are not confined to it.
The first of these comprises the three species of sand-badger
(*Arctonyx*), two of which are found in the Himalayan and Burmese
sub-regions, while the third is Tibetan. The single Oriental
species of ratel (*Mellivora*) is restricted to India, exclusive of
Ceylon; a fossil species occurring in the Siwaliks. The only other
living form is Ethiopian. Its distribution would thus seem to
indicate that the genus originated in northern India, whence it

Fig. 63.   INDIAN RATEL (*Mellivora ratel*).

migrated into Africa while there was a free communication between
the two continents, and that it only reached southern India (where
it is unknown on the Malabar coast) at a comparatively recent
epoch. The third genus, *Helictis*, comprising badger-like animals
with long bushy tails, is represented by four species, ranging from
India to China, but unknown in Ceylon. Lastly, the teledu, or
small burrowing badger (*Mydaus meliceps*) of the Malayan sub-
region, is the sole representative of its genus, and is found at con-

siderable elevations in Java, as well as in Sumatra and Borneo. No fossil representatives of either of these two genera are at present known.

Among the rodents, the grooved-toothed squirrel (*Rhithrosciurus*) is a peculiar type confined to the island of Borneo; and the pigmy squirrels (*Nannosciurus*) are represented by four Malayan species, the only other member of the genus being West African. The true squirrels (*Sciurus*), as mentioned above, attain their maximum development in the Malayan sub-region. The flying-squirrels are represented by the genera *Pteromys* and *Sciuropterus*, the former being exclusively Oriental, and the latter having one species in the eastern, and a second in the western half of the Holarctic region, in addition to being represented in a fossil state in the French Miocene. In the *Muridæ* there are no less than eleven genera—in most cases respectively represented by only a single species—peculiar to this region, while another is Oriental and Ethiopian only. Of the peculiar types, one of the most remarkable is *Chrotomys*, from the mountains of Luzon, in the Philippines, belonging to the sub-family *Hydromyinæ*, of which the typical forms are Australian[1]. From other members of the sub-family the single species of this genus differs in having three (in place of only two) pairs of molar teeth, thereby forming a link with ordinary murines. This animal, which is about the size of a rat, is easily recognised by the presence of an orange or buff line running down the middle of the back. Luzon has also yielded another rat, provisionally referred to the Australian genus *Xeromys*[2]. Another unique Oriental type is found in the long-tailed dormouse-like form from the Malabar coast known as *Platacanthomys*, which constitutes a sub-family by itself. *Phlæomys*, likewise representing a separate group of the same rank, is restricted to the Philippines, and is characterised by the molar teeth being divided into three transverse lobes. The one species is of very large size. Nearly allied is a huge, rough-haired, grey or blackish rat, from the mountains of Luzon, which may be compared in size to a small marmot, and for which the name *Crateromys* has been suggested. This differs from *Phlæomys* by

[1] *Vide suprà*, p. 40.

[2] Appendix, No. 31.

the smaller claws and more bushy tail, and also by the completely tuberculate molars. The single species of the Burmese rat-like *Hapalomys* differs from all other members of the sub-family *Murinæ* in possessing three longitudinal rows of tubercles on the lower as well as on the upper molar teeth. The one representative of the allied genus *Vandeleuria* ranges from India and Ceylon to Yunnan. The pencil-tailed tree-mice (*Chiropodomys*), of which there are three species, are restricted to the Burmese and Malayan sub-regions; and the small red rat representing the genus *Pithechirus* is known only from Sumatra and Java. With the shrew-rat (*Rhynchomys*) we revert to several peculiar forms from the mountains of Luzon, in the Philippines. This rodent, which is about the size of an ordinary rat, has the muzzle extraordinarily slender and elongated, with very feeble incisors, and it is probable that it lives on animal substances, possibly caterpillars. The two other Philippine types form the genera *Carpomys* and *Batomys*; the former with two, and the latter with a single species. They are more or less dormouse-like forms, with blunt muzzles, thick woolly fur and long and well-haired tails. Lastly, the bush-rats (*Golunda*) have one Indian and another Ethiopian representative.

An interesting instance of how the present distribution of a genus is explained by palæontological discoveries is afforded by the brush-tailed porcupines (*Atherura*), now represented by one species from the Malayan, and a second from the West African sub-region; the connecting form being one of which fossil teeth have been found in the Karnul district of Madras. From this it may be inferred that the genus was probably also represented in the Siwaliks. To the same family (*Hystricidæ*) belongs a peculiar porcupine from Borneo, constituting the genus *Trichys*.

Passing on to the ungulates, we have first to notice a peculiar group of oxen forming a section of the genus *Bos*, which is confined to this region, and characterised, in addition to certain features of the skull and horns, by the dark colour of the males, or of both males and females. Of these, the gaur (*B. gaurus*) inhabits both India and the Malayan countries, but appears never to have reached Ceylon; while the banteng (*B. sondaicus*) is confined to the countries on the east of the Bay of Bengal. Fossil representa tives of this group occur in the Indian Plistocene ; and certain generalised oxen from the Siwalik Hills and the Pliocene of southern Europe, in which the females were generally or always hornless, may have been the ancestral type.　The Indian buffalo (*B. bubalus*) is markedly distinct from the Ethiopian forms, and has ancestral representatives in the Indian Pliocene and Plisto-cene.　While abundant in Ceylon, it is probably unknown in a truly wild state to the east of the Bay of Bengal.　The Philippine buffalo, or tamarao (*B. mindorensis*) is regarded by some as a

FIG. 64.　JAPANESE SEROW (*Nemorhœdus crispus*).

cross between the last and the anoa of Celebes ; ancestral types of the latter occurring, as already mentioned, in the Siwaliks.　In the same family the short-horned goats of the genus *Hemitragus* are represented by two Indian species, one inhabiting the Himalaya and the other the Nilgiris ; the third living species being south Arabian.　One extinct species occurs in the Siwaliks and a second one in Perim Island, so that the group is essentially an Indian one ; and, as already mentioned, its present distribution can only be accounted for by a lowering of the temperature during a past epoch. Of the goat-like genera *Nemorhœdus* and *Cemas*, the former has a wide range in the region and also extends into northern China and Japan, while the latter is represented solely by the Himalayan goral; no fossil types of either being known. In its poverty of antelopes (exclusive of the widely-spread gazelles) the Oriental presents a most remarkable contrast to the Ethiopian region, although this poverty is largely a feature of the present epoch, African types being common in the Siwaliks. The sole existing forms are the four-horned antelope (*Tetraceros quadri-cornis*), the black-buck (*Antilope cervicapra*), and the nilgai (*Bos-elaphus tragocamelus*), each of which forms a genus by itself, and all of which are restricted to India, exclusive of Ceylon. Indeed, it is a remarkable feature that true antelopes and gazelles are unknown to the eastward of the Bay of Bengal; although this may be chiefly or entirely due to the countries to the eastward being unsuited to their habits. The nilgai, which has fossil representatives in the Indian Plistocene and Pliocene, is allied to the kudu group of Africa, while the four-horned antelope is a near relative of the duikers. It will be unnecessary to say anything with regard to the true goats (*Capra*) and sheep (*Ovis*) inhabiting the region, since these are found only on the north-western frontier of India, and are obviously intruders from the Holarctic region. It is, however, important to mention that extinct representatives of one, if not of both groups, occur in the Siwalik Hills.

The abundance of *Cervidæ* is one of the most noticeable features distinguishing the Oriental from the Ethiopian region; there being an equally marked difference in this respect between the former and the Holarctic area. Although the majority of the Oriental deer are now included in the genus *Cervus*, the typical, or elaphine group, as represented by the red deer and the wapiti, is entirely wanting, its place being taken by the sambur and its allies (*C. unicolor*), forming the rusine group; the swamp-deer (*C. duvauceli*), which with another species constitutes the rucervine group, and the Indian spotted deer (*C. axis*), alone representing the axine group. Rusine deer are abundant in the Indian Siwaliks, but appear to be unknown in the Pikermi beds. Although they have one Tibetan representative, the smaller deer known as muntjacs (*Cervulus*)—which are characterised by the length of the pedicles of the antlers and the shortness of the antlers themselves—form a very characteristic Oriental group, ranging over the entire region. Not improbably they are represented in the Pliocene of Europe.

The chevrotains, or *Tragulidæ*, which have already been shown to be abundant in the European Oligocene and Miocene—remains of the West African genus occurring in the latter deposits and the Indian Pliocene—are represented in the Oriental region by *Tragulus*, which dates from the Siwalik epoch, and ranges from India and Ceylon to the Philippines. Although wild camels are now everywhere unknown, it is probable that India and the Holarctic region was their original home, remains of the genus *Camelus* being found in the Pliocene of the Siwalik Hills.

The large number of species of true pigs (*Sus*) characterising the Oriental region is a notable feature, India itself being inhabited by a species (*Sus cristatus*) nearly allied to the European wild boar, while the Malayan sub-region is the home of a considerable number of species differing more or less markedly from the latter. The genus is well represented both in the Pliocene and Plistocene of India, but in neither of these formations are there any of the Ethiopian types of the family.

With the exception of the Ethiopian, the Oriental region is now the sole one where the family *Rhinocerotidæ* still exists; but there is a remarkable difference between the species inhabiting the two areas, all the three living Asiatic forms being furnished with teeth in the front of the jaws, which, as we have seen in the last chapter, are wanting in the African species. While one of the Oriental rhinoceroses (*R. sondaicus*) ranges from eastern Bengal to the Malayan islands, and a second (*R. sumatrensis*) from Assam to the same, the great Indian species (*R. unicornis*) is unknown to the eastward of Assam, as it is in Ceylon. Fossil remains of the latter are found in the Plistocene of the Narbada valley, while ancestral types both of this species and of *R. sondaicus* are met with in the Pliocene of the Siwalik Hills. It is, however, very remarkable that Ethiopian types of the genus occur not only in the last-named deposits, but likewise in the Plistocene of Madras; the total extinction of this group in India being, as in the case of

other Ethiopian types, almost impossible to account for.   One of
the two-horned extinct Indian rhinoceroses (*R. platyrhinus*) appears to have been the ancestor both of the existing *R. simus* of
Africa and *R. antiquitatis* of the Plistocene of northern Asia and
Europe ; the evolution of the latter species having not improbably
taken place in the countries lying between India and China,
whence the creature wandered northwards and westwards with the
mammoth to the Arctic tundras.   With regard to the *Equidæ*, it
will suffice to mention that species of *Equus* occur in the Plistocene of Central India and Madras, and that wild asses (of a type
markedly different from the African wild ass) occur in Sind and
Kach.   The genus, like the antelopes, is, however, totally unknown in the countries to the east of the Bay of Bengal, as it is in
Ceylon.   In the *Tapiridæ*, the Malayan tapir .(*Tapirus indicus*)
inhabits the Malay Peninsula as far north as Mergui, and also the
islands of Sumatra and Borneo.   It is important to notice that
although fossil remains of tapirs are unknown from the Pliocene of
the Siwaliks, they are met with in caverns in China.

Distinguished, among other features, from its African cousin by
the thinner and more numerous enamel-plates of its molar teeth,
the Indian elephant (*Elephas indicus*) ranges over the greater part
of the region, being found in suitable districts in India, Ceylon,
Burma, the Malay Peninsula, Cochin China, and Sumatra.   This
species is a near ally of the mammoth (*E. primigenius*); and it
may prove that both are descendants of a Siwalik species (*E.
hysudricus*), which has molar teeth of the type we should expect to
find in such an ancestral form.   If this view be correct, the
mammoth has probably wandered to northern Europe and Siberia
from the countries lying just to the east of India.   It has been
mentioned in an earlier chapter that the extinct so-called stegodont elephants (such as *E. clifti* and *E. insignis*) are mainly
confined to this region, although some of the species are found in
north China and Japan.   As these elephants form the transition
between *Elephas* and *Mastodon*, and also since the species of the
latter genus which may be regarded as the original stock of the
elephants is confined to the Indian and Malayan Pliocene, it may
be taken for granted that the elephants have been developed from
the mastodons within the limits of the Oriental region.   In the

Plistocene of the Narbada Valley in central India there occurs a species (*E. namadicus*) closely allied to the contemporary European *E. antiquus*, in both of which the molars are intermediate in structure between those of the living Indian and African species. *Elephas planifrons* of the Siwaliks, which has molars of a still more generalised type, is equally closely allied to *E. meridionalis* of the upper Pliocene of Europe; and it is quite probable that the former may be the original ancestral stock of the African elephant. It is worth mentioning that the stegodont elephants survived till the Plistocene; and also that some of the species of this group inhabiting India, as well as certain mastodons, ranged as far eastwards as Java, Borneo, China, and Japan.

FIG. 65.   WHITE-BELLIED PANGOLIN (*Manis tricuspis*).

The last mammals that we have to mention are the pangolins (*Manidæ*), which are now common to the Oriental and Ethiopian regions, and appear to be represented by an extinct genus in the European upper Oligocene. The presence of horny scales investing the whole of the body and tail serves to distinguish the pangolins from all other mammals whatsoever; and the Oriental species are further characterised by having the median series of scales on the body continued to the tip of the tail, and likewise by the presence of numerous isolated hairs between the scales of the back, as well as by the presence of small ears. Of the three Oriental species, *Manis javanica* ranges from Burma through the Malay Peninsula to Java and Borneo; *M. aurita* extends from Nipal to the Indo-Chinese sub-region; while *M. pentedactyla* is restricted to India and Ceylon. The most remarkable feature connected with the distribution of the group is, however, the circumstance that claw-bones indistinguishable from those of the giant pangolin (*M. gigantea*) of West Africa have been discovered in a cavern in the Karnul district of Madras.

In the following list the leading results of the foregoing survey are put in tabular form, the italics indicating groups or species peculiar to the region.

I. **Primates.**

    Simiidæ.

        Simia.   Borneo and Sumatra; fossil in India.
        Hylobates.   Burmese and Malayan; fossil in European Miocene.

    Cercopithecidæ.

        Macacus.   Now mainly Oriental, but occurring on the southern borders of the Holarctic region; fossil in the European and Indian Pliocene.
        Semnopithecus.   An outlying species in Eastern Tibet and one in Kashmir; fossil in Pliocene of Europe and India.
        Nasalis.   Borneo.

    Lemuridæ.

        Nycticebus.   Burmese, Malayan, and Indo-Chinese.
        Loris.   S. India and Ceylon.

    Tarsiidæ.   Elsewhere only in Celebes.
        Tarsius.   Malayan, extending into Celebes.

II. **Insectivora.**

GALEOPITHECIDÆ.

    Galeopithecus. Malayan.

TUPAIIDÆ.

    Tupaia. Indian and Malayan.

    Ptilocercus. Borneo and some adjacent islands.

ERINACEIDÆ.

    Gymnura. Burmese and Malayan.

    Hylomys. Burmese and Malayan.

SORICIDÆ.

    Soriculus. Himalayan and Indo-Chinese.

    Anurosorex. Known by one species from Assam and a second from Tibet and Pekin.

    Chimarrogale. Represented by one species from the eastern Himalaya, hills north of Burma, and Mt Kina Balu, Borneo, and a second from Japan.

III. **Carnivora.**

FELIDÆ. Very numerous in the region.

    Cynælurus. Indian and Ethiopian ; the one species being common to the two areas ; fossil in Indian Pliocene.

VIVERRIDÆ.

    Viverra. All the species, except a single Ethiopian one, are Oriental, one of these being frequently regarded as the representative of a distinct genus (*Viverricula*) ; fossil in European Oligocene and European and Indian Pliocene.

    Hemigale. Malayan.

    Linsanga. Malayan and E. Himalayan.

    Paradoxurus. An outlying species in Celebes.

    Arctogale. Burmese and Malayan.

    Arctictis. Burmese and Malayan.

    Cynogale. Malayan.

URSIDÆ.

    Melursus. India and Ceylon.

III. **Carnivora.** (*Cont.*)

PROCYONIDÆ.

    Ælurus.   Eastern Himalayan and Burmese; fossil in English Pliocene.

MUSTELIDÆ.

    Arctonyx.   Two E. Himalayan and Burmese species, and probably a third from Tibet.

    Mellivora.   One Indian and another Ethiopian species; fossil in Indian Pliocene.

    Helictis.   India to China.

    Mydaus.   Malayan.

IV. **Rodentia.**

SCIURIDÆ.

    Rhithrosciurus.   Borneo.

    Nannosciurus.   Represented elsewhere by a single West African species.

    Sciurus.   This almost cosmopolitan genus appears to attain its maximum development in the Malayan sub-region.

    Pteromys.

MURIDÆ.

    Chrotomys.   Philippines (Luzon).

    Xeromys.   Philippines (Luzon), and Australia.

    Phlæomys.   Philippines.

    Crateromys.   Philippines (Luzon).

    Hapalomys.   Burma.

    Vandeleuria.   India and Burma.

    Chiropodomys.   Burmese and Malayan.

    Pithechirus.   Sumatra and Java.

    Rhynchomys.
    Carpomys.   } Philippines (Luzon).
    Batomys.

    Golunda.   One Indian and one Ethiopian species.

HYSTRICIDÆ.

    Atherura.   One Malayan and one West African species; fossil in Indian Plistocene.

    Trichys.   Borneo.

V. **Ungulata.**

BOVIDÆ.

Bos.　The *Bibovine* group exclusively Oriental.

Hemitragus.　Two Indian species, and a third in the South Arabian sub-region of Ethiopia; fossil in Indian Pliocene.

Nemorhædus.　Largely Oriental (Himalayan, Burmese, and Malayan), but extending into northern China and Japan.

Cemas.　Himalayan.

Tetraceros.　Indian.

Antilope.　Indian; fossil in Plistocene.

Boselaphus.　Indian; fossil in Plistocene and Pliocene.

CERVIDÆ.

Cervus.　The *Rusine*, *Rucervine*, and *Axine* groups of this genus are characteristic of the region, although the first is also represented in the Austro-Malayan.

Cervulus.　Mainly Oriental, but with one Tibetan species.

TRAGULIDÆ.

Tragulus.　India, Ceylon, and Malayan sub-region; fossil in Indian Pliocene.

SUIDÆ.

Sus.　Attains its maximum specific development in the Malayan sub-region.

RHINOCEROTIDÆ.

Rhinoceros.　The three existing Oriental species differ from the Ethiopian forms in having front teeth.

TAPIRIDÆ.

Tapirus.　Malayan: elsewhere living only in the Neogæic realm, but widely distributed in a fossil state, although absent from the Siwaliks.

V. **Ungulata** (*cont.*).

    ELEPHANTIDÆ.

        Elephas. The existing Oriental elephant is widely
           different from the Ethiopian, although nearly
           allied to the Holarctic mammoth; the extinct
           Stegodont group is mainly Oriental, although
           extending into north China and Japan.

VI. **Effodientia.**

    MANIDÆ.

        Manis. Elsewhere only in Ethiopian region; fossil
           in Indian Plistocene.

The relations of peninsular India to the Himalayan area have
been already discussed at the commencement of
this chapter; while the land-connection which
appears to have existed between India and Madagascar, and thus with Africa, has been alluded to in an earlier
one. The latter connection must have ceased to exist before
the Pliocene era; and, as we have seen, the descendants of the
Siwalik mammals would appear to have made their way into
Ethiopia across Syria or Arabia. During the Pliocene, India, at
least, could not have been distinguished as a region from Ethiopia
as it exists at the present day; and even in the Plistocene the
connection between the faunas of the two areas was much more
intimate than it is now. The full reason for this gradual disappearance of the modern Ethiopian types from the Indian area
will probably never be known; but there can be little doubt that
the gradual refrigeration of the northern hemisphere with the
advent of the glacial period has been largely instrumental; the
present distribution of *Hemitragus* being only explicable on the
hypothesis of a marked lowering of the temperature over India.

    The more peculiar mammals now inhabiting the Oriental
region may be roughly arranged under five headings. The first
will include those that are common to India and some of the
countries to the west or south-west, but are, for the most part,
unknown in either Ceylon or the countries to the eastward of the
Bay of Bengal. Under this category may be included the following, viz. :—

[margin note:] Past History of the Region.

INSECTIVORA.   Erinaceus.

CARNIVORA.   Felis leo.   Ethiopian, and European Plisto-
                 cene.
             Felis chaus.   Ethiopian, and European Plistocene ;
                 ranges eastward into Burma.
             Felis caracal.   Ethiopian.
             Cynælurus jubatus.   Ethiopian.
             Hyæna striata.   Western Asia and North Africa.
             Hyæna crocuta.   Ethiopian ; occurs both in India
                 and Europe during the Plistocene. No
                 representative of the genus known in the
                 Burmese or Malayan countries.
             Canis aureus.   S. W. Asia ; also ranges into
                 Burma.
             Canis pallipes.   South Arabian.
             Mellivora.   The Indian and Ethiopian species
                 very closely allied.

RODENTIA.   Golunda.   Ethiopian ; the Indian species ranges
                 into Ceylon.

UNGULATA.   Hemitragus.   South Arabian.
             Antelopes. ⎫ None known east of the Bay of
             Equus.   ⎭   Bengal.

The second group includes such genera and species as are
common to India and Ceylon, but are unknown elsewhere. Here
we have :—

PRIMATES.   Loris.

CARNIVORA.   Felis rubiginosa.
             Melursus.

RODENTIA.   Golunda ellioti.

UNGULATA.   Bos bubalus.
             Cervus axis.
             Tragulus memimna.

EFFODIENTIA.   Manis pentedactyla.
     L.   19

The third group, which is a small one, comprises types which are confined to India; it includes

> Felis ornata.
> Boselaphus.
> Tetraceros.
> Antilope.
> Rhinoceros unicornis.

In the fourth group we have generic or specific forms common to India and the countries to the eastward of the Bay of Bengal, but unknown in Ceylon or in the countries to the west or southwest. This list comprises the following, viz. :—

PRIMATES.	Hylobates.
INSECTIVORA.	Tupaia.
CARNIVORA.	Felis tigris.
	Felis nebulosa.
	Felis marmorata.
	Ælurus.
	Helictis.
	Arctonyx.
UNGULATA.	Bos gaurus.
	Bos frontalis.
	Nemorhædus.
	Cervus porcinus.

Finally, we have (among others) the following group of genera and species confined to the countries lying immediately to the eastward of the Bay of Bengal, viz. :—

PRIMATES.	Simia.
	Nasalis.
	Nycticebus.
	Tarsius.
INSECTIVORA.	Galeopithecus.
	Gymnura.
	Hylomys.
CARNIVORA.	Mydaus.
RODENTIA.	Rhithrosciurus.
	Trichys.

UNGULATA.   Bos sondaicus.
              Tragulus javanicus.
              Tapirus.
EFFODIENTIA. Manis javanica.

Other forms might be added to several of these lists, but those included are sufficient for the purpose of showing that the present mammalian fauna of India is a complex formed by an admixture of western and eastern types.

The first group is an essentially modern one, all the generic types contained in it, with the exception of *Erinaceus* (which dates from the Miocene), being unknown before the lower Pliocene; while, if we except *Erinaceus* and *Golunda*, all occur in the Siwaliks. In the Carnivora, there is evidence of all the species except the first three being descended from Siwalik ancestors; and it is quite probable that the three species of *Felis* may trace their origin to felines which lived either in the Siwaliks or Persia during the Pliocene, in which event the lion, and not the tiger, should be regarded as the characteristic large Indian feline.

With the probable exception of *Loris*, the second group is also a modern one; all the forms save *Loris* and *Golunda* having ancestral types in either the Pliocene or Plistocene of India, and none of the genera being known before the former epoch. And it may be mentioned here that the absence of so many of the smaller types of Oriental mammals from the Siwaliks is no indication that the genera did not flourish in India during the Pliocene age, but is due to the strata being unsuited to the preservation of their remains. The remarks applicable to the second group will likewise befit the third.

On the other hand, the fourth and fifth groups appear to have less connection with the Siwaliks, while several of the types are older ones. For instance, we have no proof of the existence of oxen nearly allied to *Bos gaurus* in the Siwaliks, although such are found in the Indian Plistocene; neither is there any evidence of a Siwalik tapir. *Tupaia* is a near relative of the European Miocene *Galerix* and the Oligocene *Lanthanotherium*; as is *Gymnura* of the Oligocene *Necrogymnura*; while *Hylobates* is

represented in the European Miocene[1]. The reasons for regarding the tiger as a comparatively modern immigrant into southern
India have already been stated. A Siwalik origin is, however,
indicated for *Simia;* but concerning the other genera the palæontological history is unfortunately a blank.

The affinity between the faunas of West Africa and the Malayan
sub-region has been already alluded to ; but there are also indications of a connection between that of the latter area on the one
hand and that of southern India and Ceylon on the other, as
exemplified by the occurrence of *Nycticebus* in the Malayan sub-
region and of *Loris* in south India and Ceylon. To explain the
latter connection, Dr Blanford[2] has discussed the possible existence of a direct land-communication between the two areas ; but
this connection, as he admits, scarcely seems necessary, since in
such cases the true explanation would seem to be the survival of
old types in the tropical forest-regions. And here it may be
noticed that the Malayan types common to, or represented by
allied forms in West Africa, are such as either have representatives
in the Indian Pliocene or Plistocene, or such as we might naturally
expect to meet with there if small forms were commonly preserved.
For instance, *Simia* and *Anthropopithecus*, and *Dorcatherium* and
Tragulus, are all represented in the Siwaliks, and *Atherura*
occurs in the Madras Plistocene. This being so, what is more
likely than that lorises, linsangs, and palm-civets, of types more or
less intermediate between the existing Malayan and West African
representatives of those groups, should have flourished in India
during the Pliocene era ? *Nannosciurus*, again, should certainly
be expected to occur in the Indian Pliocene. On this point I
think Dr Wallace[3] was on the right track when, in writing of the
Malayan sub-region, he observed that " Here alone, in the
Oriental region, are found the most typical equatorial forms of
life-organisms adapted to a climate characterised by uniform but
not excessive heat, abundant moisture, and no marked departure
from the average meteorological state throughout the year. These

[1] If my memory serves me right, I have been shown teeth from the upper
Oligocene Phosphorites of France closely resembling those of *Hylobates*.

[2] *Manual of Geology of India*, 1st Ed. pt. 1. p. lxviii.

[3] *Geographical Distribution of Animals*, vol. 1. p. 335.

favourable conditions of life only occur in three widely separated districts of the globe—the Malay Archipelago, Western Africa, and equatorial South America. Hence, perhaps, it is that the tapir and trogons of Malacca should so closely resemble those of South America ; and that the great anthropoid apes and crested horn-bills of Western Africa should find their nearest allies in Borneo and Sumatra."

In addition to the resemblances between the mammalian fauna of the Malayan sub-region and that of West Africa, there are, however, equally well-marked differences between the former and that of Ethiopia in general. Among Burmese and Malayan types wanting in Africa, we have especially to note *Tapirus*, *Gymnura*, *Tupaia*, *Hylobates*, and *Ælurus*. From the small size of their representatives it would be unfair to argue anything from the absence of the last four of these from the Siwaliks, but the case is very different with regard to *Tapirus*, which ought surely to have been found did it exist there. As this genus is equally wanting in the Pikermi and Persian Pliocene, while it occurs in that of France, Germany, England and China, it is a fair inference that it has reached the Malayan countries by a route lying north of India. And the occurrence of *Ælurus* in the English Pliocene suggests that the same may be the case with that genus. If this be so, it is not an improbable hypothesis that the other genera men-tioned, all of which have representative types in the European Tertiaries, may have migrated eastwards by a similar route ; and in this connection it is especially noteworthy that such of the genera in question as enter India at all, occur only in the eastern or southern districts.

With regard to the date of the separation of Ceylon from India, the numerous species of mammals common to the two areas show that this must have taken place at a very recent date, comparatively speaking, although, as aforesaid, at a period when several of the mammals now inhabiting southern India had not yet occupied that portion of their distributional area.

When discussing the possibility of a former land-connection across the Bay of Bengal between Ceylon and southern India on the one hand, and the Malayan countries on the other, Dr Blanford was careful to point out that the ocean-bed afforded no evidence in favour of such a line of communication. This feature, together with certain marked differences between the mammals of the two areas, appears to afford a conclusive argument that these countries have never been much more closely connected than they are at present. Had any more extensive connection existed, we should surely expect to find antelopes, gazelles, and perhaps asses, in the more open districts of upper Burma; while the Bay of Bengal would scarcely have formed such a sharp line limiting the eastward range of wolves, foxes, hyænas, and the other mammals mentioned in the list on page 289, as it actually does. That list is confined to existing types, but if fossil forms had been included, *Hippo-potamus* might have been added, since the extinct Oriental representatives of that genus do not range further east than Burma (whither they evidently migrated down the river-valleys from northern India), no species being known from the Tertiaries of China, Japan, or the Malayan islands. These circumstances, together with the depth of the sea in the Bay of Bengal, seem to disprove the suggestion of Dr Wallace[1] that a continuous tract of land formerly connected Borneo and the rest of Malaysia with the central peak of Ceylon, and extended eastwards to Hainan.

Within the limits of the present volume it would be quite impossible to give a detailed description of the mammalian faunas of the Malay Peninsula and Islands; and I have accordingly selected that of the Bornean group, as an example of what may be called the typical Malayan sub-region, as distinct from Java, which differs markedly in its fauna from Borneo and Sumatra. The chief reason for selecting Borneo is that its fauna has been carefully worked out by Mr A. H. Everett[2] and Mr C. Hose[3], from whose papers the following list of mammals (exclusive of bats), with some emendations and additions, has been compiled; species, like the rat, mouse, and buffalo, which have obviously been introduced, being omitted. Mr Everett includes within the Bornean group the island of Paláwan, and states that the group may be defined "by a line which starts from a point immediately to the west of

Malayan Sub-region.

---

[1] *Op. cit.* p. 359. [2] Appendix, No. 14.
[3] *Descriptive Account of the Mammals of Borneo*, Diss, Norfolk, 1893.

St Julian Island in the Tambelan Archipelago, and being drawn south of the Great Natuna (Bungoran Island), passes northward of Labuan and thence follows the 100-fathom line, so as to embrace Balábac, Paláwan (Paragua), the Calamianes, and the Cuyo Islands, and, returning along the same line of soundings on the southern side of Paláwan, is drawn immediately to the islands of Cagayan Sulu, and Sibutu,—whence it is continued through the Makassar Straits south of the Paternoster, Lauriot (Laset Ketjil), and Solombo islets, and in a north-westerly direction through the Karimata Strait back to the island of St Julian." In the following list the genera and species peculiar to the group are printed in italics, those which are confined to the Paláwan sub-group having that name placed after them in brackets. The distribution of the other forms has been indicated as far as practicable.

PRIMATES. Simia satyrus. Sumatra.

Hylobates leuciscus. Java to Philippines.

,, *muelleri.*

Semnopithecus maurus. Malay Peninsula and Java.

,, *chrysomelas.*

,, *cruciger.*

,, *hosei.*

,, *everetti.*

,, *rubicundus.*

,, *frontatus.*

Nasalis larvatus.

Macacus arctoides. Burma, China, E. Tibet, etc.

,, nemestrinus. Burma, Malay Peninsula, Sumatra, and Java.

,, cynomolgus. Burma to Philippines.

Nycticebus tardigradus. Burma to Philippines.

Tarsius spectrum. Sumatra, Java, and Banka.

INSECTIVORA. Chimarrogale himalayica. E. Himalaya and hills north of Burma.

Crocidura fuliginosa. E. Himalaya.

,, *fœtida.*

,, *doriæ.*

,, indica.

.. *hosei.*

INSECTIVORA (*cont.*).

Gymnura rafflesi[1]. S. Tenasserim, Malay Peninsula, and Sumatra.

Hylomys suilla. Burma, Malay Peninsula, and
Sumatra.

Ptilocercus lowi. Also in some neighbouring islands.

Tupaia *murina.*

,, javanica. Java and Malay Peninsula.

,, *minor.*

,, *gracilis.*

,, *melanura.*

,, ferruginea. Burma, Malay Peninsula, Sumatra, and Java.

,, *splendidula.*

,, tana. Sumatra, Natuna Islands.

,, *dorsalis.*

,, *picta.*

,, *montana.*

Galeopithecus volans. Malay Peninsula, S. Tenasserim, Siam, Sumatra, and Java.

CARNIVORA. Felis planiceps. Malay Peninsula and Sumatra.

,, *badia.*

,, temmincki. E. Himalaya, Burma, Malay
Peninsula, and (?) Sumatra.

,, bengalensis. India to Philippines.

,, marmorata. E. Himalaya to Sumatra.

,, nebulosa. E. Himalaya to Formosa.

Viverra tangalunga. Malay Peninsula, Sumatra,
Philippines, and Celebes.

Linsanga gracilis. Java and (?) Sumatra.

Paradoxurus leucomystax. Malay Peninsula and
Sumatra.

,, hermaphroditus. India to Java and
Sumatra.

,, philippinensis. Philippines.

Arctogale leucotis. Sikhim, Burma, Malay Peninsula, and Sumatra.

[1] Mr Jentink regards the Bornean form as a distinct species (*G. alba*).

Carnivora *(cont.)*.

Hemigale hardwickei.   Malay Peninsula and Sumatra.

,,   *hosei*.

Arctictis binturong.   E. Himalaya, Burma, Siam,
Malay Peninsula, Sumatra, and Java.

Helictis *everetti*.

Cynogale bennetti.   Malay Peninsula and Sumatra.

Herpestes brachyurus.   Malay Peninsula.

,,   semitorquatus.   Sumatra[1].

(?) Canis rutilans.   Malay Peninsula, Sumatra, and
Java.

Ursus malayanus.   Arakan, Tenasserim, Malay
Peninsula, Java, and Sumatra.

Mydaus meliceps[2].   Java and Sumatra.

Mustela flavigula.   India to China.

,,   nudipes.   Malay Peninsula and Sumatra.

Lutra sumatrana.   Malay Peninsula, Sumatra, and
Java.

,,   cinerea.   India to Java and China.

Rodentia.   Sciuropterus pulverulentus.   Malay Peninsula.

,,   horsfieldi.   Malay Peninsula and Java.

,,   setosus.   Sumatra.

,,   genibarbis.   Java.

,,   *nigripes* (Paláwan).

Pteromys nitidus.   Malay Peninsula and islands.

,,   *phæomelas*.

Rhithrosciurus macrotis.

Sciurus bicolor.   E. Himalaya to Siam and
(?) Celebes.

,,   prevosti.   Malay Peninsula, Sumatra, and
Celebes.

,,   hippurus.   Malay Peninsula and islands.

,,   *pryeri*.

---

[1] Jentink, *Notes Leyden Museum*, vol. XVI. p. 210 (1894).
[2] The form from Calamianes has been separated as *M. marchei*.

Rodentia (*cont.*).

Sciurus *brookei.*

    „   tenuis.   Malay Peninsula and islands, to Siam.

    „   *lowi.*

    „   *jentinki.*

    „   notatus.   Malay Peninsula, Sumatra, Java, etc.

    „   insignis.   Malay Peninsula, Sumatra, and Java.

    „   *hosei.*

    „   *everetti.*

    „   *steerei* (Paláwan).

    „   laticaudatus.   Malay Peninsula.

    „   soricinus.   Java and Sumatra.

Nannosciurus exilis.   Malay Peninsula and Sumatra.

    „   *whiteheadi.*

    „   *melanotis.*

Mus *infraluteus.*   Sumatra.

    „   muelleri.   Sumatra.

    „   *sabanus.*

    „   neglectus.   Philippines.

    „   jerdoni.   E. Himalaya, Tenasserim, Java.

    „   *alticola.*

    „   ephippium.   Sumatra, Philippines.

    „   *margarettæ.*

    „   *raja.*

    „   *ochraceiventer.*

    „   *whiteheadi.*

    „   *bœodon.*

    „   *baluensis.*

Chiropodomys *major.*

    „   *pusillus.*

Hystrix *crassispinis.*

    „   *pumila* (Paláwan sub-group).

    „   muelleri.   Sumatra.

Trichys guentheri.

UNGULATA.   Bos sondaicus. Burma, Malay Peninsula, Java, and Bali.

,,   moellendorffi[1] (Paláwan sub-group).

Cervus unicolor[2], var.   Probably introduced.

Cervulus muntjac.

Tragulus napu. S. Tenasserim to Java and Sumatra.

,,   *nigricans* (Paláwan sub-group).

,,   javanicus. Malay Peninsula to Cochin China.

Sus vittatus.   Java, Sumatra, Amboyna, Batjian.

,,   verrucosus.   Java, Ceram.

,,   *barbatus*[3].

,,   longirostris.   Java.

Rhinoceros sumatrensis.   Assam to Siam, Malay Peninsula, and Sumatra.

Tapirus indicus.   S. Tenasserim, Malay Peninsula, and Sumatra.

Elephas indicus.   Probably introduced.

EFFODIENTIA.   Manis javanica (Paláwan sub-group).

In this list the genera *Nasalis*, *Trichys*, and *Rhithrosciurus* are peculiar to the group, while *Ptilocercus* is almost so; and, including the latter, there are no less than fifty-one species peculiar to Malaysia. Of these a very large number are common to Sumatra or the Malay Peninsula, or both together, while a smaller number occur in Java. There are but six species common to peninsular India, among which the Indian elephant and the sambar may have been introduced; but there are ten which are found in the Eastern Himalaya or Assam. The most remarkable among these is the Himalayan water-shrew (*Chimarrogale himalayica*), which is only

[1] Founded on a skull from the island of Busuanga, in the Calamianes, which probably indicates only a race (? introduced) of the buffalo.

[2] The so-called *C. equinus* is regarded by Dr Blanford as not specifically distinct from *C. unicolor*. This being so, it is probable that the Bornean form described by Mr C. Hose (*Ann. Mag. Nat. Hist.* ser. 6, vol. XII. p. 206) as *C. brookei* is also a variety.

[3] *Sus ahænobarbus*, Huet, from Paláwan, and *S. calamianensis*, Heude, from the Calamianes, are identified by Dr Nehring (*Sb. Ges. Naturf. Berlin*, 1894, pp. 190, 191) with this species.

found in the Eastern Himalaya, the mountains north of Burma, and Mount Kina Balu, in North Borneo; a musk-shrew (*Crocidura fuliginosa*), common to the Eastern Himalaya and Borneo, being likewise unknown elsewhere. These two instances alone are sufficient to prove that Borneo must have been in immediate connection with the lands to the north-west within the period during which the living species of mammals have come into existence; while the restriction of the water-shrew to the mountains seems likewise to imply a former lowering of the temperature of the whole region sufficient to enable the creature to pass from the one area to the other, or perhaps rather to have allowed of its existence in the intermediate lowlands, whence it migrated to its present isolated haunts. That Borneo was connected with the mainland during the Pliocene epoch is proved by the occurrence in that island of the Siwalik *Mastodon latidens*, the tooth figured on page 173 being of Bornean origin.

The large number of species common to Borneo, Sumatra, and the Malay Peninsula also shows that these three areas must have been very recently in connection; but the excessive number of peculiar Bornean forms seems to indicate that the former island, with the adjacent islets, was the first to be isolated. Even so, however, the extraordinarily large percentage of distinctive types is most remarkable. Regarding the relationship of the Paláwan sub-group to Borneo, Mr Everett writes that "although the general facies of the mammalian fauna of the sub-group is clearly Bornean, it is to be noted that no species appears to be peculiar to the group as a whole, a fact which suggests the inference that closely connected as Borneo has undoubtedly been with Balábac and Paláwan, and isolated as they have been together from the mainland of Asia, there has also been much isolation of Borneo and Paláwan *inter se*."

From Sumatra and Borneo, which have so much in common, and in a somewhat less degree from the Malay Peninsula, Java differs very remarkably as regards its mammalian fauna; a large number of typically Malayan forms being absent, while others as characteristically Indian are present. In the first place, the orangs (*Simia*), common to Borneo and Sumatra, are absent; and the elephant and tapir are likewise wanting, the former certainly existing in a wild state in Sumatra, although it has been considered that its occurrence in Borneo is due to human introduction[1]. Although the Javan rhinoceros (*R. sondaicus*), as we have seen, is common to eastern Bengal, Burma, and Java, its reputed occurrence in either Borneo or Sumatra does not appear to rest upon any solid basis of fact[2]; while the Sumatran species (*R. sumatrensis*), which is common to Borneo, is wanting from Java. It has, indeed, been stated[3] that certain teeth from the Plistocene of Borneo are referable to the last-named species, but the molars of both kinds are so alike that it is almost, if not quite, impossible to distinguish between them. A noteworthy circumstance is that whereas there is no Siwalik species allied to *R. sumatrensis*, yet *R. sondaicus* is almost indistinguishable from the Siwalik *R. sivalensis*, and is thus proved to be a very ancient Indian type. As another instance of the distinctness of the mammalian fauna of Java from that of Borneo and Sumatra, we may take the case of the banteng (*Bos banteng*), which is wanting[4] from both those islands, but is present in Java, the Malay Peninsula, and Burma. Again, the genus *Hemigale*, of which the type species is common to the Malay Peninsula, Sumatra, and Borneo, is quite unknown in Java. That the latter island was directly connected with the mainland is, of course, proved by such species of existing mammals as are common to the two areas ; but if further evidence be needed, it is to hand in the fossil mammals which have been obtained from Pati-Ajam, in Java[5]. These comprise *Elephas trigonocephalus, E. bombifrons, E. clifti, E. namadicus, E. hysudricus, Sus hysudricus, Bos sivalensis*, and *Cervus lydekkeri*. With the exception of the first and last (which may prove not to be distinct likewise), all these are Indian forms, *E. namadicus* belonging to the Plistocene, while the others pertain to the Siwalik fauna ; and it may be added that

[1] Dr Wallace states that *Ursus malayanus* is absent from Java, but according to Dr Blanford this is incorrect.

[2] See Jentink, *Notes Leyden Museum*, vol. XVI. p. 231 (1894).

[3] Busk, *Proc. Zool. Soc.* 1869, p. 409.

[4] In the *Fauna of British India—Mammalia*, p. 490, Dr Blanford gives Borneo, and perhaps Sumatra, as part of the habitat of the banteng, but the animal is omitted from fauna of the former by Mr Everett.

[5] K. Martin, *Sammlungen Geol. Reichsmuseums in Leiden*, vol. IV (1887).

the first three belong to the group of stegodont, or intermediate elephants.

In endeavouring to explain the relationship of the Javan fauna to that of the rest of the Malayan sub-region, Dr Wallace[1] was first of opinion that Java, which was evidently isolated before Sumatra and Borneo, had a brief land-connection with the Siamese peninsula, independently of those two islands. This view, however, was subsequently abandoned[2], and the following hypothesis proposed. From the evidence of certain Tertiary rocks in Java believed to be of Miocene age, it is considered probable that at the epoch in question that island " would have been at least three thousand feet lower than it is now, and such a depression would probably extend to considerable parts of Sumatra and Borneo, so as to reduce them all to a few small islands. At some later period a gradual elevation occurred, which ultimately united the whole of the islands with the continent. This may have continued till the glacial period of the northern hemisphere, during the severest part of which a few Himalayan species of birds and mammals may have been driven southward, and ranged over suitable portions of the whole area. Java was then separated by subsidence, and these species became imprisoned there ; while those in the remaining part of the Malayan area again migrated northward when the cold had passed away from their former home, the equatorial forests of Borneo, Sumatra, and the Malay Peninsula being more especially adapted to the typical Malayan fauna—which is there developed in rich profusion. A little later the subsidence may have extended further north, isolating Borneo and Sumatra, but probably leaving the Malay Peninsula as a ridge between them as far as the islands of Banka and Billiton. Other slight changes of climate followed, when a further subsidence separated these last-named islands from the Malay Peninsula, and left them with two or three species which have since become slightly modified. We may thus explain how it is that a species is sometimes common to Sumatra and Borneo, while the intervening island (Banka) possesses a distinct form[3]."

[1] *Geographical Distribution of Animals*, vol. I. p. 359.
[2] *Island Life*, p. 360.
[3] As exemplified in the case of the birds of the genus *Pitta*.

Although not taking into account the relationship between the fauna of Borneo and that of the Paláwan sub-group, this may be accepted as, on the whole, a very probable explanation of the facts of the case. It may be added that while the Javan rhinoceros (*R. sondaicus*) is, as already stated, closely allied to the Siwalik *R. sivalensis*, the nearest ally of *R. sumatrensis* appears to be the extinct *R. schleiermacheri* of the European Miocene, thus affording one more instance of affinity between the typical Malayan fauna and that of the middle Tertiaries of Europe.

Recent investigations on the mammals from some of the small chain of islands lying to the south-west of Sumatra, such as Nias and Sipora in the Mentawi group, together with Christmas Island[1], lying still further to the south, have shown that they differ very markedly from those of the larger islands. From Sipora, in addition to bats, Mr O. Thomas[2] records the following species, of which those peculiar to the island are printed in italics :—

Nicobars, Mentawi, and Christmas Islands.

Semnopithecus *potenziani*.
Macacus nemestrinus.   Widely spread.
Tupaia ferruginea, var. hypochrysa.   Java.
Paradoxurus, sp.
Pteromys nitidus.   Widely distributed.
Sciuropterus *lugens*.
    ,,   aurantiacus.   Banka.
Sciurus *melanogaster*.
    ,,   *fraterculus*.
Mus *siporanus*.
    ,, raja.   N. Borneo.

From Christmas Island the same writer[3] records a peculiar species of fruit-bat (*Pteropus natalis*), a variety of a widely-spread musk-shrew (*Crocidura fuliginosa*), and two peculiar rats (*Mus macleari*, and *M. nativitatis*), remarkable for their large size. Regarding the bearings of the faunas of these islands on the general problem of distribution, Mr Thomas writes that the collec-

[1] This must not be confounded with the island of the same name in Polynesia.

[2] *Ann. Mus. Civ. Genova*, Ser. 2, vol. XIV. pp. 660—672 (1895).

[3] *Proc. Zool. Soc.* 1887, pp. 511—514, and 1888, pp. 532—534.

tion from Sipora "does not show the very slightest special relationship to Sumatra, and therefore lends weight to the view that the Mentawi chain is the remnant of a long peninsula or island, similar in shape to, but separate from the Malay Peninsula or Sumatra. Further than this I cannot at present go, mainly because we know so little of the small terrestrial mammals of the other islands of the chain, those of the Nicobars being almost unknown, and those of Simalu, Sibiru, and Pagi entirely, while in Nias and Engaño the collections consist mainly of bats. Still the few indications there are, such as the relations to each other of *Pteropus nicobaricus, modiglianii*, and *natalis*, and of *Mus siporanus* and *macleari*, show that the mammals, like the other animals, present a general similarity throughout the chain the whole way from the Nicobars to Christmas Island."

Hitherto the Philippine Islands (exclusive of Paláwan, Calamianes, etc., which are classed with the Bornean group [1]) have been regarded as forming a portion of the Malayan sub-region ; but the discovery of a very peculiar mammalian fauna in the mountains of Luzon [2] clearly proves their right to form a sub-region apart. This mountain fauna, which it is highly probable may also prove to be existent in Mindanao, is evidently a very ancient one, showing certain indications of affinity with that of Australia; while the plain fauna is of a more modern and generally Oriental type. Curiously enough, the indications of affinity with the fauna of Celebes are by no means strongly marked. Unfortunately, there is at present no knowledge of the palæontological history of the mammalian fauna of the group. The following species of mammals, exclusive of bats, have been recorded from this sub-region [3]; the names of such genera and species as are peculiar being printed in italic type.

PRIMATES.   Hylobates leuciscus.

Macacus cynomolgus.   A distinct race [4].

Tarsius *philippinensis* [5].

Nycticebus tardigradus.

---

[1] *Vide suprà*, p. 294.   [2] See Thomas, Appendix, No. 31.

[3] In addition to the paper cited above, see Bourns and Dear, Appendix, No. 11.   [4] Often separated as *M. philippinensis*.

[5] Meyer, *Abh. Mus. Dresden*, 1894, Art. 1, p. 1.

Insectivora.   Galeopithecus volans[1].
                Tupaia *everetti*.

Carnivora.   Viverra tangalunga.   Ranges to Celebes.
                Paradoxurus philippinensis.   Also Bornean.
                Felis bengalensis.

Rodentia.   Sciurus *philippinensis*.
                    „   *mindanensis*.
                    „   *samarensis*.
                    „   *cagsi*.
                Nannosciurus *concinnus*.
                Chrotomys *whiteheadi*. ⎫
                Xeromys *silaceus*.   ⎬ Mountains of Luzon.
                Phlæomys *cumingi*.
                    „   *pallidus*[2].
                Crateromys *schatenbergi*. ⎫
                Rhynchomys *soricoides*.   ⎪
                Carpomys *melanurus*.   ⎬ Mountains of Luzon.
                    „   *phæurus*.   ⎪
                Batomys *granti*.   ⎭
                Mus *luzonicus*.
                    „   *chrysocomus*.   ⎫
                    „   *neglectus*.   ⎬ Mountains of Luzon.
                    „   *ephippium*.   ⎭

Ungulata.   Bos *mindorensis*.
                Cervus *philippinus*.
                    „   *alfredi*.
                Sus celebensis, var.[3]   Elsewhere only in Celebes.

With the exception of the *Tarsius*, which is now regarded as a peculiar species, the Primates are all wide-ranging forms, as is also the case with *Galeopithecus volans*, *Viverra tangalunga*, and *Felis bengalensis*. A relationship with Borneo is indicated by the *Paradoxurus* and two species of *Mus*; while the pig is typically a Celebean form. The tamarao (*Bos mindorensis*) has its nearest ally in the anoa of Celebes, but, as mentioned in an earlier chapter,

---

[1] The Philippine race has been separated as *G. philippinensis*.
[2] Doubtfully distinct.
[3] Equal *S. marchei*, Huet.

L.   20

it is suggested that the animal in question is really a hybrid between the latter and the Indian buffalo. The two species of deer are small forms allied to the race of the sambar inhabiting Java and Borneo ; the first in the list being uniformly coloured, while the second is spotted with white at all ages. The so-called *Cervus marianus* of Luzon appears to be inseparable from *C. philippinus*.

It is remarkable that the six genera peculiar to the group all belong to the *Muridæ*, and that five of these are known only from the mountains of Luzon. Moreover it is quite probable that the species from the latter locality referred to the Australian genus *Xeromys* ought really to be generically distinguished. It is among the *Muridæ* that evidences of Australian affinities are alone exhibited. As these murines have already been noticed on page 277 it will be unnecessary to allude to them further in this place.

Next to the peculiar rodents of the mountains, the most remarkable feature about the fauna of the typical Philippines is the absence of such a number of the most characteristic Malayan genera of mammals. Among the Primates the deficiency of orangs (*Simia*) is perhaps not very remarkable, but the total lack of langurs (*Semnopithecus*) and the presence of only a single species of *Macacus* and *Hylobates* are most noteworthy. The macaque is, however, distributed over all the islands of the group, and differs from other forms of its species in its extremely light coloration, so that it is scarcely likely to have been introduced by human agency. Of Malayan genera which are absent, special note may be taken of *Linsanga, Arctogale, Arctictis, Cynogale, Herpestes*, the wild dogs of the sub-genus *Cyon, Ursus, Tragulus*[1], and *Elephas*. Almost equally well-marked peculiarities are exhibited by the bird fauna.

Apart from the tamarao, which has yet to be proved entitled to rank as a valid species, the lowland mammalian fauna of the Philippines is such as might have well reached the group by means of a narrow connection of limited duration with some portion of the Malay countries; say, for instance, with Borneo by way of Paláwan. That such a connection must have been comparatively recent is indicated by the identity of several of the Philippine

---

[1] Represented by *T. nigricans* in Paláwan.

species with Malayan forms and the absence of any peculiar genera save *Phlæomys*. The absence of such a number of Malayan types indicates, however, either that the connection must have been exceedingly brief, or that a large number of species formerly inhabiting the islands have been destroyed by submergence. On the other hand, the presence in the group of a considerable proportion of widely spread continental genera of birds suggests a more free communication with China than at present exists; such communication having not improbably taken place by way of Formosa. The mountain fauna of Luzon doubtless indicates an earlier type of colonisation.

Note.—Since the foregoing was written two papers have appeared on the Fauna of the Natuna Islands; viz. O. Thomas, *Novitates Zool.* Vol. II. pp. 26—28, and Thomas and Hartert, *l.c.* pp. 409—429.

CHAPTER IX.

THE HOLARCTIC REGION.

By far the most extensive of all the zoological regions of the globe is that which is equivalent to the whole of the Palæarctic and the greater portion of the Nearctic region of Messrs Sclater and Wallace, the one to which Dr Heilprin (after a suggestion of Professor A. Newton) applied the name Holarctic. In defining this region, Dr Heilprin separated from it a Sonoran "transitional region " in the Western Hemisphere, and a similar Mediterranean or Tyrrhenian tract in the Eastern. Of these the former is now accepted as a definite region ; but our knowledge of the distribution of species in the Eastern Hemisphere is either too imperfect, or the interdigitation of the two faunas is too complete to admit of the full definition of a Mediterranean region. Accordingly, without prejudice as to what it may be possible to accomplish in this direction in the future, the Mediterranean area is provisionally included in the Holarctic region. It is, however, important to observe that the reservation by Dr Heilprin of the two transitional tracts already named justifies the use of the term Holarctic even if both such tracts be raised to the rank of separate regions ; and there is accordingly no necessity for the adoption of Dr Blanford's term Aquilonian as equivalent to the restricted Holarctic. Used in the sense here indicated, the Holarctic region will comprise all that portion of Arctogæa lying north of the

Sonoran region in America, and of the Ethiopian and Oriental regions in the Old World. The whole area is extra-tropical; and, as Dr Heilprin remarks, "no other region can compare with the Holarctic in the manifold variety of its physical characteristics. Every form of terrestrial configuration, or condition of soil and climate that may be represented in any other region, is also represented here, and on an imposing scale. From the ice-bound fields of the far north to the burning desert-wastes of Turkestan on the south, and from the deep forest-grown lowlands to mountain-summits soaring thousands of feet above the level of perpetual snow, we pass through all those various gradations of climate which respectively characterise the Frigid, Temperate, and Torrid zones. Densely covered forest-tracts, supporting, as in the north, a sombre growth of pine and other coniferous trees, or, as in the south, a vegetation of almost tropical luxuriance, alternate with broadly open grass or pasture lands (*tundras* of Siberia, American prairies and plains), which in some cases support over enormous areas only a very scanty vegetation, and in others display a profuse variety of vegetable productions. It is in this region that, in addition to a most bountiful development of desert tracts, we meet with the most elevated table-land (the Central Asian), and, at the same time, with the greatest expanse of lowland on the surface of the globe, the great plain of Siberia and north-eastern Europe."

Although the essential unity of the greater portion of the Nearctic and Palæarctic regions has long been fully recognised by the American zoologists, several attempts to bolster up the alleged distinctness of these artificial divisions have recently been made in England[1], one proposal being to recognise an Arctic or Boreal circumpolar province cut off from both areas, although this is practically begging the whole question.

If I rightly understand his view, Dr C. H. Merriam[2], who is an ardent advocate for the zoological unity of the more northern portions of both hemispheres, would distinguish a Boreal circumpolar region common to both hemispheres; while in America he recognises south of this a Transition region, followed still more to the south by the Sonoran. In the Old World he would have an

---

[1] Appendix, Nos. 28 and 35.
[2] *Ibid.* No. 19, pp. 24, 63.

analogue of the Sonoran—practically equivalent to the Mediterranean sub-region of European writers—but says nothing about
an analogue of the Transition; from which I infer that he
would use the term Boreal really as practically equivalent to the
Holarctic, if the Sonoran and Mediterranean areas were subtracted.
In the New World the Boreal, exclusive of the Transitional, is
divided into an Arctic and a Coniferous Forest Boreal zone, the
latter being frequently spoken of as the Boreal " zone," in contra-
distinction to the Boreal circumpolar "region"; the Arctic zone
including the tract beyond the limit of trees. The distinction
between the Boreal and Transition areas is certainly not of
regional value; and as the term Boreal is used in several senses,
it had better give place to the earlier Holarctic.

As has been partially indicated in earlier chapters, and as will
be more fully noticed in the sequel, there is undoubtedly a marked
distinction between the mammals of North America as a whole
and those of Europe and northern Asia, but this has been considerably exaggerated by including the Sonoran region in the old
Nearctic, and is overshadowed by the number of genera and
species common to the two areas and unknown elsewhere. Could
a Mediterranean region be satisfactorily defined, the homogeneity
of the mammalian Holarctic fauna would be still more apparent;
but this, from the great mingling of northern and southern types
which has taken place in the Old World, is, I think, impracticable.
As has been already mentioned, it is probable that the western
and eastern halves of the Holarctic region have never had more
than a comparatively small area of communication by way of
Bering Strait, and, therefore, the further south we travel in the
two areas the more distinct do the faunas become; while only
such forms as are capable of withstanding a certain degree of cold
have ever been able to cross at all. It may be added that the
evidence for the unity of the Holarctic region is by no means
solely dependent upon the mammalian fauna, but is supported by
many other groups of animals. To take an instance from the
insects, I may quote from Mr W. F. Kirby[1], who writes as follows:
" Had I been dealing with Lepidoptera only, I would certainly

[1] *Journ. Linn. Soc.—Zool.* 1873, p. 432.

have united Dr Sclater's Palæarctic region and Nearctic region ; for although the species of North American Rhopalocera are seldom identical with those of northern Asia and Europe, still the genera are the same with scarcely an exception, except a few representatives of South American genera, which have no more right to be considered Nearctic species than the similar chance representatives of African[1] forms in North Africa or south-west Europe, or of Indian forms in south-east Europe, have to be con-sidered Palæarctic species."

On the other hand, North America differs remarkably from the eastern half of the Holarctic region as regards its land molluscs. Thus the Rev. A. H. Cooke[2] writes, that " no district in the world of equal extent is so poor in genera, while those which occur are generally of small size, with scarcely anything remarkable either in colouring or form.   The elongated land-shells (*Clausilia*, *Buli-minus*) so characteristic of Europe are entirely wanting, but a few *Bulimulus*, of Neotropical origin, penetrate Texas, and from the same sources come a few species of *Glandina* (as far north as South Carolina), *Holospira* (Texas), and *Helicina*." Probably this poverty is largely due to the unsuitableness of the greater part of North America to molluscan life ; aided by the circumstance that land-molluscs are just the creatures that would have been unable to pass over from Asia by way of Bering Strait.   The batrachians, again, which differ most remarkably in their distribution from mammals, are not indicative of the unity of the Holarctic region, the American types being very different from those of the Eastern Hemisphere.

According, however, to Mr F. E. Beddard[3], " the earth-worms offer the best evidence of any group in favour of the Holarctic region."

Although during the Plistocene era—even subsequently to the passing away of the glacial period—elephants, rhino-ceroses, and hippopotami abounded over the greater part of Europe, while species of the two former groups ranged as far north as Siberia, and macaques

Characteris-tics of the Mammalian Fauna.

---

[1] The author obviously means Ethiopian.
[2] *Cambridge Natural History*—Mollusca, p. 339 (1895).
[3] Appendix, No. 5, p. 80.

were found in France and England, yet the Holarctic region is now characterised by the absence of all these animals, save a few species of *Macacus* on its southern borders ; and if a Mediterranean region could be satisfactorily defined, even these, as well as hyænas and certain other southern types, would likewise be excluded. The fruit-bats constituting the family *Pteropodidæ* are likewise practically wanting in this region ; and Effodientia are quite unknown. On the other hand, carnivores, such as wolves, foxes, bears, martens, weasels, and the glutton, are abundant ; while the rodents are specially represented by such types as the marmots, beavers, voles, and picas ; and the ungulates comprise the bisons, nearly all the sheep, the true goats (absent in the western half of the region), and all the typical deer.

Referring in more detail to the peculiar generic types common to the region as a whole, we have, first of all, among the Insectivora the typical or true shrews (*Sorex*)—which belong to that section of the *Soricidæ* characterised by having the tips of the teeth stained brownish-red—in the main characteristic of this region, although in America they extend southwards into the Sonoran. In the allied family of the *Talpidæ*, the mole-shrews (*Urotrichus*), which are near relatives of the European desmans, are represented solely by one Japanese, and a second North American species, although the latter is frequently separated generically as *Neurotrichus*.

The peculiar genera of Carnivora are but few, although certain groups are either confined to, or very strongly represented in, the region. For instance, among the *Felidæ* the true lynxes—which although generally included in the genus *Felis*, are by some regarded as entitled to form a genus by themselves—are absolutely confined to this region, where they range as far south as Spain. The bears (*Ursus*), also, are very strongly represented ; brown-coloured species being peculiar to the Holarctic area, as is the very distinct polar species to its Arctic portions. The peculiar sea-otter, the sole representative of the genus *Latax*, has a distribution very similar to that of the mole-shrews ; this animal occurring on the coasts of Kamschatka and the Kurile Islands, as well as on those of the Aleutians. The wolverene (*Gulo*), which is likewise the only member of its genus, has a more extended range, being found throughout the forest regions of northern Europe, Asia, and North America, while in the Plistocene era it ranged as far south as England. Although marine mammals are for the most part omitted in this work, the walruses (*Trichechidæ*), which now have a circumpolar distinction, can scarcely be omitted, since these animals never wander far from land. Remains of the existing forms have been disinterred from the peat of the English fens, while tusks of fossil species have been discovered in the Pliocene Crag of the east of England, and also in the corresponding deposits of Belgium.

Passing on to the rodents, we have in the *Sciuridæ* the ground-squirrels or chipmunks (*Tamias*) practically confined to the region. Although in America they also extend into the Sonoran, in Europe they are unknown in the Mediterranean area. The pouches in the cheeks for storing food and the alternate dark and light longitudinal bands down the back serve to distinguish the chipmunks from other squirrels. Fossil remains of the genus, probably belonging to existing species, occur in the Plistocene of Europe and North America. The family of the beavers (*Castoridæ*) seems always to have been mainly confined to the Holarctic region, one of the two existing species (*Castor fiber*) being European, and dating from the English Plistocene, while the other (*C. canadensis*) is North American. Whereas, however, the latter ranges southwards into the Sonoran region, the former is unknown in the Mediterranean sub-region. Fossil species of this genus occur in the upper Tertiaries of Europe and North America, where the extinct *Chalicomys* is likewise met with; but no beavers are known from either the Siwaliks or the Pikermi beds. Although there are few generic types of *Muridæ* peculiar to the whole region, such as there are are important. Foremost of these are the great tribe of the voles (*Microtus*[1]), constituting the typical representatives of a sub-family nearly allied to the cricetines, but distinguished by the two longitudinal rows of tubercles on the crowns of the molar teeth being modified into alternating triangular prisms, and likewise by these teeth being generally of the hypsodont type. In the Old World they are found from the Arctic zone to Asia Minor, while

---

[1] Commonly known by the later name of *Arvicola*.

in America they enter the Sonoran region. As fossils, they appear to be first known from the upper Pliocene Crag of England; and were thus probably evolved within the Holarctic region from the more generalised cricetines at a comparatively late epoch. Nearly allied are the lemmings (*Myodes*), which are, however, a more northern type, unknown in the Mediterranean sub-region, and likewise in the Sonoran. Still more northern, and indeed circumpolar in its range, is the banded lemming, which alone represents a genus (*Cuniculus*) distinguished from the last by the absence of external ears, the short and thick fur on the feet, the rudimentary first toe of the fore feet, and the elongation of the two middle claws of the same. The second of the two families peculiar to the region is that of the picas, or tailless hares (*Lagomyidæ*), comprising small hare-like animals with short ears, of which all the living forms are included in the single genus *Lagomys*. While the majority of the picas are confined to the highlands of Central Asia, some are found on the first snowy range of the Himalaya, and both south-east Europe and the Rocky Mountains severally possess a representative. Fossil forms are common in the middle and upper Tertiaries of Europe, as far south as Sardinia. Although the hare-tribe (*Leporidæ*) have an almost cosmopolitan distribution, the majority of species of *Lepus* are inhabitants of the Holarctic region, Central Asia being especially rich in representatives of the genus.

Among the *Bovidæ*, the bisons, which form a well-marked group of the genus *Bos*, may be regarded as now characteristic of the region, although the American *Bos americanus* ranges into the Sonoran. In addition to the European *B. bison*, which was formerly spread over the greater part of Europe and during the Plistocene extended into Arctic America, there is also the somewhat aberrant yak (*B. grunniens*) of Tibet. In the Plistocene the range of the group was somewhat more extensive, remains of an extinct species having been found in deposits of that age in Texas; and there is likewise another from the Pliocene of northern India. The sheep constitute a group mainly characteristic of the Holarctic region; their headquarters being the Central Asian plateau, where they are more numerous than anywhere else in the world, although one species (*Ovis vignei*) impinges on the north-western frontier of the Oriental region, while the single North American form also enters the Sonoran. In a fossil state it is possible that sheep occur in the Indian Siwaliks, but elsewhere they are first known from the Forest-bed of the Norfolk coast, belonging to the early part of the Plistocene epoch. More nearly allied to the sheep than to the goats, the musk-ox (*Ovibos*) of Arctic America is now extinct in the Old World, but as it is abundant in the Plistocene of Europe and Asia, where it ranged as far south as England, it is surely entitled to rank among the forms common to the western and eastern halves of the Holarctic region. Nearly allied extinct species occur in the Plistocene formations of the United States. Among the *Cervidæ* there are three types common to the entire region. The first, or Elaphine group, includes the typical members of the genus *Cervus*, as represented by the red deer (*C. elaphus*) of the Old World and the wapiti (*C. canadensis*) of North America; this group being characterised, among other features, by the general presence of both a brow- and a bez-tine to the antlers, although the latter is wanting in the North African variety of the red deer, and also in the Tibetan *C. thoroldi*. The alliance between the wapiti and some of the forms inhabiting Central Asia is so close as to render it. doubtful whether they are really anything more than varieties of a single species. The single species of elk (*Alces*) is common to both halves of the region; and the same is also the case with the reindeer (*Rangifer*), which although now not found to the south of Europe, ranged during the Plistocene era into the south of France. The genera or groups which may be regarded as characteristic of the entire Holarctic region may be tabulated as follows, those which are practically peculiar being printed in italics :—

Insectivora.

Soricidæ.	*Sorex.*	In America ranges into Sonoran.
Talpidæ.	*Urotrichus.*	Japan and N. America.

Carnivora.

Felidæ.	Felis, the true lynxes solely Holarctic.	
Mustelidæ.	*Latax.* } One species common to both	
	Gulo. } hemispheres.	
Trichechidæ.	*Trichechus.*	

Rodentia.

SCIURIDÆ.	*Tamias.* Ranges into Sonoran.
	Arctomys.
	Spermophilus.
CASTORIDÆ.	*Castor.* ⎫
MURIDÆ.	*Microtus.* ⎬ Also Sonoran.
	Myodes.
	Cuniculus.
LAGOMYIDÆ.	*Lagomys.*
LEPORIDÆ.	*Lepus*, the greater number of species Holarctic.

Ungulata.

BOVIDÆ.   Bos, the bison group chiefly Holarctic, although the American species reaches the Sonoran.

Ovis, just touches Oriental, and also reaches Sonoran.

Ovibos, now extinct in the Old World, where it was common in the Plistocene.

The arguments for the unity of the Holarctic region are, however, by no means confined to the case of genera, for there are a number of species which either have a circumpolar range, or which are represented by closely allied forms in the opposite hemisphere. It is true, indeed, that many of these are now more or less exclusively Arctic in their distribution, but some range a long distance to the south; while during the Plistocene epoch this was the case with the majority. The following list includes the more important species which are either common to the eastern and western halves of the region or have representative forms in the two hemispheres; those which are strictly Arctic having the letter P appended :—

1. Common lynx (*Felis lynx*).   Canadian lynx (*F. canadensis*).   P.
2. Wolf (*Canis lupus*).
3. Fox (*Canis vulpes*).
4. Arctic fox (*Canis lagopus*).   P.
5. Brown bear (*Ursus arctus*).

 6. Polar bear (*Ursus maritimus*).   P.
 7. Sea-otter (*Latax lutris*).
 8. Pine-marten (*Mustela martes*).   American marten (*M. americana*).
 9. Weasel (*Mustela vulgaris*).
10. Wolverene (*Gulo luscus*).
11. Walrus (*Trichechus rosmarus*).
12. Arctic vole (*Microtus rutilus*).   P.
13. Common lemming (*Myodes lemmus*).   American lemming (*M. obensis*).
14. Banded lemming (*Cuniculus torquatus*).   P.
15. European beaver (*Castor fiber*).   American beaver (*C. canadensis*).
16. Mountain hare (*Lepus timidus*[1]).
17. Bison (*Bos bison*).   American bison (*B. americanus*).
18. Kamschatkan sheep (*Ovis nivicola*).   Bighorn (*O. canadensis*).
19. Musk-ox (*Ovibos moschatus*).
20. Central Asian deer (*Cervus eustephanus*).   Wapiti (*C. canadensis*).
21. Elk (*Alces machlis*).
22. Reindeer (*Rangifer tarandus*).

If the Plistocene be taken into consideration there may be added the mammoth (*Elephas primigenius*) and the horse (*Equus caballus*), remains of both of which have been discovered in Eschscholtz Bay, where those of the European bison also occur.

In this list it may be noticed that the North American lynx is so closely allied to the European form that it has been a question whether it is really more than a local variety. Although the American wolf has been separated as a distinct species, it is now generally identified with the European form; the same being also the case with the so-called cross-fox of North America, which, together with another form from the Himalaya, and a third from North Africa, may be considered a mere variety of the ordinary

---

[1] This name is commonly applied to the English hare, although it properly belongs to the more northern species.   The so-called Polar hare (*L. glacialis*) of Arctic America appears to be only a variety.

fox. Much discussion has taken place with regard to whether any of the bears of North America are distinct from the common brown bear (*Ursus arctus*), with its many Asiatic varieties. According, however, to one of the latest memoirs on this subject[1], all these forms appear mere varieties of the latter, the so-called cinnamon and black bears presenting more distinctly marked differences than does the grizzly. The American marten is so nearly related to the European marten and the Asiatic sable that it is almost impossible to point out valid characters by which the three forms can be specifically distinguished. In the case of the walrus, the Pacific form is distinguished by certain external features from the one inhabiting the Atlantic coasts, although these are scarcely of sufficient importance to be ranked as specific. In regard to the Arctic vole, it may be mentioned that although it is typically a polar form, yet it is represented in southern Europe by the bank-vole (*Microtus glareolus*) and in the United States by Gapper's vole (*M. gapperi*), both of which may be regarded as southern climatic offshoots from the northern stock. Among the other species of rodents it will suffice to mention that the European and American beavers are merely distinguished from each other by the relative lengths of the nasal bones of the skull. And it may be added that the Kamschatkan wild sheep is so closely related to one race of the bighorn, or Rocky Mountain sheep, that it is very questionable whether the two are really entitled to specific distinction. The same is also the case with the two deer mentioned in the list. Reference has already been made to the circumstance that the musk-ox is extinct in the eastern division of the region. A few of the American forms, such as the bear, beaver, bison, and bighorn, enter the limits of the Sonoran region.

Although, as will be shown immediately, there are a large number of generic types respectively confined to its eastern and western divisions, the lists given above, especially the one relating to the species, are amply sufficient to demonstrate the essential unity of the Holarctic region. None of the other zoological regions have anything like the number of common or representa-

---

[1] A. E. Brown, *Proc. Ac. Philad.* 1894, pp. 119, 129. Merriam, *P. Biol. Soc.* Washington, vol. x. pp. 65—83 (1896), regards the N. American bears as forming several distinct species.

tive species which characterise the two divisions of the present one. It must, moreover, be remembered that in the case of the other regions we have taken the whole of the peculiar generic types into consideration, although many of them are confined to small portions of such regions, whereas in the present instance only those which range over nearly the whole area have been mentioned. If, for instance, we were to take the genera respectively confined to the Indian and Malayan areas of the Oriental region, it would be found that the distinction between the two areas would be nearly or quite as marked as are the two great divisions of the Holarctic, while in the matter of species the differences between the two would be far greater. There would accordingly be stronger grounds for making an Indian and a Malayan region than there are for the separation of a Palæarctic and a Nearctic.

Having now discussed the leading mammalian types characteristic of the Holarctic region as a whole, it remains to notice those confined to its eastern division, after which such as are restricted to its western half will be taken into consideration. And here it is advisable to repeat that since the communication between eastern Asia and North America by way of Bering Strait appears always to have been of very limited extent, as we proceed south in the two great continental areas the difference between their faunas appears more and more strongly marked. Accordingly, if it were possible definitely to establish a Mediterranean region, the distinction between the faunas of the eastern and western divisions of the Holarctic would be much less than is the case under the arrangement here followed.

Mammals of the Eastern Holarctic Region.

Passing by the bats entirely, no notice will be taken of groups which, like the hedgehogs (*Erinaceus*), are spread over several of the regions of Eastern Arctogæa, since these are not in any way characteristic of the eastern division of the Holarctic region[1]. The first mammal that presents itself for notice is accordingly the water-shrew, which is the sole representative of the genus *Crossopus*, and belongs to that division of the *Soricidæ* in which the

[1] Such types have been discussed when considering the fauna of Eastern Arctogæa, *suprà*, p. 181.

teeth are stained red; the especial characteristic of the genus being the thickly-fringed feet and tail. The water-shrew is a typical Holarctic mammal, ranging as far east as the Altai mountains and unknown in the Mediterranean sub-region. In the second division of the same family, or that in which the teeth are white, there is a peculiar shrew from the Kirghiz steppes nearly allied to the

FIG. 66. RUSSIAN DESMAN (*Myogale moschata*).

widely-spread musk-shrews (*Crocidura*), but constituting a genus (*Diplomesodon*) by itself. To the same sub-family belongs the Tibetan water-shrew (*Nectogale*), which is likewise the solitary representative of its genus, and is closely allied to *Chimarrogale*, which has, as elsewhere stated, one Oriental and one Japanese species. Both these types are accordingly closely connected with the Oriental region, although nothing is known of their past history. Far more characteristic of the eastern half of the region under consideration are the two species of desman (*Myogale*), which are aquatic insectivores, with long trunk-like snouts, somewhat intermediate between the shrews and the moles. Of the two living species, the smaller is confined to the region of the Pyrenees, extending as far north as the Department of the Landes, while the other is now restricted to south-eastern Russia, although its fossilised remains have been discovered in the Plistocene Forest-bed of the east of England. Extinct species occur in the Miocene and upper Oligocene of the Continent. A slate-coloured insectivore, with the external form of a shrew but the skull of a mole, inhabiting Eastern Tibet, constitutes the genus *Uropsilus*; and the more mole-like creature known as *Scaptonyx* is likewise from the same locality. In any case, these two animals only just enter the border of the region, so that they cannot be regarded as characteristic Holarctic types; and, indeed, the Moupin district of Eastern Tibet is included by Dr Wallace in the Oriental region, although it is assigned by others to the Holarctic. Although two species occur to the south of the Himalaya, the true moles (*Talpa*) are very characteristic of the eastern Holarctic region, the common species ranging from England to Japan, and dating from the epoch of the Norfolk Forest-bed; while fossil species, some of which have been separated as *Protalpa*, range through the Tertiaries of Europe to the epoch of the upper Oligocene. By some writers *Talpa moschata* of Eastern Tibet is distinguished as *Scaptochirus*.

Among the Carnivora the widely spread Old World genera *Genetta*, *Herpestes*, and *Hyæna* enter the Mediterranean sub-region, but are unknown elsewhere in the Holarctic area at the present day. The *Ursidæ* include a most remarkable generic type known as *Æluropus*, represented by the parti-coloured bear of Tibet, in which the cheek-teeth present a decided approximation to the extinct *Hyænarctus* and also resemble those of the panda (*Ælurus*). This genus is another of the border-forms between the Holarctic and Oriental regions. On the other hand, the badgers (*Meles*) are very characteristic of the area under consideration, ranging from England through the rest of Europe to Japan and China, where one species enters the Oriental region, being found as far south as Hongkong. Remains of extinct badgers occur in the lower Pliocene of Persia.

The list of more or less peculiar generic types among the rodents is relatively large. Foremost among these stands a very large flying-squirrel (*Eupetaurus*), inhabiting the regions north of Kashmir and Tibet, and distinguished from all other members of the family to which it belongs by its tall-crowned (*hypsodont*) cheek-teeth. In the dormouse family (*Myoxidæ*) the squirrel-tailed species, which is the sole representative of the genus *Myoxus* in its restricted sense, and the common dormouse, alone constituting *Muscardinus*, are exclusively European; fossil forms occurring through the upper and middle Tertiaries.

In the *Muridæ* the hamsters (*Cricetus*), if regarded as generically distinct from their allies the white-footed mice (*Sitomys*) of the New World, are absolutely characteristic of the eastern division of the Holarctic region, where they range over a large portion of Europe and northern Asia. Extinct species are abundant in the European Tertiaries. Two genera of the mole-like rodents, having the dentition of voles, but approximating in the form of their body and limbs to the moles, constitute a sub-family which is also restricted to this area. Of these mole-voles, the first genus (*Ellobius*) is represented by one species from Russia and another from Afghanistan, while the second (*Siphneus*) includes several forms from North and Central Asia; fossil species of the latter having been described from the Plistocene of the Altai and the Pliocene of northern China. The great mole-rat (*Spalax typhlus*) of southern Europe, Persia, Mesopotamia, Syria and Egypt, represents the only genus among the *Spalacidæ* which is confined to the present area. The *Dipodidæ*, on the other hand, contain several characteristic eastern Holarctic generic types. The most aberrant of these are the rat-like animals constituting the genus *Sminthus*, of which one species inhabits eastern and northern Europe and Central Asia, while a second is found in Kashmir, and a third in the Kansu district of China. Of the more typical forms, the true jerboas (*Dipus*), characterised by having only three toes, are exclusively Holarctic, ranging from Egypt into Central Asia, where they always frequent desert districts. Of the genera with five toes, *Euchoretes* is represented by a single long-snouted and long-eared species from the neighbourhood of Yarkund; and *Platycercomys*, which differs by its flattened and lancet-shaped tail, includes several species, extending from Siberia to Nubia, so that the genus just enters the Ethiopian region. It is represented in the Plistocene of northern Asia. Lastly, there is the genus *Alactaga*, of which there are several species, mostly inhabitants of Northern and Central Asia. The typical *A. jaculus* extends, however, into Persia and southern Russia, while in the Plistocene it ranged as far west as Germany; and another species is an inhabitant of Afghanistan. Although mainly an Ethiopian

FIG. 67.   HEAD OF SPANISH IBEX (*Capra pyrenaica*).

and Neotropical family, the *Octodontidæ* have an Holarctic representative in the gundi (*Ctenodactylus*), which inhabits the borders of the Sahara in the neighbourhood of Tripoli. The extinct *Pellegrinia*, of the Sardinian Pliocene, also belongs to the same family.

Among the artiodactyle ungulates there are six genera, either entirely or mainly confined to the eastern Holarctic region. Of these, the true goats—that is to say, the members of the genus *Capra*, as distinct from the Oriental and Arabian *Hemitragus*—are almost exclusively Holarctic, although *C. walie* inhabits the Abyssinian highlands, and *C. siniatica*, of Palestine and upper Egypt, may also enter the confines of the Ethiopian region. All the goats, it may be observed, are essentially mountain-dwelling animals, and the occurrence of the same species of ibex (*C. sibirica*) in both the Altai and Himalaya is a clear proof of the former prevalence of colder conditions, as without these the animal could not have passed from the one range to the other. The sheep also—in spite of the existence of outlying North American species, and a variety of one Central Asian form (*Ovis vignei*) which enters the north-western confines of India—are mainly characteristic of the area under consideration, attaining their greatest numerical development, and also their maximum size, in the highlands of Central Asia. The range of the genus includes almost the whole of the eastern Holarctic region, the mouflon (*O. musimon*) inhabiting the Corsican islands, and the aberrant arui (*O. tragelaphus*) being found in northern Africa. It is possible that fossil sheep occur in the Indian Siwaliks (where remains of a goat allied to the Himalayan markhor are also met with), and a large species is definitely known from the Norfolk Forest-bed. The next on our list is the remarkable goat-like antelope from the hills to the north of the Assam known as the takin (*Budorcas*), which is allied to the Oriental *Nemorhædus*, and is therefore probably an immigrant into the region from the southward. Allied to this group is the chamois, or gemse (*Rupicapra*), now confined to the higher mountain-ranges of Europe from the Pyrenees to the Caucasus, but which during the Plistocene epoch ranged over many of the lowlands. Among the true antelopes, the addax (*Addax*), an ally of the oryx group, is an exclusively Mediterranean type, inhabiting North Africa and Syria, where there are also representatives of other genera which are typically Ethiopian. More thoroughly Holarctic are the two peculiar but allied genera *Saiga* and *Pantholops*, each represented by a single living species. The saiga is now confined to the steppes of western Asia and Eastern Europe, but during the Plistocene epoch extended as far westwards as Germany, France, and England. The chiru, as the representative of the second genus is called, is, on the other hand, an exclusively Tibetan form; and it is believed that a fossil species occurs in the later Tertiary formations of the same area, where, curiously enough, a rhinoceros also existed. Although gazelles (*Gazella*) have representatives in both the Oriental and Ethiopian regions, they are mainly characteristic of the desert districts of the eastern Holarctic region, being especially numerous in North Africa, Syria, and parts

FIG. 68.   MUSK-DEER (*Moschus moschiferus*).

of Central Asia. An inhabitant of the cooler regions of Asia, where it extends from the south of Siberia to Kashmir and Cochin China, the musk-deer (*Moschus*) represents a peculiar sub-family of the *Cervidæ*, confined to the region under consideration. A second species has been described from Kansu, in north-western China, and it is not improbable that the genus is also represented in the Indian Siwaliks. Although agreeing with the musk-deer in the absence of antlers and the presence of long tusks in the upper jaw of the males, the Chinese water-deer (*Hydropotes*), from the valley of the Yang-tsi-kiang, belongs to the more typical *Cervidæ*. Another genus (*Elaphodus*) more nearly allied to the muntjacs is also Asiatic, being represented by one species from near Ningpo, in China, and by a second from Moupin, in eastern Tibet. Of the true deer (*Cervus*) there are two groups confined to the area under consideration. Firstly, there is the elaphurine group, represented solely by the aberrant David's deer (*C. davidianus*), of northern China; and, secondly, we have the damine group, of which the fallow deer (*C. dama*) of the Mediterranean countries and the Persian fallow deer (*C. mesopotamicus*) are the living forms. Allied types occur in the East Anglian Forest-bed, and the gigantic extinct Irish deer (*C. giganteus*) must likewise be included in the group, all the members of which have the antlers more or less palmated.

As regards the camels (*Camelus*), there is some difficulty in arriving at a satisfactory conclusion, since although a feral race of the Asiatic *C. bactrianus* is met with in the deserts bordering Kashgar, it is now pretty well ascertained that really wild camels exist nowhere in the world. Still, however, as fossil species of the genus are met with in the Pliocene of the Siwalik Hills (on the borders of the Oriental and Holarctic regions) and in the Plistocene of Algeria, it is probable that the group is of Holarctic origin[1].

The foregoing survey may be summarised as follows : the names of such genera or groups as are mainly or exclusively confined to the eastern division of the Holarctic region being printed in italic type.

Insectivora.

SORICIDÆ.	*Crossopus.*
	Diplomesodon. C. Asia.
	Nectogale. Tibet.
TALPIDÆ.	*Uropsilus.* } E. Tibet.
	Scaptonyx. }
	Talpa. Also enters Oriental.
URSIDÆ.	*Æluropus.* Tibet.
MUSTELIDÆ.	*Meles.* Enters E. Oriental.

[1] See page 281.

Rodentia.

SCIURIDÆ.	*Eupetaurus.* Tibetan.
MYOXIDÆ.	*Myoxus.*
	Muscardinus.
MURIDÆ.	*Cricetus.*
	Ellobius.
	Siphneus.
SPALACIDÆ.	*Spalax.*
DIPODIDÆ.	*Sminthus.*
	Dipus.
	Euchoretes. Central Asian.
	Platycercomys. Enters Ethiopia.
	Alactaga.
OCTODONTIDÆ.	*Ctenodactylus.* North African.

Ungulata.

BOVIDÆ.	*Capra.* An outlying Ethiopian species.
	Ovis. One N. American species, and one from Central Asia entering Oriental.
	Budorcas. Tibetan.
	Rupicapra. European.
	Addax. Mediterranean.
	Saiga. Central Asian.
	Pantholops. Tibetan.
	Gazella. A large proportion of the species E. Holarctic.
CERVIDÆ.	Cervus. The Elaphurine and Damine groups exclusively E. Holarctic; the former Central Asian, and the latter Mediterranean.
	Elaphodus. E. Tibet and China.
	Hydropotes. Chinese.
	Moschus. Asiatic.
CAMELIDÆ.	(?) *Camelus.*

With the possible exception of the *Camelidæ*, none of the families in the foregoing list are peculiar to the area in question—a feature presenting a marked contrast to the lists of the characteristic mammalian genera of the Ethiopian and Oriental regions given above. The total number of genera which can in any sense be considered as peculiar to the eastern half of the Holarctic region does not exceed thirty ; and among these *Uropsilus, Scaptonyx, Elaphodus,* and *Hydropotes* can only be regarded as intruders from the Oriental region ; while *Ctenodactylus* and *Addax* are manifestly Ethiopian types. Indeed, if a Mediterranean region were established, the whole of these, and probably also the true Tibetan forms, would have to be removed from the lists. Apart from the absence of peculiar families, the list bears no comparison as regards numbers with that of the mammalian genera distinctive of the Ethiopian region ; while, with the aforesaid deductions, it is also considerably inferior to that of the Oriental region. All this confirms the conclusions already drawn as to the inadvisability of regarding the area under consideration as a separate zoological region.

The Pliocene and earlier Tertiary faunas of this area having already been considered in connection with Eastern Arctogæa in general in an earlier chapter, we may pass on to a brief review of the Plistocene mammals of the eastern division of the Holarctic region, preparatory to the consideration of the sub-regions into which the latter is divided. The Plistocene period may be taken in England to commence with the Forest-bed of the Norfolk coast, which overlies the topmost of the Pliocene Crag series, and is itself overlain by the glacial deposits. To a later epoch of the same period belong the brick-earths and gravels of our river valleys, as well as the cavern-deposits ; many of these being either of post- or inter-glacial age.

During the Plistocene period two very remarkable differences from the existing state of things have to be noticed. In the first place, the eastern Holarctic region was at that time very much less distinctly differentiated from either the Ethiopian or the Oriental than is the case at the present day ; macaques, hyænas, the lion, rhinoceroses, hippopotami, and elephants abounding in Europe even as far north as England. The second point is the curious mixture of remains of existing species of mammals respectively characteristic of hot and cold climates met with in many parts of England and France. For instance, the glutton, rein-deer, Arctic fox, and musk-ox are animals whose presence seems indicative of a more or less decidedly Arctic climate; many of the voles, picas, jerboas, and susliks, together with the saiga antelope, appear to point with equal force to the prevalence of steppe-like conditions; while the hippopotamus and spotted hyæna seem as strongly in favour of a subtropical state of things. Nevertheless, remains of several of these groups have been found in such close association as to leave no doubt that the animals lived and died hard by where they are now buried. Much evidence on this point has been collected by Sir H. H. Howorth[1], who writes as follows: "Cuvier, whose prejudices were the other way, was long ago con-strained to write of the remains of reindeer found with those of the mammoth and rhinoceros in the cave at Breugue: ' Il ne faut pas douter qu'il [the rhinoceros] n'ait été enseveli avec lui [the reindeer] à Breugue. Ses os y étaient pêle-mêle avec ceux de ce grand quadrupède, enveloppés dans la même terre rouge, et revêtus en partie de la même stalactite.' In the high-level gravels of the Thames valley the mammoth and woolly rhinoceros occurred with the *Elephas antiquus*, while in the low-level gravels the *Rhinoceros leptorhinus* and the hippopotamus occurred with the bison and the musk-ox[2], together with a worked flint. The lemming and the reindeer occurred with the lion and hyæna at Bleadon, the lemming with the lion and hyæna at Wookey-Hole. The reindeer and the grizzly bear were associated with the hippopotamus at Cefn, and the *E. antiquus* with the mammoth at Durdham Down. The hippopotamus and the *E. antiquus* have been found with the reindeer and bison in Kirkdale Cave; the hippopotamus with the wild boar, the rein-deer, the mammoth, and the *E. antiquus* at Brentford. Lartet says that in France remains of the hippopotamus have been found in one cave, that of Arcy, in which the reindeer has also occurred, accompanied by a worked flint. At St Acheul and in the Somme valley the same two animals have occurred together, and also at Levallos, in the Seine valley. At Viry Noureuil, near Chauny

---

[1] *The Mammoth and the Flood* (London, 1887), pp. 114, 115.

[2] The author uses the term musk-sheep for this animal; a few other verbal alterations have been made in the quotation.

(Aisne), the mammoth and *Rhinoceros antiquitatis* have occurred with the hippopotamus, the reindeer, and the musk-ox. At Bicêtre, close to Paris, the lion is associated with northern voles, a marmot, a lizard, and a snake. At Montmorency the mole and the hedgehog have occurred with the hamster, the suslik, and the pica. In Auvergne, M. Pomel has found an elephant and the woolly rhinoceros with a cat, a suslik, and a hamster, together with snakes, lizards, frogs, and with shells such as are still found in the district. In Germany it is the same. At Westeregeln the lion and the spotted hyæna, the mammoth and rhinoceros were found with the marmot, the suslik, the lemming, the pica, and the rein-deer. At Thiede, the mammoth, woolly rhinoceros, the horse, the ox, the reindeer, the Arctic fox, the lemming, and the pica; and so we might continue throughout the majority of the German caverns."

Many attempts have been made to reconcile these apparently contradictory facts; one of the older views being that while the tropical types of mammals lived during warm interludes, they migrated southwards with the incoming of colder conditions to give place to the more Arctic fauna. The associations mentioned above render it, however, perfectly certain that such an explanation is not the right one. On the other hand, it must be remembered that there is yet much to be learnt about the effects of climate on mammals; and the mammalian fauna of the Tibetan plateau shews that many types of animals formerly regarded as more or less essentially tropical or sub-tropical are capable of withstanding a winter climate of extreme severity. Thus in parts of Tibet, as well as in Kashmir, langurs and macaques may be seen leaping among the snow-clad branches of pines. Still it is perhaps diffi-cult to understand how two such animals as the hippopotamus and the reindeer could have inhabited the same locality contem-poraneously[1].

In spite of this association of Arctic and sub-tropical forms, there appears, however, to be evidence of a northern and southern

---

[1] The present writer is not prepared to accept the view of Mr A. H. Keane (*Ethnology*, Cambridge Geographical Series, p. 65, 1896) that the reindeer has only recently become adapted to a northern habitat. Among other circum-stances, its remains are unknown from the Forest-bed.

type of Plistocene fauna, England being apparently somewhere near the border-line where the two met, and, at times at least, overlapped each other. Probably all or nearly all of the living European mammals were in existence at the same time ; but most of these need not be referred to here, attention being concentrated on those which are either extinct, or are now inhabitants of other regions or districts. The following list includes the more important of these forms ; those which are decidedly northern types being indicated by a *, and those which appear essentially southern with a †. The scientific names of extinct species and genera are printed in italic, and of those still living in ordinary type.

Primates.

† Barbary Ape.		Macacus inuus.	
†	,,	,,	*pliocenus.*	England.
†	,,	,,	*suevicus.*	Switzerland.
†	,,	,,	*tolosanus.*	France.

Carnivora.

† Lion.   Felis leo.
† Kafir Cat.   Felis caffra.
† Sabre-tooth.   *Machærodus latidens.*
† Spotted Hyæna.   Hyæna crocuta.
† Striped Hyæna.   ,,   striata.
  Cave Bear.   Ursus *spelæus.*
* Arctic Fox.   Canis lagopus.
  European Wild Dog.   ,,   (Cyon) *europæus.*
      ,,   Hunting Dog.   Lycaon *anglicus.*   England.
* Wolverene.   Gulo luscus.

Rodentia.

† Maltese Squirrel.   *Leithia melitensis* [1].
* Giant Beaver.   *Trogontherium cuvieri.*
* Northern Vole.   Microtus rutilus.
  Sardinian Pica.   Lagomys *sardus.*

---

[1] Originally described as a gigantic dormouse, but shown by the writer in a communication to the *Proceedings of the Zoological Society*, 1895, p. 860, to be allied to the *Sciuridæ*.

Ungulata.

Aurochs.   Bos taurus, var. *primigenius.*
* Musk-Ox.   Ovibos moschatus.
† Barbary Sheep.   Ovis tragelaphus.
† Spanish Ibex.   Capra pyrenaica.
English Gazelle.   Gazella *anglica.*
† Saiga Antelope.   Saiga tartarica.
Irish Deer.   Cervus *giganteus.*
† Hippopotamus.   Hippopotamus amphibius.
† Pentland's Hippopotamus.   ,,   *pentlandi.*
* Woolly Rhinoceros.   Rhinoceros *antiquitatis.*
Megarhine   ,,   ,,   *megarhinus.*
† Leptorhine   ,,   ,,   *leptorhinus.*
† Etruscan   ,,   ,,   *etruscus.*
* Elasmothere.   *Elasmotherium sibiricum.*
* Mammoth.   Elephas *primigenius.*
† Straight-tusked Elephant.   ,,   *antiquus.*
† Southern Elephant.   ,,   *meridionalis.*
† Dwarf Elephants. $\left\{\begin{array}{l} \\ \\ \end{array}\right.$,,   *melitensis.* $\left.\begin{array}{l} \text{Maltese} \\ \text{Islands.} \end{array}\right.$
                                ,,   *mnaidriensis.*
† African Elephant.   ,,   *africanus.*

In this list the Barbary ape is now confined to North Africa
and Gibraltar.   The lion, although now restricted to Africa, India,
Persia, and Mesopotamia, ranged during the historic period into
Thessaly ; while the Kafir cat is solely African.   The sabre-
toothed tiger of the caves was the last survivor of a genus common
in the Pliocene, which in the Plistocene is unknown further north
than Cromer.   The spotted hyæna, whose remains are so abundant
in the English caves, is, as we have seen in an earlier chapter, now
restricted to southern Africa; while the striped species, which
dates from the upper Pliocene, now ranges from north Africa to
India.   The cave-bear, a gigantic extinct species, was distinguished
from the brown bear by the more complex structure of its molar
teeth.   Of the *Canidæ*, the European wild dog has its nearest living
ally in the Altai, the other species of the group being Oriental;
while the European hunting-dog, which is known only by a single
jaw from the Glamorganshire caves, appears to have been closely

related to the living Cape species.   Among the rodents, it is only necessary to mention that the giant beaver (*Trogontherium*) represents a distinct genus ranging from the Norfolk Forest-bed to Siberia; and also that the Maltese squirrel (*Leithia*) was restricted to the islands from which it takes its name.   In the ungulates, the aurochs[1] was the gigantic ancestor of the domestic cattle of the present day, but is unknown living in a wild state.   The arui, or Barbary sheep, is now restricted to north Africa; while the Spanish ibex is confined to the mountains of the Iberian penin-sula, its fossil remains occurring in the Gibraltar caves.   The Irish deer, distinguished by its great size and widely-spreading antlers, was an ally of the fallow-deer, with which it is connected by means of another extinct species or variety (*C. ruffi*); and it may be mentioned that there are species of extinct deer from the Forest-bed, which it is unnecessary to name in this place.   The latter deposits are the source of the known remains of the English gazelle.   The common hippopotamus, which dates from the upper Pliocene of Italy, is now exclusively confined to Ethiopian Africa, but in the Plistocene is known to have wandered as far north as Yorkshire.   Pentland's hippopotamus is a smaller species from Italy and the Mediterranean islands, where there may be a second still smaller form.   Of the rhinoceroses, *R. antiquitatis*, which is exclusively Plistocene, ranged from Central Europe to Siberia; its remains being dredged abundantly, in common with those of the mammoth, from the Dogger Bank, in the North Sea.   The relation-ship of this species to the extinct Indian *Rhinoceros platyrhinus* and the living African *R. simus* has been alluded to in a previous chapter.   The other three European species of the genus, which, like the last, were two-horned and devoid of front teeth, date from the Pliocene, and form a group differing remarkably in dental characters from *R. antiquitatis*.   While two of these species were southern types, the third accompanied the mammoth and woolly rhinoceros in Siberia.   A near ally of the rhinoceroses, the huge Siberian *Elasmotherium*, differed remarkably in the structure of its cheek-teeth, which are tall-crowned, and shew some approximation to those of the horses.

---

[1] The European bison is frequently miscalled the aurochs.

From a distributional point of view, the European Plistocene elephants are of especial interest. Foremost and best-known of these is the mammoth (*Elephas primigenius*), which, as stated in an earlier chapter, was a very near ally of the existing Indian species, although distinguished—as we know from the evidence of specimens preserved in the frozen soil of Siberia—by its coat of woolly red hair, among which were intermingled long bristly black hairs. Curiously enough, traces of this woolly coat have been detected in the Indian elephant, so that it is probable that this species originated in some part of Asia where the climate is colder than is that of India. Regarding the range of this species, Professor Boyd Dawkins[1] remarks that "the mammoth is very abundant in the caverns and river-deposits of Britain and of France, and is known to have ranged over the Pyrenees into Spain, from the discovery of specimens in the zinc-mines of Santander. It has been proved by Prof. E. Lartet and Dr Falconer to have lived in the neighbourhood of Rome when the volcanoes of central Italy were active, and poured currents of lava and clouds of ashes over the [site of the] imperial city. It is common in northern and southern Germany, but it has not been found in Europe north of a line passing through Hamburg, or in any part of Scandinavia or Finland. It occurs in the auriferous gravels of the Urals; and in Siberia, as is well known, it formerly existed in countless herds, being buried in the morasses in large numbers, in the same manner as the Irish elks at the bottom of the Irish peat-bogs. The admirable preservation of some of the carcases is undoubtedly due to their having been entombed directly after death, and then quickly frozen up, a process which need not necessarily imply climatal conditions unlike those of the present time in Siberia." That the mammoth ranged across Bering Strait into Arctic America, is proved by the discovery of its remains in the frozen soil of Eschscholtz Bay; but in the greater part of North America it was replaced by the closely-allied *E. columbianus*. In eastern Europe there existed a variety or species known as *E. armeniacus*, of which the molar teeth still more closely resemble those of the Indian elephant than do those of the typical form. The straight-

---

[1] *Early Man in Britain* (London, 1880), p. 106.

tusked elephant (*E. antiquus*) is a more southern Plistocene type, of which the molars are to a certain extent intermediate between those of the living Indian and African species. Still more southerly in its distribution is the gigantic southern elephant (*E. meridionalis*), of which the remains are found in the upper Pliocene of Italy, as well as in the Plistocene Forest-bed of Norfolk, and equivalent strata at Dewlish, in Dorsetshire. The Maltese Islands were the habitat during the Plistocene epoch of the two or three species of dwarf elephants, which appear to have been nearly allied to the existing African species, but whose size was diminished by the smallness of the areas where they flourished. Lastly, the African elephant, which is now restricted to Ethiopian Africa, has left evidence of its existence during the Plistocene epoch in Algeria, Spain, and Sardinia.

The fauna of the Forest-bed period, among which the mammoth, megarhine rhinoceros, and Irish deer are wanting, is, as already stated, of pre-glacial age, and, on the whole, indicative of a fairly warm climate, although there is some evidence that the musk-ox then ranged as far south as England. At the close of this epoch, the southern elephant, together with a small bear known as *Ursus arvernensis*, appear to have become extinct. Soon after, glacial conditions made their appearance, causing much disturbance and migratory movements among the original southern pre-glacial fauna, and bringing an incursion of northern forms like the reindeer, Arctic fox, wolverene, and musk-ox, as well as of species from the eastern steppes such as the Saiga antelope and the Kirghiz jerboa (*Alactaga*), together with mountain animals like the chamois, the ibex, and the marmot, into the lowlands of south-western Europe. Among the northern forms that then spread themselves southward were the mammoth and the woolly rhinoceros, which at this epoch appear to have attained their maximum development.

Unfortunately, there is much uncertainty as to the part played by the glacial epoch in the extermination of the large mammals characterising Plistocene Europe. By most English geologists the brick-earths of the Thames valley, which contain remains of rhinoceroses and elephants in abundance, as well as those of monkeys more sparingly, are regarded as of post-glacial age ; but Prof. von

Zittel[1] considers them pre-glacial, or more probably inter-glacial. If they are either inter- or post-glacial, it is clear that the cold was not the exterminating cause; and it is quite possible that many or all were killed off by man, although this could scarcely be the case with the Siberian fauna.

Be this as it may, there is good evidence that when northern forms, such as the reindeer, wolverene, and banded lemming, had once obtained an entrance into central and southern Europe, they remained there for a considerable time, since they were present during the latter portion of what is known as the palæolithic epoch. With the advent of the present climatic conditions came in the present woodland fauna of central Europe, constituting what has been termed the squirrel- or bison-epoch; and from that date, when animals became domesticated, man has exercised a large influence on the fauna.

It is important to notice that, in spite of the mingling of northern and southern types in England, France, and Germany, to which allusion has already been made, there seems to have been a distinction between the northern and the Mediterranean fauna throughout the whole of the later Plistocene epoch, such forms as the Barbary sheep and the fallow-deer being essentially southern, although the hippopotamus, as we have seen, extended as far north as Yorkshire.

Although the later Plistocene fauna was spread not only over Europe, but also through North and Central Asia, a number of the characteristic European types, such as the hippopotamus, ibex, chamois, fallow-deer, cave-bear, and wild cat, were wanting in Asia. In that area forms are met with which are still characteristic of the same districts. As examples may be noticed : the Mongolian gazelle (*Gazella gutturosa*), the Himalayan ibex (*Capra sibirica*), the Persian wild goat (*Capra ægagrus*), the argali (*Ovis argali*), the musk-deer (*Moschus moschiferus*), the tiger (*Felis tigris*)—of which the remains have been found even within the Arctic Circle—together with a number of smaller forms, such as *Siphneus aspalax*, *Ellobius talpinus*, *Spalax typhlus*, *Sminthus vagans*, *Tamias asiaticus*, and *Mustela zibellina*. Here, then,

[1] Appendix, No. 36, p. 189.

are clear indications of a Central Asiatic sub-region as far back
as the Plistocene epoch.

The foregoing brief survey of the Plistocene mammals of the
eastern division of the Holarctic region enables
certain deductions to be drawn as to geographical
changes which have taken place in the area since
that epoch. *Geographical changes since the Plistocene.*

In the first place, the occurrence of remains of the tiger in the
New Siberian, or Liakov Islands, lying far within the Arctic
Circle, indicates the union of those islands with the Siberian mainland; and this greater extension of the land at the north-eastern
extremity of Asia would naturally lead to the conclusion that there
was also a land-connection with Alaska across Bering Strait.
That such was really the case is proved by the discovery during
the voyage of H. M. S. "Blossom," in the years 1825–28, of
remains of the horse, mammoth, and bison, in the frozen soil of
Eschscholtz Bay, Kotzebue Sound, Alaska[1]; this evidence being
confirmed by the occurrence of the musk-ox in the European
Plistocene, as it is also by the number of species of mammals still
common to the more northern parts of the two hemispheres. And
here it may be mentioned that, according to the researches of the
Russian geologists, Siberia, instead of being covered like northern
Europe with a continuous ice-sheet during the glacial epoch,
had only a number of comparatively small glaciers, so that the pre-
glacial fauna was able to exist here at a time that it could not live
in Europe. Still the frozen condition of the subsoil, and the
formation of ground-ice in the rivers, rendered the preservation of
the carcases of mammoths, rhinoceroses, bison, and musk-oxen an
easy matter.

Passing to south-western Europe, the occurrence of remains of
the African elephant in Sicily and Spain, together with the
presence of small allied species in the Plistocene of Malta, and
likewise of remains of the Barbary sheep and Barbary ape in
southern Europe, indicates a free land-communication between
Europe and Africa, both by way of the Straits of Gibraltar, and
likewise between Italy, Sicily, and Tunis; Malta being then also in

[1] Beechey's *Voyage to the Pacific and Behring's Straits in H.M.S.*
"*Blossom,*" Vol. II.

connection with the mainland. Probably also it was by one or both of these routes that the hippopotamus and spotted hyæna passed between Europe and Africa, as it is scarcely likely that the former animal, at least, travelled round by way of Turkey and Syria. Writing of the Maltese islands, Leith-Adams[1] observes that "although from their smallness the islands furnish only scant evidences of the complicated and extensive oscillations of level to which the original area has been subjected from first to last, nevertheless the data I have furnished are at the least suggestive, and, in conjunction with the fossilised remains, seem to lead to the belief, that in the first place there was an upheaval of a large tract of land in this portion of the Mediterranean at some period towards or after the close of the Miocene epoch. In the second place, that during the Quaternary [Plistocene] period, the whole, or at least all excepting perhaps the tops of the Benjemma heights and Gozo hills, were again submerged ; and, thirdly, that a re-elevation of the land took place, ending in the present insular fragments. Perhaps in the first case there was a connection or contemporaneity in the upheaval of the Miocene beds of Malta, Sicily, Italy, Candia, the Red Sea, Egypt, Arabia, Cerigo, Azores, Algeria, Southern France, and Spain. Thus the islands of the inland sea may represent portions of a land area now occupied more or less by water. When this area began to sink is not apparent, but the fact that the same elephant and hyæna now living in Africa existed in Sicily, shews that there was a land-connection between the two at a very recent epoch."

Regarding the nature of the former connection between Italy, Sicily, and Malta, Dr Wallace[2] writes that a comparatively shallow sea or submerged bank incloses Malta and Sicily, and "that on the opposite coast a similar bank stretches out from the coast of Tripoli, leaving a narrow channel, the greatest depth of which is 240 fathoms. Here, therefore, is a broad plateau, which an elevation of about 1,500 feet would convert into a wide extent of land connecting Italy with Africa ; while the same elevation would also connect Morocco with Spain, leaving two extensive lakes to repre-

---

[1] *The Nile Valley and Malta* (London, 1870), p. 211.
[2] *Geographical Distribution of Animals*, Vol. I. p. 201.

sent what is now the Mediterranean Sea, and affording free communication for land animals between Europe and North Africa."

Probably the dwarf elephants of Malta were developed from a
larger form, closely allied to or identical with the African elephant,
after the separation of the island itself from the mainland.   With
regard to Leith-Adams' idea of the subsequent submergence of
Malta, it is pretty certain that this could not have been complete,
since that island is inhabited by a large species of weasel (*Mustela
africana*) common to Egypt, and perhaps the south of Italy[1];
this animal being doubtless a survivor from the old fauna of the
Plistocene land connecting Italy, Sicily and Malta with northern
Africa.

In north-western Europe there are equally conclusive evidences
of the connection of the British Islands with the Continent during
the Plistocene epoch.   On many parts of the English coasts there
occur, for instance, submerged forests dating from a comparatively
recent epoch, which, when exposed during exceptionally low tides,
are seen to contain the stumps of trees in their original upright
position, and with their roots still implanted in the soil.   Forests
of this kind are found near Torquay and Falmouth, as well as on
several parts of the Welsh coasts and in Holyhead harbour; the
submergence which has taken place in the case of the one at
Falmouth being estimated at upwards of 70 feet.   Again, the pre-
glacial Norfolk Forest-bed, so often alluded to in the foregoing
pages, affords evidence of an extensive submergence on the east
coast of England; this being supplemented by the Dogger Bank
in the North Sea, from which, as already mentioned, such numbers
of remains of the mammoth, as well as those of the woolly rhinoceros and other mammals, have been dredged.   Additional evidence
in favour of the same subsidence is afforded by the numerous
ancient river-channels and valleys found in many parts of Britain,
which are situated at depths of from one to two hundred feet
below the present level of the land, and frequently cut right across
the existing drainage lines, so as to connect valleys now completely distinct.   These ancient channels, which are now completely
choked with sand, mud, or gravel, have only been revealed by the

---

[1] See Thomas, *Proc. Zool. Soc.* 1895, pp. 128—131.

aid of the borer, but their evidence is, nevertheless, unimpeach-
able.

From these and other lines of evidence, we learn that during
the Plistocene epoch not only was England connected with France
across the English Channel, but that the land extended up the
North Sea at least as far as the Dogger Bank; the Ouse, the
Thames, the Rhine, and perhaps the Elbe originally uniting to
form one mighty river, discharging far up in the North Sea.
During a portion of this period Ireland was in connection with
the British Islands; and it has been suggested by Leith-Adams[1]
that the connection was with Scotland, owing to the circumstance
that, with the exception of the cave-bear, all the living and extinct
Irish mammals have been recorded from Scotland, while a number
of the English Plistocene mammals appear never to have reached
the latter country. On the other hand, Dr R. F. Scharff[2], from a
study of the freshwater fishes and molluscs, is of opinion that
" Ireland was in later Tertiary times connected with Wales in the
south and Scotland in the north; whilst a freshwater lake occupied
the present central area of the Irish Sea. The southern connection broke down at the beginning of the Plistocene period, the
northern connection following soon after. There is no evidence
of any subsequent land-connection between Great Britain and
Ireland." Since the above was written Dr Scharff (*Mem. Soc.
Zool. France*, vol. VIII. pp. 436—474, 1895) has further developed
his views on the origin of the Irish fauna. He concludes that
all the Irish mammals reached the island in the early Plistocene
(Forest-bed); such British forms as are unknown in Ireland being
considered to have reached Britain later, when Ireland was
isolated.

Further reference to the former connection or connections
between Britain and the Continent will come more conveniently
later.

The generic and specific mammalian types common to the
eastern and western divisions of the Holarctic region
having been already referred to, we may at once
proceed to the consideration of those characteristic

Western
Division of the
Region.

[1] *Proc. Roy. Irish Acad.* Ser. 2, Vol. III. (1883).
[2] Appendix, No. 25.

of the western division.　And here it may be mentioned in respect to the two areas, that whereas many generic types of animals were unable to pass from the one to the other owing to the high latitude of the strip of connecting land, yet in other cases the geographical limits of the range of certain genera in the Old World form also an important factor in the case.　As stated a few paragraphs back, many of the characteristic European Plistocene mammals, such as the hippopotamus, the fallow-deer, and the cave-hyæna, never extended into the Asiatic portion of the Holarctic region, so that these and many other forms never could have had an opportunity of crossing Bering Strait, even had they been capable of existing in such a high latitude.

Excluding bats and seals, the following genera of mammals will be found confined to the western half of the Holarctic region, although some of these range southwards into the Sonoran.　There are also certain genera which appear to be typically Sonoran, whose range includes part of the western Holarctic region, but these are best considered in the light of intruders from the south[1].

Among the shrews of the western Holarctic there are two species, viz. *Sorex palustris*, of the Rocky Mountains, and *S. hydrodromus*, of Unalaska Island, which differ from all their allies in the presence of long fringes of hair to the feet, although they resemble ordinary species of the genus in the characters of their dentition and tail.　In consequence of these differences these aquatic shrews have been referred by some writers to a separate genus, under the name of *Neosorex*; although such distinction is considered by Dr Merriam unnecessary.　There is, however, one genus of Insectivora (*Condylura*), represented by the star-nosed mole, absolutely characteristic of this area.　Allied in structure and habits to the Old World moles, which are totally wanting in America, this animal takes its name from the presence of a star-like ring of fleshy appendages at the extremity of the muzzle.

The Carnivora include no peculiar genera[2]; but the Rodentia,

---

[1] It may be well to mention here that the majority of American zoologists regard as genera a number of groups to which the present writer would not be disposed to grant more than sub-generic rank.

[2] *Mephitis*, *Taxidea*, etc., appear to be of Sonoran origin.

which, as in the eastern division, are very numerous, comprise one family, as well as several genera, restricted to this area. In the *Sciuridæ* the marmot-like genus *Cynomys* ranges into the Holarctic, but is considered by Dr Merriam as chiefly characteristic of the Sonoran region; and the same is the case with the white-footed mice (*Sitomys*)—of which there is but a single Holarctic representative, while the Sonoran species are very numerous—and also with the wood-rats (*Neotoma*), of which a sub-genus is restricted to the Holarctic. The family peculiar to the region is that of the *Haplodontidæ*, or sewellels, represented by two species of the genus *Haplodon*, from the districts west of the Rocky Mountains. Closely allied to the squirrels, these rodents are distinguished from the latter by the absence of postorbital pro-cesses to the frontal bones of the skull, the depressed skull, and the rootless, or hypsodont, cheek-teeth; all these characters indi-cating a more specialised type. In the *Muridæ*, the voles of the genus *Phenacomys* connect the more typical members of the group with cricetines like the wood-rats (*Neotoma*). Several species have been described. A more southern type is the single representative of the allied genus *Synaptomys*, distinguished by its grooved upper incisors; its molar teeth resembling those of the lemmings, while its skull is of the same structure as in the true voles. According to Dr Merriam, this animal is restricted to the southern part of the Holarctic area, or what he terms the Transi-tion region. In the same great family the well-known aquatic musk-rat, or musquash (*Fiber*), may be considered an Holarctic type, since it is found in the "barren-grounds" on the borders of the Arctic sea, although it ranges southwards into the Sonoran. Closely allied to the voles, with which it agrees in the characters of the skull and teeth, this animal differs by the long, compressed, nearly naked, and reticulate tail; the naked-soled feet being partly webbed, and the whole body adapted to an aquatic mode of life. Its fossil remains occur in the Plistocene of the United States. It is noteworthy, as a negative characteristic of the Holarctic area, that no members of the exclusively New World family *Geomyidæ* are found within its limits. On the other hand, in the family *Dipodidæ*, the jumping-mice of the genus *Zapus*, of which several species are recognised by North American zoologists, are solely

Holarctic, the typical *Z. hudsonianus* dating from the Plistocene epoch. A distinctive feature of the western Holarctic region is the absence of true porcupines (*Hystrix*), their place being taken by the Canadian porcupine (*Erethizon*), which belongs to the same sub-family as the South American porcupines, although distinguished, among other characters, by its short and non-prehensile tail. It is a native of the wooded portions of Canada and the United States, and its remains have been discovered in a cave in Pennsylvania.

FIG. 69.   ROCKY MOUNTAIN GOAT (*Haploceros montanus*)·

Among the ungulates, the remarkable animal known as the Rocky Mountain goat, which alone represents the genus *Haploceros*, and differs from all other ruminants by the extreme shortness of the cannon-bone in both the front and hind limbs, is exclusively an inhabitant of the western Holarctic region. The same is now the case with the musk-ox (*Ovibos*), but as this animal ranged over Europe and northern Asia during the Plistocene, it can scarcely be regarded as distinctive of the western area.   Of other peculiar New World ungulates, the prong-buck (*Antilocapra*) and certain deer of the genus *Cariacus* are found within the Holarctic region, but the former seems to be essentially a Sonoran type, while the latter, although probably also of Sonoran origin, occurs through Central and South America.

The Tertiary genera of mammals peculiar to North America may be best considered in the chapter devoted to the Sonoran region, to which they for the most part belong; and this portion of the subject may be accordingly concluded by tabulating the existing genera or groups peculiar to the area under consideration. These will stand as follows, viz.:—

Insectivora.

        SORICIDÆ.   Sorex.   The sub-genus or genus *Neosorex*.

        TALPIDÆ.   *Condylura*.

Rodentia.

        HAPLODONTIDÆ.   *Haplodon*.

        MURIDÆ.   *Phenacomys*.

                    Synaptomys.   Confined to the southern portion of the area.

                    Fiber.   Enters Sonoran.

        DIPODIDÆ.   *Zapus*.

        HYSTRICIDÆ.   *Erethizon*.

Ungulata.

        BOVIDÆ.   *Haploceros*.

Even if we add to the above certain other sub-generic types, such as the spruce-squirrels (*Tamiasciurus*) and the bushy-tailed wood-rats (*Teonoma*), and likewise take into account the number of Old World types (in many cases widely-distributed ones) that are absent, it can scarcely be urged that such an assemblage is sufficient to constitute a zoological region by itself.   Those of my readers desirous of consulting lists of the species inhabiting the Arctic and Boreal zones of Dr Merriam, will find them in his memoir[1].

       [1] Appendix, No. 19, pp. 24, 25.

In America, probably owing to the north and south trend of the mountain-ranges, the glacial period has had an even more marked effect than in the Old World. On this subject Dr Merriam[1] writes that "not only are the pre-Plistocene animals and plants now represented imperfectly and in greatly reduced numbers, but the areas at present inhabited by their descendants, except in the case of the Boreal forms, are insignificant in comparison with their former extent. It should be remembered that the refrigeration of the glacial epoch has only in part disappeared. In earlier Pliocene times, characteristic representatives of subtropical faunas and floras existed northwards over much of the United States and Canada, and in still earlier times reached the Arctic circle. During the advance of cold in the glacial epoch these forms were either exterminated or driven southward into the narrow tropical parts of Mexico and Central America. The retreat of cold at the termination of this period was not complete, and our continent has never regained its former warmth. Hence the expelled species were not permitted to advance more than a short distance into the region formerly occupied by them, and the tropical species have been held back, and at the present day are not found except along the extreme southern confines of our territory [the United States]. For example, peccaries in early Plistocene times ranged northward over a large part of western North America, while at present they are restricted to parts of Texas and Louisiana below the Red River of the south ; and capivaras, tapirs and other tropical forms whose fossil remains have been found in many parts of the United States have not been able to return. The same is true of plants, for the palms, tree-ferns, and numerous other tropical types that formerly ranged over much of our country are now either altogether extinct or exist only in the tropics.

"The llama and many plants now inhabiting the Andes may be looked upon as representing a class of cases in which Boreal forms were driven so far south that they actually reached the great mountain-system of South America and spread southward over its elevated plateaus and declivities to the extreme end of the continent in Patagonia and Tierra del Fuego."

[1] Appendix, No. 19, p. 44.

Coming to the consideration of sub-regions, we have first of
all the Arctic sub-region, which corresponds to the
Boreal sub-region of Dr Heilprin, and the Arctic
zone of the Boreal region of Dr Merriam, and is of
circumpolar extent. According to the former writer, in the Old
World it may be defined as the tract lying to the north of a line
starting from about the 66th parallel of latitude on the Norwegian
coast, and passing south-eastwards to the coast of eastern Asia
in about the 50th parallel, thus including the greater part of
Kamschatka, and Amurland. In America, according to Dr
Merriam's map, after running just inside the shores of Newfound-

Arctic Sub-region.

FIG. 70.   MUSK-OX (*Ovibos moschatus*).

land and Labrador, the boundary line bends southwards after
passing Cape Chudleigh to coincide with the southern shore of
Hudson Bay, and then takes a north-westerly direction so as
to include within the sub-region only a narrow strip on the north-
eastern coast of Alaska, and a somewhat broader one on the
north-western shore of the same. In the Old World the boundary
line coincides approximately with the northern limit of the cultiva-
tion of cereals, and also with that of the southern migrations of
the reindeer; but in America certain reindeer (which are regarded

by the American zoologists as specifically distinct from the circumpolar "barren-ground" variety) extend considerably further to the
south. For the most part of its extent, the mammals inhabiting
this sub-region are few in number, a large proportion of them
having a circumpolar range. Among them may be included the
Arctic fox, polar bear, wolverene, the ermine or stoat, the eastern
and western species of lemming (*Myodes*), the banded lemming
(*Cuniculus torquatus*), the Arctic vole (*Microtus rutilus*), Parry's
suslik (*Spermophilus empetra*), the musk-ox, and the reindeer;
several of these being restricted to the sub-region. The sea-otter
(*Latax*) frequents the shores of Alaska and Kamschatka, but also
ranges as far south as the Kurile Islands and California, so that it
is not confined to the sub-region. During the Plistocene epoch,
as we have seen, such animals as the mammoth, horse, bison, and
tiger were inhabitants of this tract; the latter animal being still
found in eastern Siberia. Towards Amurland and the Kamschatkan peninsula, the fauna becomes somewhat less scanty; the
large Kamschatkan sheep (*Ovis nivicola*) being here met with, as
well as a true deer, and the brown bear.

Of other groups of animals inhabiting the more typical portions
of this region, there may be noticed among the birds the ptarmigan
(*Lagopus*), the snowy owl (*Nyctea scandiaca*), the Greenland falcon
(*Falco candicans*), the eider-duck (*Somateria mollissima*), as well as
various species of divers (*Colymbus*) and guillemots (*Uria* and
Lomvia), together with the little auk (*Mergulus alle*). Dr Heilprin
writes that "Captain Markham observed the footprints of the
polar hare in the snow-bound ice in latitude 83° 10′, and the
antlers of a reindeer were picked up by the officers under Sir
George Nares, in latitude 82° 45′ (Grinnell Land). A skeleton of
the latter animal, recently picked by wolves, was also obtained in
latitude 80° 27′. Traces of the rock-ptarmigan (*Lagopus rupestris*)
have been met with as far north as latitude 83° 6′, and the snow-
bunting (*Plectrophanes nivalis*) in latitude 82° 33′. The reptile-
fauna is very limited, no serpent, apparently, passing beyond the
sixty-seventh parallel of latitude, and no lizard above the seventieth.
The fishes, which include the common perch and pike, are mainly
salmonoids. Insects are fairly numerous, and even in the far
north the number of species is considerable."

During the Plistocene the region within the Arctic circle enjoyed a decidedly less rigorous climate than it at present possesses. In Baron von Toll's recent expedition to the New Siberian Islands[1], where, as previously stated, remains of the tiger have been obtained, it was discovered "that under the perpetual ice, in a freshwater deposit, which contained pieces of willow and bones of post-tertiary mammals (the mammoth-layer) were complete trees of *Alnus fruticosa*, fifteen feet long, with leaves and fruit. It was thus evident that during the mammoth-period tree-vegetation reached the seventy-fourth degree of latitude, and that its northern limit was at least three degrees further north than it is now."

The next sub-region is the European, which may be taken to include all that part of Europe lying between the Arctic sub-region in the north, and the line of the Pyrenees and Alps, continuing eastwards along the northern shore of the Black Sea to the Caucasus and the Caspian Steppes. This area includes the typical fauna of the eastern Holarctic region, among its more or less characteristic mammals being (in the north) the elk—also ranging into America—, the red deer (unknown in America, but represented by a variety in North Africa), the roe, the bison, the chamois, the Alpine ibex, the typical variety of the brown bear, the badger, the wolverene (in the north), the Alpine marmot (*Arctomys marmotta*), the dormouse, hamster, mole, and hedgehog; several of these being, however, common to the Arctic and Central Asian sub-regions. The desmans (*Myogale*) are restricted to this sub-region; and the same was probably the case with the aurochs (*Bos taurus*, var. *primigenius*), the ancestral stock of our domestic cattle. Finally, the Caucasus is the home of two or three peculiar species of goats (*Capra cylindricornis* and *C. caucasica*) known as ture.

It will be unnecessary, even if this could be accomplished, to give a complete list of the mammalian fauna of this sub-region, but it is essential to refer to the comparative poverty of the fauna of the British Islands as compared with that of the Continent. The following list includes all the mammals (exclusive of bats) known to have inhabited the British Islands within the historic

European Sub-region.

---

[1] See *Knowledge*, 1895, p. 106.

period ; those which are now extinct having an asterisk prefixed
to them, while such as occur in Ireland have the letter I added.
Those that have been introduced by man have a † before them.
The list stands as follows, viz.:—

    Hedgehog.   Erinaceus europæus.   I.
    Mole.   Talpa europæa.
    Common Shrew.   Sorex araneus.
    Lesser   „   „   minutus.   I.
    Water-Shrew.   Crossopus fodiens.
    Wild Cat.   Felis catus.
  *Wolf.   Canis lupus.   I.
    Fox.   „   vulpes.   I.
    Pine-Marten.   Mustela martes.   I.
    Polecat.   „   putorius.
    Stoat.   „   erminea.   I.
    Assogue.   „   hibernica.   I.
    Weasel.   „   vulgaris.
    Badger.   Meles taxus.   I.
    Otter.   Lutra vulgaris.   I.
  *Brown bear.   Ursus arctus.   I.
    Squirrel.   Sciurus vulgaris.   I (? introduced).
  *Beaver.   Castor fiber.
    Dormouse.   Muscardinus avellanarius.
    Harvest-Mouse.   Mus minutus.
    Wood-Mouse.   „   sylvaticus.   I.
    Yellow-necked Mouse.   „   flavicollis.
    Common Mouse.   „   musculus.   I.
  † Black Rat.   „   rattus.   I.
  † Brown Rat.   „   decumanus.   I.
    Common Field-Vole.   Microtus agrestis.
    Bank-Vole.   „   glareolus.
    Water-Vole   „   amphibius.
    Common Hare.   Lepus europæus.
    Mountain Hare.   „   timidus.   I.
  † Rabbit.   „   cuniculus.   I.
 *?Wild Cattle.   Bos taurus.
    Red Deer.   Cervus elaphus.   I.

† Fallow Deer.   Cervus dama.   I.
Roe Deer.   Capreolus caprea.
*Wild Boar.   Sus scrofa.   I.

The total number in this list is only 28, out of which at least four are introduced.   With the exception of the recently-described assogue[1], which is intermediate between the stoat and the weasel, and is peculiar to Ireland, the whole of these mammals are common to the Continent.   As shown in an earlier portion of the present chapter, during the Plistocene epoch Britain possessed a fauna apparently identical with that of the Continent; and there must accordingly be some good reason for its present poverty in mammalian life as compared to the latter area.   The difference is accounted for by Dr Wallace, through the occurrence of one or more periods of subsidence, which took place during the close of, or subsequent to, the Glacial epoch; after which England again became united to the Continent, when its present fauna entered, the period of connection being, however, of comparatively short duration, and thus permitting of the passage of only a moiety of the continental forms, or those which happened at the time in question to be inhabiting the districts nearest to the connecting line.   Only a certain number of the mammals which thus crossed into Britain have ever succeeded in reaching Scotland; and it is from this country, if we accept the views of Dr Scharff, referred to above, that Ireland appears to have received its still more impoverished mammalian fauna.

It will be seen that the foregoing hypothesis attributes the clean sweep supposed to have been made of the original British fauna to the effects of submergence, and not to the ice-sheet.   On the other hand, Mr G. W. Bulman[2], who doubts whether the submergence has been sufficient for this, attributes such extermination as he believes to have taken place solely to the effects of an ice-sheet.   And he further believes that a number of the original British mammals survived in the southern and south-western counties of England, whence they re-populated Britain on the disappearance of the ice-sheet, without there having been any

---

[1] See Thomas, *Natural Science*, Vol. VI. p. 377 (1895).

[2] Appendix, No. 12.

subsequent connection with the Continent. The difficulty connected with this explanation is that it apparently necessitates a pre-glacial or early glacial age for the mammaliferous deposits of the Thames valley, which are almost certainly inter-glacial or post-glacial. The whole subject of glaciation is, however, so complicated and involved, that it is almost impossible to form workable theories as to the exact mode of the repopulation of Britain after the changes which took place during the glacial epoch.

In contrast to the British Isles, which are eminently of the continental type, may be cited Iceland, lying near the border-line between the Arctic and European sub-regions, which is as markedly oceanic in its character. Beyond an occasional ice-borne polar bear, Iceland possesses only the Arctic fox, and a mouse, which has been stated to be a peculiar species; the fox having doubtless been originally introduced from the north on floating ice.

According to the scheme of Dr Heilprin, the next sub-region on the list is that of Central Asia, which includes the countries bounded on the west by the European, and on the north by the Arctic sub-region, and extends eastwards as far as Mantchuria and China proper, being bordered on the south by the Kuenlun and Nanshan mountains[1]. A large portion of the western districts of this tract are open steppes or deserts; and in such tracts several peculiar types of rodents, such as the Kirghiz jerboa (*Alactaga*) and the Yarkand jerboa (*Euchoretes*), are met with, while the saiga antelope (*Saiga*), and the Mongolian gazelle (*Gazella gutturosa*) are likewise characteristic types. Susliks (*Spermophilus*), marmots (*Arctomys*), and picas are very abundant; and the place of the European wild cat is occupied by Pallas's cat (*Felis manul*), the tiger being also sparingly found in the western districts, where the ounce is likewise met with. In part of this sub-region the red deer is replaced by a variety or species known as *Cervus xanthopygus*, while Yarkand is the home of a variety of the Kashmir stag (*C. cashmirianus*), and the Thian-Shan possesses the very fine and wapiti-

Central Asian Sub-region.

---

[1] Dr Heilprin included the Tibetan plateau in this sub-region.

like form described under the name of *C. eustephanus*; all these deer being mostly inhabitants of forest-districts. The Tatarian roe (*Capreolus pygargus*), inhabiting suitable localities in the mountains forming the watershed between the Russian and Chinese empires and Turkestan, is also generally regarded as specifically distinct from its western ally. The sub-region is also the chief home of the magnificent sheep known as argali, among which the splendid Pamir sheep (*Ovis poli*) ranges from the Pamirs to the Altai, while the true argali (*O. ammon*)—if the Tibetan *O. hodgsoni* be really distinct—is also restricted to this sub-region, where it is now confined to northern Mongolia, although it formerly inhabited the Altai. The ibex of the Altai is, however, identical with the Himalayan and Tibetan *Capra sibirica*.

Although included by Dr Heilprin in the preceding, the area typified by the Tibetan plateau is regarded by Dr Blanford[1] as constituting a sub-region by itself.

Tibetan Sub-region.

Typically this region is bounded on the north by the ranges of the Kuenlun, Altyn Tag, and Nanshan, and extends eastwards to China proper, while to the west it must be taken to include Ladak and the upper Indus valley as far as Gilgit[2]. To the south it extends to the main chain of the Himalaya. The following list of mammals is given by Dr Blanford as distinctive of this sub-region; the names of such species and genera as are entirely or mainly confined to the area being printed in italics.

Insectivora.

    Crocidura aranea.
    Nectogale elegans.

Carnivora.

    Paradoxurus *laniger.*
    Canis lupus, var. *laniger.*
      ,,   vulpes, var. flavescens.
      ,,   *ferrilatus.*
      ,,   deccanensis, var.
    Mustela foina, var.
      ,,   *larvata.*

[1] *Proc. Zool. Soc.* 1893, p. 449.
[2] See Blanford, *Fauna of British India*, Mammalia, p. v.

Carnivora (*cont.*).

Mustela *canigula*.
,, alpina, var. *temon*.
,, erminea.
Meles *leucura*.
,, *albogularis*.
Æluropus melanoleucus.
Ursus *pruinosus*.

Rodentia.

Eupetaurus cinereus.
Arctomys *himalayanus*.
,, *robustus*.
Mus *sublimis*.
Microtus *blythi*.
,, *strauchi*.
,, *przevalskii*.
Siphneus fontanieri.
Lagomys *curzoniæ*.
,, *rutilus*.
,, *erythrotis*.
,, *melanostomus*.
,, *ladacensis*.
Lepus *oiostolus*.
,, *hypsibius*.

Ungulata.

Equus hemionus, var. *kiang*.
Bos *grunniens*.
Ovis *hodgsoni*.
,, vignei, var.
,, *nahura*.
Capra sibirica.
Pantholops hodgsoni.
Budorcas taxicolor.
Gazella *picticaudata*.
Cervus *affinis*.
,, *thoroldi*.
Moschus moschiferus.

This list includes all the species inhabiting the plateau at elevations exceeding 12,000 feet. Dr Blanford writes that "many of the forms named only inhabit small portions of the area, and whilst *Bos grunniens*, *Pantholops hodgsoni*, and *Gazella picticaudata*, with several rodents, appear to be peculiar to the high plateaus above 14,000 feet, the two species of *Cervus* are probably found in brushwood at a rather lower elevation in the more broken region of Eastern Tibet, where the rainfall is heavier and the vegetation more abundant.

"As was printed in the paper in the *Geological Magazine*[1], there is, so far as I am aware, no equally peculiar mammalian fauna to be found in any continental area of equal extent, and for a parallel it is necessary to turn to some island like Celebes, that has long been isolated from all surrounding lands."

This, however, is not all, for there occur at Hundes, on the Tibetan plateau, mammaliferous strata yielding, among other remains, bones of a rhinoceros, and of an antelope which is apparently generically identical with the chiru (*Pantholops*), now inhabiting the same area. The isolation and development of this most peculiar fauna is intimately connected with the date of elevation of the Himalaya. After pointing out that both the fossil chiru and the fossil rhinoceros appear to have inhabited the area when it had attained something approaching its present enormous elevation, Dr Blanford[2] writes as follows : "Bearing in mind that the isolation of the Tibetan plateau is far less perfect as regards mammals than that of any island, and that some of the forms— the Carnivora especially—found in Tibet are evidently very recent immigrants, it is a reasonable conclusion that the peculiar fauna of the Tibetan plateau has been distinct from that of neighbouring countries since middle Tertiary times.

"But what has caused the isolation of the Tibetan fauna? Why in this one continental tract is there a generic and specific differentiation of the mammalia, of which no other example exists? There is only one character in which Tibet is different from other continental areas, its great height. This alone renders the climate of Tibet so different from that of other parts of

[1] Decade 3, Vol. ix. p. 161 (1892).
[2] Geol. Mag. *op. cit.* p. 165.

Central Asia, which are equally cold and barren. It seems a reasonable inference that the elevation of the Tibetan plateau dates back to middle Tertiary times.

"It is of course probable that the elevation was gradual; and although the area may have been sufficiently high at the close of the Miocene period to produce a difference in climatal conditions, the greater part of the upward movement may have been post-Miocene, and a great part post-Pliocene."

Bordering as it does upon the tropics, where it abuts against the Oriental region, the Mantchurian sub-region is not easy to define, since the intermingling of Holarctic and Oriental types is very strongly marked on its southern confines. Starting somewhere about the Amur river, it may, however, be taken to include the Japanese islands, Mantchuria, Corea, and northern China; its southern limit being placed approximately in the latitude of Fuchau. Westwards it may be taken to include Moupin, in Eastern Tibet, although this district is referred by Dr Wallace to the Oriental region.

From all the other sub-regions, with the exception of the Mediterranean, the Mantchurian is distinguished by the presence of monkeys belonging to the genera *Macacus* and *Semnopithecus*, some of these occurring in Japan and others in Eastern Tibet. Of the latter, one (*Semnopithecus roxellanæ*) is peculiar, and the other is identified by Mr H. O. Forbes with the widely-spread Oriental *Macacus arctoides*. Among the Carnivora, the Oriental genus *Helictis* enters this sub-region, one species occurring in the neighbourhood of Shanghai; while Japan is the home of a peculiar long-haired dog (*Canis procyonides*), which is frequently separated generically under the name of *Nyctereutes*, although it unquestionably pertains to the typical genus. Perhaps, however, the most characteristic mammals are the deer. Foremost among these are a group of small deer belonging to the genus *Cervus*, and distinguished from the red deer group by the invariable absence of a bez-tine to the antlers, each of which has but four points. These deer are further characterised by the coat of the adult being spotted in summer with white, but uniformly brown in winter, and also by the black lateral margins to the white blaze on the hindquarters. The species include the Japanese deer (*C. sica*), common to Japan and North China, the larger Mantchurian deer (*C. mantchuricus*), and Dybowski's deer (*C. dybowskii*) from the upper Ussuri district of Mantchuria, in the neighbourhood of Vladivostock.   Elsewhere the group is represented in Formosa, and also in the Caspian provinces of Persia.   In addition to these, there are the hornless Chinese water-deer (*Hydropotes*), and the two species of tufted deer (*Elaphodus*); the latter being closely allied to the Oriental muntjacs.   What is known of the palæontological history of the southern portion of this area indicates that during the Pliocene epoch its mammalian fauna was closely allied to that of the Siwalik Hills, thus showing that at this time there was no distinction between the Oriental and Holarctic regions, which even now grade imperceptibly into one another in this district.

The remains of fossil elephants from Japan[1] are referable to *Elephas clifti, insignis*, and *namadicus*, of which the two first are common to the Siwaliks, while the third occurs typically in the Plistocene Narbada beds of India.   From the known distribution of these elephants, it is probable that Japan was connected with the mainland during the Pliocene epoch by way of the Corean peninsula, although Dr Wallace is of opinion that its latest connection was to the north.   Of existing animals common to Japan and the mainland, allusion has already been made to *Cervus sica*; and another common type is the giant salamander *Megalobatrachus*.   The latter genus is represented in a fossil state in the Miocene of Baden, and as it is closely allied to the North American *Cryptobranchus*, there is clear evidence of the eastern migration of this ancient type, of which the two survivors are respectively confined to China and Japan on the one hand, and North America on the other.   Further evidences of affinity between the fauna of Japan and North America are afforded by the circumstance that one species of the mole-like genus *Urotrichus* is confined to the former islands, while the other is an inhabitant of the northwestern districts of the latter continent.   The sea-otter (*Latax*) is likewise common to the coasts of Japan, the Kurile Islands, and Kamschatka, and the Pacific shores of North America.   More remarkable, however, is the fact that a North American scincoid

[1] See Naumann, *Palæontographica*, Vol. XXVIII. Art. 1 (1881).

lizard (*Eumeces quinquelineatus*) is represented in Japan by a form (*E. marginatus*) so closely allied that the two were long considered inseparable, although they are now regarded as distinct[1]. All these facts are indicative that Japan was formerly joined to both Corea and Kamschatka, whence land was continued across Bering Strait to unite the Old World with Alaska.

Although, as already stated, the Mediterranean or Tyrrhenian sub-region has strong claims to be regarded as representing a region by itself, it may be more conveniently considered here than later on in the chapter. In addition to such parts of Africa and Arabia as lie to the north of the Ethiopian region, this sub-region includes Spain, those parts of Europe situated south of the Alps, together with Turkey, Asia Minor, Persia, Baluchistan, and Afghanistan. Whether Kashmir should be regarded as an aberrant outlier of this region, I am not yet satisfied. Although gerbils (*Gerbillus*[2]) are also found in the Oriental and Ethiopian regions, their distribution in the Holarctic is very nearly coincident with the limits of the present sub-region.

Whereas to the north of the Mediterranean Sea a large proportion of the mammals are more or less typically Holarctic, in North Africa and Syria those with an Ethiopian facies are met with, and an Oriental element makes its appearance in the eastern districts of the sub-region. Even in Africa, however, some of the forms have an Oriental facies, the Barbary ape (*Macacus inuus*) belonging to a genus whose home is now in the Oriental region, and which is totally unknown in the Ethiopian. As a wanderer from the Ethiopian region, mention may first be made of a species of jumping-shrew (*Macroscelides*) met with in Barbary, while among the octodont family of the rodents, the gundi, forming the sole representative of the genus *Ctenodactylus*, has its nearest allies in Ethiopia, although it is confined to North Africa. The Barbary ape, although occurring on the rock of Gibraltar, where it may have been introduced, is otherwise confined to North Africa.

---

[1] See Boulenger, *Cat. Lizards, Brit. Mus.* Vol. III. p. 369.

[2] Many writers separate certain species as *Meriones*, but as the two groups are connected by *G. indicus* (see Lataste, *Proc. Zool. Soc.* 1884, p. 88), such distinction seems superfluous.

In the Carnivora, the striped hyæna, which is also an inhabitant of India, is widely spread in this sub-region, ranging through western Asia to northern Africa. The common genet (*Genetta vulgaris*), which belongs to a genus otherwise exclusively Ethiopian, is mainly confined to this region, inhabiting southern France, Spain, Turkey, North Africa, and Palestine. A nearly similar distribution characterises the common mungoose, or ichneumon (*Herpestes ichneumon*), which frequents southern Spain, Asia Minor, North Africa, and Palestine. The large weasel (*Mustela africanus*) common to Egypt, Malta, and perhaps the south of Italy has been already referred to in an earlier part of this chapter. In addition to *Ctenodactylus*, the rodents possess another and more widely-spread generic type confined to this sub-region in the form of the great mole-rat (*Spalax typhlus*), whose range includes southeastern Europe, Persia, Mesopotamia, Syria, and Egypt. In the same order the common porcupine (*Hystrix cristata*), although ranging into West Africa, is found but little, if at all, to the north of the present sub-region, where it is common to northern Africa and southern Europe.

Among the ungulates the addax antelope (*Addax nasomaculata*), although allied to Ethiopian types, is solely Mediterranean, its home being North Africa and Syria. More closely allied to the Ethiopian fauna are certain hartebeests of the genus *Bubalis*, the smaller of which (*B. mauritanica*) is common to North Africa, Syria, and Arabia, while the second (*B. major*) inhabits Tunis. The same is the case with the Beatrix antelope (*Oryx beatrix*) of Western Arabia and Bushire. In gazelles, this sub-region is remarkably rich, doubtless from the number of sandy or desert tracts it contains. Algeria is the habitat of the three species known as *Gazella loderi*, *G. kevella*, and *G. rufina*, while *G. dorcas* ranges through Egypt, Algeria, Syria, Palestine, and a part of Asia Minor, and *G. subgutturosa* roams from Persia through Afghanistan and Turkestan. The aberrant sheep known as the arui (*Ovis tragelaphus*) is now restricted to North Africa; and the mouflon (*O. musimon*), although its fossil remains have been found on the Continent, appears to be now restricted to Corsica. Another species peculiar to the sub-region is the Armenian sheep (*O. gmelini*) of eastern Persia and Asia Minor, represented by a closely-allied form in Cyprus. Of the goats, the Spanish ibex (*Capra pyrenaica*) is restricted to the mountains of Spain; while the Sinaitic ibex (*C. sinaitica*) represents the group in Palestine and upper Egypt. Among the *Cervidæ*, the two species of fallow-deer were originally confined to this area, the common *Cervus dama* being a native of the Mediterranean countries, while the Persian *C. mesopotamicus* is found in the mountains of Luristan, in Mesopotamian Persia. In North Africa the ordinary red deer is represented by a variety distinguished by the invariable absence of a bez-tine to the antlers. A connection with the Tibetan sub-region is afforded by the wild asses inhabiting the desert-plains between the Red Sea and the Indus, since both these and the Tibetan form are but varieties of a single species (*Equus hemionus*). Lastly, Ethiopian affinities are exhibited by the occurrence of a species of hyrax (*Procavia*) in Syria. In the early part of the present century the hippopotamus still inhabited lower Egypt, while, as we have seen, the lion, which is now common in parts of Persia, was found within the historic period in Thrace. At a still earlier date, both these animals, as well as the spotted hyæna, extended as far north as England.

On the whole, therefore, the fauna of this sub-region is a very mixed one; and this fact, together with the difficulty in defining its boundaries, suggests the need of further deliberation before the area is raised to the rank of a separate region. The former connections between southern Europe and Africa having been alluded to in an earlier part of the chapter, require no further notice in this place.

Very difficult to determine is the position which should be given to the valley of Kashmir, since its fauna exhibits such a mingling of Oriental and Holarctic **Kashmir.** types that it might almost be as well assigned to one region as the other. Holarctic affinities are, however, exhibited by the occurrence of a species of the red deer group, *Cervus cashmirianus*, and likewise by one of the rodent genus *Sminthus*, of which the second species inhabits northern and eastern Europe, and the third Kansu, in western China. A variety of the brown bear is also indicative of Holarctic affinities, and this is still more markedly the case with the spiral-horned goat known as the markhor (*Capra falconeri*), of which one variety inhabits the Pir Panjal range, on the south side of the valley, while the others are found in the districts to the north and west of Kashmir. The musk-deer, again, is another essentially Holarctic type. On the other hand, the occurrence of a langur (*Semnopithecus*) and a macaque (*Macacus*) points to a connection with the Oriental fauna; and a Kashmir mungoose (*Herpestes auropunctatus*) is identical with one from India. There are, however, none of the exclusively Oriental genera in Kashmir; and this fact, coupled with the absence of all deer of the sambar-group, leaves little doubt that the valley really belongs to the Holarctic. Whether it should be regarded as pertaining to the Mediterranean sub-region, or as forming a distinct sub-region by itself, must be reserved for future consideration.

Passing to the western division of the Holarctic region, the tract lying to the southward of the circumpolar Arctic sub-region, designated by Dr Merriam the Boreal zone of his Boreal region, may be conveniently termed the Canadian sub-region. Its northern boundary is, of course, identical with the southern limits of the Arctic sub-region, and the area includes the greater part of the Dominion of Canada, although a long strip runs down the line of the Rocky Mountains, and another along the Pacific coast; into the United States. Indeed, Dr Merriam includes in this sub-region all the higher plateaus of Wyoming and Colorado, so that the sub-region embraces a number of small disconnected areas on its south-western extremity, and it is consequently impossible to define its limits by description. It may be stated, however, that on the eastern side of the continent the sub-region extends from Hudson Bay to the middle of Lake Michigan, while on the western coast it stretches from near the extremity of Alaska to San Francisco; a big loop extending northwards of Montana nearly to latitude 55°.

The mammalian fauna of the Canadian sub-region is that of the western division of the Holarctic region generally, and includes those forms mentioned on page 344. According to Dr Merriam, the following genera from this sub-region do not range further south than the undermentioned Transition zone; namely, *Condylura*, *Urotrichus*, *Gulo*, *Latax*, *Arctomys*, *Haplodon*, *Phenacomys*,

Myodes, *Cuniculus*, *Zapus*, *Erethizon*, *Lagomys*, *Cervus*, *Alces*, *Rangifer*, and *Haploceros*. On the other hand, the following, which are as clearly of northern origin, penetrate as far south as the Sonoran region, which some of them enter. These are *Sorex*, *Mustela* (only the members of the sub-genus *Putorius*), *Ursus*, *Fiber*, *Microtus*, *Castor*, *Tamias*, *Bos*, and *Ovis*.

Between the Canadian sub-region of the Holarctic and the Sonoran region is interposed a tract whose fauna contains a mixture of Canadian and Sonoran forms, **Transition Zone.** and it is consequently termed by Dr Merriam the Transition zone. Under this somewhat indefinite title the area may best be left. It is described by the author just cited as follows[1] : "The humid division of this zone, known as the Alleghanian fauna, covers the greater part of New England (except Maine and the mountains of Vermont and New Hampshire), and extends westerly over the greater part of New York, southern Ontario, and Pennsylvania, and sends an arm south along the Alleghanies, all the way across the Virginias, Carolinas, and eastern Tennessee, to northern Georgia and Alabama. In the Great Lake region this zone continues westerly across southern Michigan and Wisconsin, and then curves northward over the prairie-region of Minnesota, covering the greater parts of North Dakota, Manitoba, and the plains of the Saskatchewan ; thence bending abruptly south, it crosses eastern Montana and Wyoming, including parts of western South Dakota, and Nebraska, and forms a belt along the eastern base of the Rocky Mountains in Colorado and northern New Mexico, here as elsewhere occupying the interval between the Upper Sonoran and Canadian zones.

" In Wyoming the Transition zone passes broadly over the well-known low divide of the Rocky Mountains, which affords the route of the Union Pacific railway, and is directly continuous with the same zone in parts of Colorado, Uta, and Idaho, skirting the Canadian boundaries of the Great Basin all the way around the plains of the Columbia, sending an arm northward over the dry interior of British Columbia, descending along the eastern base of the Cascade Range and the High Sierra to the southern extremity

---

[1] Appendix, No. 19, p. 30. In this extract the word Canadian has been substituted for Boreal.

of the latter, and occupying the summits of the Coast Ranges in California and of many of the desert ranges of the Great Basin.

"The Transition zone, as its name indicates, is a zone of overlapping of Canadian and Sonoran types. Many Canadian genera and species here reach the southern limits of their distribution, and many Sonoran genera and species their northern limits. But a single mammalian genus (*Synaptomys*) is restricted to the Transition zone, and future research may show that it inhabits the Canadian region also."

FIG. 71.   MUSQUASH (*Fiber zibethicus*).

The writer adds, however, that there are a considerable number of species—mostly rodents—restricted to this zone. The following Canadian genera, namely, *Condylura, Urotrichus, Ursus, Arctomys, Tamias, Fiber*[1], *Zapus, Erethizon, Cervus,* and *Ovis,* almost or completely disappear in this zone; while the following intruders from the Sonoran, namely *Scalops, Bassariscus, Spilogale, Perognathus, Thomomys, Geomys, Cynomys* and *Antilocapra* do not range further north, several of them, indeed, only intruding into the zone in a small area in the west.

[1] Penetrates the Sonoran along the lines of streams where cool currents of air are carried down.

CHAPTER X.

THE SONORAN REGION.

Limits—Characteristics of Mammalian Fauna—Extinct Groups of Mammals characteristic of Western Arctogæa—Distinctness of the Region—Dual Origin of Groups.

As stated in the introductory chapter, wherever one zoological region of the globe has no definite physical barrier by which it is separated from the next well-marked region, there must always occur an intermediate tract where the characteristic types of the faunas of the two regions inosculate and intermingle. That this is the case with that area of North America denominated the Sonoran region has been indicated at the close of the preceding chapter, and the existence of the Transition zone, which seems, on the whole, to pertain to the Holarctic region, unfortunately prevents the Sonoran from being defined with the precision which would be possible had this area a high mountain-barrier on its northern frontier.

In a map of the small dimensions of the one accompanying this volume it is impossible to show with any attempt at accuracy the complex nature of the $\qquad$ Limits. northern boundary of this region, which will, however, be found accurately laid down in the map illustrating Dr Merriam's memoir[1].

According to the latter writer, "the Sonoran region as a whole stretches across the continent from the Atlantic to the Pacific, covering nearly the whole country south of latitude 43°, and reaching northward on the Great Plains and Great Basin to about latitude 48°. It is invaded from the north by three principal intrusions of Canadian[2] forms along the three great mountain-

---

[1] Appendix, No. 19.
[2] Boreal in the original.

systems already mentioned [Alleghanies, Rocky Mountains, and Cascade and Sierra Nevada ranges]; while to the southward it occupies the great interior basin of Mexico, and extends into the tropics along the highlands of the interior. It covers also the peninsula of lower California, the southern part of which seems entitled to rank as an independent subdivision."

Later it is stated that the region "may be divided by temperature into two principal transcontinental zones, Upper Sonoran and Lower Sonoran; and each of these in turn may be subdivided into arid and humid divisions."

The proposal to form a separate region for such an insignificant area as the southern extremity of California seems unnecessary, although its fauna may differ considerably from that of the typical Sonoran.

Omitting mention of the bats, the mammalian genera characteristic of the Sonoran region may now be taken into consideration. Commencing with the Insectivora, the *Soricidæ* are represented by the peculiar genus *Notiosorex*, which is closely allied to the Oriental *Soriculus*, but has only 28 in place of 30 teeth. Of this genus the two species do not range north of this region, although they also enter Central America[1]. The short-tailed, or earless shrews (*Blarina*), with either 32 or 30 teeth, are also mainly Sonoran, although ranging northwards into the Holarctic, and southwards into Guatemala. In the *Talpidæ* the three species of the mole-like genus *Scalops*, characterised by having 36 teeth, webbed hind feet, and a short and nearly naked tail, are mainly Sonoran, although passing into the Transition zone. On the other hand, the two species of *Scapanus*, distinguished by the possession of 40 teeth, and the hairy tail, have a distribution very similar to *Blarina*, although they do not enter Central America.

In the Carnivora the raccoon-family (*Procyonidæ*) is very strongly represented, although none of the genera are absolutely peculiar to the region. The genus *Bassariscus*—a near ally of the true raccoons, and possessing two species—is nevertheless mainly Sonoran, although it ranges into the Transition zone of the

---

[1] *Teste* Dobson.

Holarctic and also into Central America. The true raccoons, on the other hand, cannot be regarded as distinctive of the region, since they range from South America into the Canadian sub-region of the Holarctic; and the coatis (*Nasua*) are now highly characteristic of the Neogæic realm. Indeed Dr Merriam considers both genera as intruders from the latter realm, but this can scarcely be regarded as the correct view. The family is represented in the two halves of the northern hemisphere (in the eastern by *Ælurus*), in both of which it dates from the Pliocene, and, as it is unknown in South America till the Plistocene or late Pliocene, it is evidently one of northern origin; the American forms having probably attained their maximum development in the Sonoran region. Much the same is the case with regard to the skunks among the *Mustelidæ*; these being probably an original Sonoran type which has spread northwards into the Holarctic region and southwards into the Neogæic realm. Of these, the single species of climbing skunk (*Spilogale*) is mainly Sonoran, although it also enters the Transition zone of the Holarctic, and likewise Central America. Of the other members of the group, the typical skunks (*Mephitis*) range from Hudson Bay to Guatemala; while the allied genus *Conepatus* is found from Texas to Patagonia. In the same family the American badgers (*Taxidea*), although ranging well into the Holarctic, are regarded by Dr Merriam as of Sonoran origin. These badgers, it may be observed, differ from the true badgers of the Old World by the form and characters of their cheek-teeth, the last upper molar being proportionately much smaller.

Turning to the rodents, the well-known prairie-marmots (*Cynomys*), which occupy a position intermediate between the true marmots and susliks, are regarded by Dr Merriam as of Sonoran origin, although they extend into the Holarctic. In the *Muridæ* the peculiar cricetine genus *Rhithrodontomys*—which, together with the allied South American *Rhithrodon*, differs from the other members of the sub-family to which it belongs by its grooved upper incisors—appears to be restricted to the region under consideration. The white-footed mice (*Sitomys*), although distributed over the whole of the New World, seem to attain their maximum specific development in the Sonoran, to which the two sub-genera

Onychomys and *Oryzomys* are restricted.   Yet their near alliance to the Old World hamsters indicates that the group must have had a northern origin, although the genus may have attained its present distinctive features within the Sonoran area. The genus *Sigmodon*, which differs from the last in the pattern of the molar teeth, and is represented solely by the rice-rat, does not range north of the Sonoran region, although extending into South America as far as Ecuador.   The wood-rats (*Neotoma*), in which the molars simulate the prismatic appearance of those of the voles, are also largely Sonoran, although they extend into the Canadian

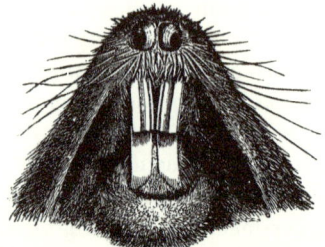 

Fig. 72.   FACE OF *Geomys bursarius*, SHOWING GROOVED UPPER INCISORS AND OPENINGS OF CHEEK-POUCHES.

Fig. 73.   FACE OF *Thomomys talpoides*, SHOWING SMOOTH UPPER INCISORS AND OPENINGS OF CHEEK-POUCHES.

sub-region of the Holarctic, where they are represented by a distinct sub-genus (*Teonoma*).   The round-tailed musk-rat of Florida (*Neofiber*) is an exclusively Sonoran type, although it is regarded by Dr Merriam merely as a sub-genus of *Microtus*. Highly characteristic of the region are the pouched rats, constituting the genera *Geomyidæ*.   Of these, the typical genus *Geomys*[1] extends northwards into the Transition zone and southwards into Central America; while the nearly-allied *Thomomys*, in which the upper incisor teeth are smooth instead of grooved, penetrates into the Canadian sub-region of the Holarctic, although unknown

[1] Subdivided into eight genera by Merriam, *North American Fauna*, Part VIII., Washington (1895).

in Central America. Both these genera are represented in the Pliocene of the Sonoran area. In the same family the three genera of kangaroo-rats known as *Dipodomys*, *Perodipus*, and *Microdipodops* appear to be confined to the region ; and the same is the case with the allied genus *Heteromys*, although *Perognathus* passes northwards into the Transition zone.

FIG. 74. UNDER SURFACE OF LEFT FORE-FOOT OF *Geomys*.

In the Ungulata, the deer belonging to the peculiar American genus *Cariacus* are very abundant in the Sonoran region (where those of the typical genus *Cervus* are entirely wanting), although they also range into the Canadian sub-region of the Holarctic, and extend right through South America. Since, however, they are wanting in the earlier Tertiary deposits of the latter area, as they are at all epochs in the Old World, there can be little hesitation in regarding them as essentially Sonoran types. Even more decidedly is this the case with the prongbuck (*Antilocapra*), the sole type of the family *Antilocapridæ*, which is distinguished from the *Bovidæ* by the horn-sheaths of the males being branched and periodically shed from their bony supports. Although the prongbuck pene-trates a considerable distance into the Canadian sub-region of the Holarctic, its true home is the prairie-district of the Sonoran lying to the westward of the Mississippi. Possibly a small deer-like animal from the Tertiaries of the same area known as *Cosoryx*, may have been the ancestral stock of the prongbuck. Lastly, the peccaries (*Dicotyles*), which are now chiefly South American, appear to have been originally Sonoran types which have migrated southwards ; their fossil remains being common in the Tertiaries of the United States, whereas they are unknown in South America before

FIG. 75.   HEAD OF MALE MULE-DEER (*Cariacus macrotis*).

the Plistocene. Their near affinity to the earlier Tertiary pigs of the Old World indicates that at a more remote date they spread from a more northerly starting-point.

With regard to the armadillo (*Tatusia*) found in the Sonoran, this is clearly a very recent immigrant from the Neogæic realm ; and although opossums (*Didelphys*) were abundant in North America during the early portion of the Tertiary epoch, it is not improbable that the same explanation will hold good for their existing Sonoran representatives.

The following list includes such exclusively New World genera of mammals (apart from bats) which are represented in the Sonoran area; those which may be regarded as more or less nearly confined to this region being printed in italics. To appreciate fully the significance of this list, reference must, however, be made to the series of Holarctic genera given on p. 360, which

FIG. 76.   HEAD OF MALE PRONGBUCK (*Antilocapra americana*).

are more or less completely restricted to the Canadian sub-region of that great region, and the intervening Transition zone. The Sonoran list is as follows, viz. :—

Insectivora.

SORICIDÆ.

Notiosorex.   Also Central America.

Blarina.   Enters Canadian sub-region of Holarctic.

Insectivora *(cont.)*.

    TALPIDÆ.

        Scalops.   Enters Transition zone.

        Scapanus.   Enters Canadian sub-region of Holarctic.

Carnivora.

    PROCYONIDÆ.

        Bassariscus.   Enters Transition and Central America.

        Procyon.   N. to S. America.

        Nasua.   Also South American.

    MUSTELIDÆ.

        Spilogale.   Enters Transition and Central America.

        Conepatus.   Texas to Patagonia.

        Mephitis.   Extends into Canadian sub-region and Central America.

        Taxidea.   Enters Holarctic.

Rodentia.

    SCIURIDÆ.

        Cynomys.   Extends into Holarctic.

    MURIDÆ.

        Rhithrodontomys.

        Sitomys.   The whole of America.

        Sigmodon.   Southwards to Ecuador.

        Neotoma.   Ranges into Holarctic.

        Neofiber.

    GEOMYIDÆ.

        Geomys.   Extends into Transition zone and Central America.

        Thomomys.   Ranges into Canadian sub-region.

        Dipodomys.

        Perodipus.

        Microdipodops.

        Perognathus.   Ranges into Transition zone.

        Heteromys.

Ungulata.

    ANTILOCAPRIDÆ.

        Antilocapra.   Ranges into Canadian sub-region.

Ungulata (*cont.*).
    CERVIDÆ.
        Cariacus.   Greater part of America.
    DICOTYLIDÆ.
        Dicotyles.   Also South American.

Edentata.
    DASYPODIDÆ.
        Tatusia.   South American.

Marsupialia.
    DIDELPHYIDÆ.
        Didelphys.   South American.

Although the Transition zone undoubtedly forms an unsatisfactory item in regard to the distinctness of the Sonoran region, yet when we look at the difference of its mammalian fauna as a whole from that of the Canadian sub-region of the Holarctic, and the close similarity between the latter and the fauna of northern Europe and Asia, there can be but little hesitation in regard to the acceptance of Dr Merriam's view that the Sonoran is a valid zoological region of the Arctogæic realm.

In a previous chapter the groups of mammals, both living and extinct, confined to the eastern division of the Arctogæic realm have been already noticed, while in the present one reference has been made to such existing types as are restricted to the western half of the same realm. It now remains to consider briefly some of the leading extinct groups which are found only in the latter area ; and the consideration of these comes most appropriately here, seeing that the majority of these peculiarly American types are of Sonoran origin, a large number of their remains having been obtained from the States of New Mexico, Kansas, Nebraska, and Dakota, which lie within that region, or from Colorado, Wyoming, and Montana, which are situated within the Transition zone.

Extinct Groups of Mammals characteristic of Western Arctogæa.

Although, in common with the higher Primates, lemuroids are now quite unknown in North America, they were well represented there during the Puerco epoch of the lower Eocene by three families. The first of these—the *Chriacidæ*—includes animals

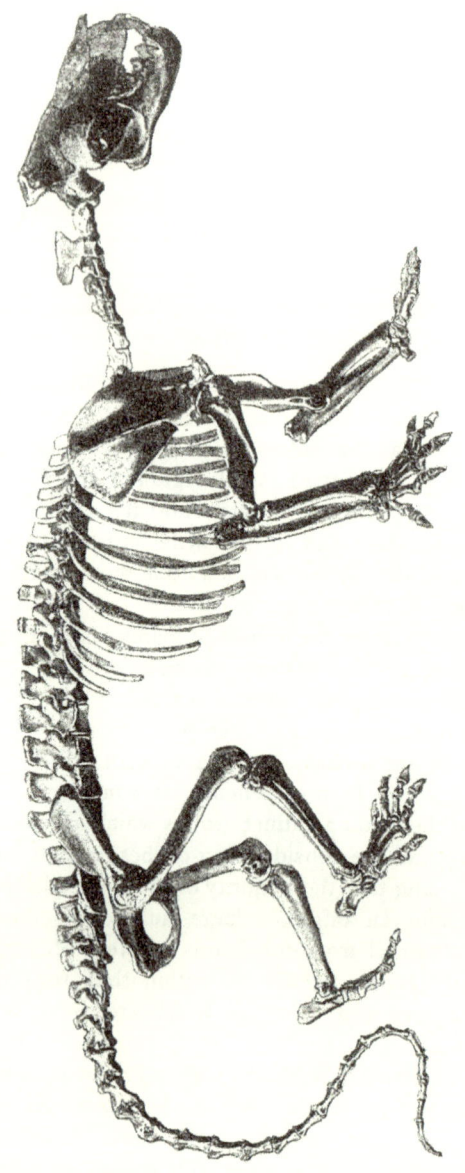

FIG. 77. SKELETON OF *Patriofelis ferox*. ⅟₁₆ nat. size.

having the same number of teeth as the allied Tertiary European family *Adapidæ*, but all characterised by their more primitive structural features.   Indeed these early lemuroids appear to present considerable resemblances to the creodont Carnivora, and differ from all the other members of the sub-order to which they belong by the great elongation of the bony symphysis connecting the two branches of the lower jaw at the chin.   Several other genera, in addition to the typical *Chriacis*, are assigned to this family.   The second group is that of the *Anaptomorphidæ*, which is represented in the Puerco Eocene by a genus known as *Indrodon*, and in somewhat higher beds by the typical *Anaptomorphus*.   Although in other respects coming closer to existing types than is the case with the *Chriacidæ*, the present family is broadly distinguished by the tritubercular structure of the upper molar teeth.   The third peculiar North American family of the lemuroids is that of the *Mixodectidæ*, typically represented by *Mixodectes* of the Puerco Eocene.

Among the extinct creodont Carnivora there are two families apparently restricted to the Tertiaries of North America, namely, the *Miacidæ* and the *Mesonychidæ*, the former of which presents such strongly marked affinities to the modern Carnivora that it is frequently assigned to that group.   The second family, on the other hand, as represented typically by the genus *Mesonyx* of the Uinta or lowest Oligocene, is characterised by the very simple structure of the whole series of cheek-teeth, which are not unlike the pre-molars of some of the higher carnivores.   One of the species of the typical genus attained dimensions as large as those of a bear.   In the widely distributed family *Hyænodontidæ*, an exclusively North American genus is *Patriofelis*, which is regarded as a specialised offshoot from *Oxyæna*.

Among the ungulates there are several extinct families confined to North America.   In the group forming a transition between the pigs and the ruminants there is first of all the family of the oreodonts (*Cotylopidæ*), which make their appearance in the middle Oligocene, and continue to the Miocene and lower Pliocene[1].   These ungulates, which were allied to the genus

[1] By American geologists the term Oligocene is not generally used, so that the whole of the Tertiary strata are classed as Plistocene, Pliocene, Miocene,

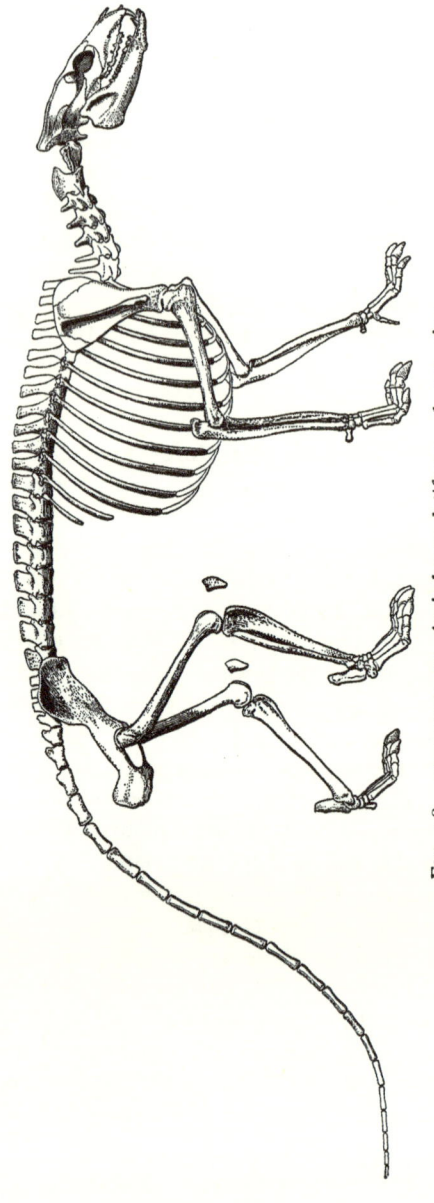

FIG. 78.   SKELETON OF *Agriochœrus latifrons*. $\frac{1}{10}$ nat. size.

Ancodus, common to the Tertiaries of both hemispheres[1], and
were represented by a large number of generic types, have
crescentic columns to the short-crowned cheek-teeth, the upper
molars usually carrying four such columns; while the lower canine
is approximated to the incisors, its usual form and function being
assumed by the first pre-molar. The last upper pre-molar is
simpler than the molars; and while the feet have usually four toes
each, in the typical genus *Cotylops* a rudiment of the thumb is
retained in the front pair, as in *Ancodus*. In *Cotylops* and most
of the other genera the molars of the upper jaw have but four
columns, but in *Protoreodon* there are five;—a feature serving to
connect the family with the *Anthracotheriidæ*, from which group
the oreodonts are probably descended. A nearly-allied but more
specialised family is that of the *Agriochœridæ*, as represented by
the genus *Agriochœrus*[2], of the upper and middle Oligocene, in
which the toes were developed into claws, instead of being
incased in hoofs. Here it may be mentioned that while the
peccaries (*Dicotylidæ*) are now exclusively New World types, and
the pigs (*Suidæ*) restricted to the Old World, the Tertiaries of both

and Eocene. Introducing the former term, the series may be approximately
classified as follows, viz. :

PLISTOCENE.		Equus Beds.	*Equus, Elephas primigenius.*
PLIOCENE	(Upper).	Blanco Series.	*Pliauchenia.*
	(Lower).	Loup-Fork.	*Protohippus, Hipparion.*
MIOCENE.		Deep River.	*Anchitherium*, First Mastodons.
OLIGOCENE	(Upper).	John Day.	*Miohippus, Ancodus.*
	(Middle).	White River.	*Agriochœrus, Titanotherium.*
	(Lower).	Uinta.	*Amynodon, Mesonyx.*
EOCENE	(Up. and Mid.).	Bridger.	*Pachynolophus, Palæosyops.*
	(Lower).	Wahsatch.	*Hyracotherium, Coryphodon.*
	(Lowest).	Puerco.	*Neoplagiaulax, Polymastodon.*

In this series the Deep River beds are identified with the European
Miocene (*suprà*, p. 117) by the presence of *Anchitherium*, and the first appear-
ance of *Mastodon*; while the existence of *Ancodus* in the John Day and Upper
White River beds correlates them with the Upper and Middle Oligocene,
Hyracotherium and *Coryphodon* serving to identify the Wahsatch beds with the
Lower Eocene.

[1] *Suprà*, p. 161.

[2] Equal *Artionyx*.

hemispheres contain intermediate types such as *Hyotherium* and *Chœrohyus* which are apparently the ancestral stock of both families.

FIG. 79.   FRONT VIEW OF RIGHT HIND FOOT OF *Agriochœrus.*

Another very peculiar type of North American Tertiary ungulates is represented by *Protoceras*, from the upper division of the White River Oligocene, which forms a family (*Protoceratidæ*) by itself.   In these creatures the feet approximate to the ruminant type; but the skull, as shown in the accompanying illustration,

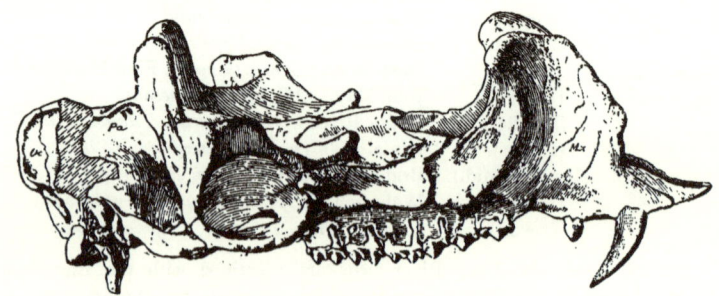

FIG. 80.   SKULL OF *Protoceras*, WITHOUT THE LOWER JAW.

has at least two pairs of large bony processes, probably covered in life with horns, and a pair of large upper tusks, in both of which respects it exhibits a curious parallelism with the perissodactyle

ungulates.   No trace of these singular artiodactyles has hitherto
been detected in the Old World.

The *Camelidæ* seem to have been primarily a North American
family, which originated in the Sonoran region, and of which one
branch (*Lama*) subsequently migrated south, while the other
(*Camelus*) crossed Bering Strait into the Old World.   In the
upper Pliocene there occurs *Pliauchenia*, with only three lower
pre-molars, and in the lower Pliocene Loup-Fork beds *Procamelus*
with four of these teeth ; while the earliest representative of the
family is *Leptotragulus* of the Uinta Oligocene.

In the perissodactyle section of the same order the family
Titanotheriidæ is mainly North American, although, as stated on
page 107, a representative of the typical genus *Titanothèrium* has
been discovered in the Tertiaries of the Balkans.   *Titanotherium*
includes huge rhinoceros-like animals, with low-crowned molar

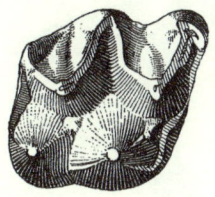

FIG. 81.   RIGHT UPPER MOLAR TOOTH OF *Palæosyops*.

teeth of the type of those of *Chalicotherium*, and frequently having
the nasal region of the skull surmounted by large bony protuberances.   The genus is characteristic of the Uinta and the lower
division of the White River Oligocene.   An earlier type of the
same family is typified by the smaller and less specialised hornless
animals from the Bridger Eocene, known as *Palæosyops*, which,
together with certain allied forms, constitute a peculiar sub-family
confined to America.   This family, like the camels, appears therefore to have originated in the Sonoran region, whence a few
representatives wandered eastwards into the Old World.

In the generalised ungulate sub-order termed Amblypoda,
of which the lower Eocene coryphodons were the earliest
representatives, North America possesses an absolutely peculiar

family in the *Uintatheriidæ*. These were huge, somewhat elephantine ungulates, with five short toes to each foot, long tusk-
like canine-teeth in the upper jaw, and the skull surmounted with
three pairs of large bony protuberances; their molar teeth being a
specialised form of the *Coryphodon* type. They occur in the
middle division of the Bridger, or the one above the zone yielding
Coryphodon, and may accordingly have been the descendants of
that genus. These uintatheres appear to have been restricted to
the "Bad Lands" of the Sonoran region and adjacent districts of
the Transition zone.

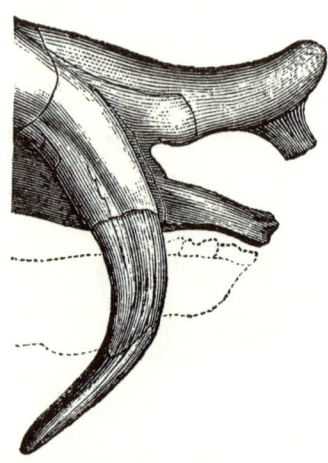

Fig. 82.   EXTREMITY OF SKULL OF *Uintatherium*, TO SHOW UPPER TUSKS.

Passing by certain other forms of less interest, attention may
be directed to a peculiar group of aberrant mammals forming
the Tillodontia, which appear to be mainly North American, and of
which the serial position cannot be precisely determined. They
are restricted to the Eocene, and seem to combine the characters
of the modern Ungulata, Rodentia, and Carnivora. In the genus
Anchippodus, forming the type of the family *Anchippodontidæ*, the
skull approximates to that of a bear, the cheek-teeth are of an
ungulate type, and there is a pair of large chisel-like incisors
(preceded by a small functionless upper pair) in each jaw, very
similar to those of the rodents and hyraces. A second family,

Psittacotheriidæ, is represented by the genera *Psittacotherium* and *Calamodon*, in which the cheek-teeth grew permanently, instead of developing roots.

The following list exhibits the chief families or minor groups of characteristically North American mammals, which are either entirely wanting or but sparingly represented in the Old World; those that are extinct being indicated by a †, and the absolutely characteristic forms being printed in italics.

Primates.

† *Anaptomorphidæ.   Anaptomorphus.*
† *Mixodectidæ.   Mixodectes.*
† *Chriacidæ.   Chriacis.*

Carnivora.

Procyonidæ.   Represented in the Old World only by Ælurus.
† *Miacidæ.   Miacis, Didymictis.*
† *Mesonychidæ.   Mesonyx.*
† Hyænodontidæ.   *Patriofelis.*

Rodentia.

Haplodontidæ.
Geomyidæ.

Ungulata.

Dicotylidæ.   Represented by ancestral types in Tertiaries of E. Hemisphere.
† *Cotylopidæ.   Cotylops, Mesoreodon, Protoreodon.*
† *Agriochœridæ.   Agriochœrus.*
† *Protoceratidæ.   Protoceras.*
Camelidæ.   † *Pliauchenia,* † *Procamelus,* † *Leptotraguius.*
Antilocapridæ.   Antilocapra, † *Cosoryx.*
† Titanotheriidæ.   *Palæosyops, Limnohyops, Telmatotherium.*
† *Uintatheriidæ.   Uintatherium.*

Tillodontia.

† *Psittacotheriidæ.   Psittacotherium, Calamodon.*
† *Anchippodontidæ.   Anchippodus.*

In an earlier chapter[1] a list has been given of the leading
mammalian families common to the two divisions
of Arctogæa, and since in the foregoing chapter it
has been shown that many of the peculiar American families
are more or less intimately related to some of those common to
the two areas, it is manifest that throughout the Tertiary period
eastern and western Arctogæa must have had a land-connection
towards the north, so that there was an interchange of the fauna
of the more northern districts. Those American types which
penetrated as far south as what is now the Sonoran area would,
however, naturally tend to become isolated, and thus develop
into the families which may be regarded as characteristic of that
region. So far, therefore, from this area being merely a part of a
so-called Nearctic region, there are indications that it was differentiated from the Holarctic at a time when the existing zoological
regions of the eastern half of the Arctogæic realm were still unde-
fined. Indeed from the community of the Pliocene fauna of
southern Europe, Asia Minor, Persia, northern India, and south
China, it seems probable that the only divisions of the Arctogæic
realm that could have been attempted would have been into (1) a
Sonoran region, (2) a Holarctic region, comprising the northern
districts of America, Asia, and Europe, (3) what may be termed a
Mediterraneo-Oriental region, including southern Europe, north
Africa, and the whole of southern Asia; and (4) a Malagasy
region, which would then, or perhaps somewhat earlier, have
included Ethiopian Africa.

As to the amount of interchange which took place during
Tertiary times between the mammals of the eastern
and western divisions of Arctogæa, and as to whether
similar generic types may have been developed independently in
the two areas, it is almost impossible to arrive at any satisfactory
conclusion. The suggestion that *Equus* has thus been independently evolved in the two areas, has been already mentioned[2], and
this idea receives support from some very remarkable observations
recently made on the invertebrates inhabiting certain European
and North American caves.

Distinctness of the Region.

Dual Origin of Groups.

[1] *Suprà*, p. 176. [2] *Suprà*, p. 168.

With a quotation from Mr G. H. Carpenter's interesting paper[1] on this subject, the present volume may be fitly closed. After describing. the inhabitants of the Mitchelstown Cave, in Ireland, the author writes that the spring-tail (*Lipura*) "is hardly to be separated from a species found in the caves of Carniola, and the *Sinella* is almost identical with one inhabiting the caves of North America ; while the spider is apparently the same as a cave-dweller from the Mediterranean district of southern France, which probably occurs in the North American caverns also.   Had we to do with animals of the upper fauna, these results, though highly interesting, would not be without parallel in species already known....   But the occurrence of cave-dwelling species with so wide a range is a truly remarkable phenomenon.   The caves cannot be of any great geological age.   Any possible geographical connection which would permit the migration of subterranean animals between southern Europe and Ireland, or between Ireland and North America, seems altogether out of the question within any period during which the fauna can have been specifically identical with that of the present day.   The only conclusion is that from ancestors, presumably of the same genus, which took to an underground life in such widely-separated localities, the similar conditions of the caves have evolved descendants so similar that, when compared, they cannot or can hardly be specifically distinguished from each other.   Should the identifications stand the test of a comparison of types, we shall have proof that the independent development of the same species, under similar conditions, but in widely-distant localities, has taken place.   It must be granted, however, that cave-conditions are so marked and exceptional, that it might not be safe to argue from them as to what may have occurred in the upper world."

Although the author of this passage is perfectly correct in his statement that there is a vast difference between cave-life and open-air life, yet if animals which appear to belong to one and the same species can be proved to have had a dual origin in the one case, it can scarcely be considered impossible that similar instances may occur in the other.   And if such dual origins exist among

---

[1] *Irish Naturalist*, Vol. IV. pp. 25—35 (1895).

species, there is surely no reason why they should not occasionally occur in the case of genera. It would, therefore, seem by no means improbable that the species of the genus *Equus* which inhabited the eastern and western halves of the northern hemisphere during the close of the Tertiary period may have been evolved from the closely-allied but separate ancestral equine stocks.

The matter does not, however, by any means end here. In an earlier chapter[1] it has been shown that the same species of a genus of fish (*Galaxias*) occurs in countries so remote from one another as New Zealand and Australia on the one hand, and Patagonia on the other. With the evidence of the cave-animals before us, is it absolutely impossible that these apparently identical fishes can have been evolved independently of each other? Should this be so, it will engender increased caution in drawing any inferences as to former land-connection from the evidence of single animals. But such instances of independent evolution, if they do occur, must be of extreme rarity, and will in no case interfere with deductions drawn from the presence of a number of allied species or genera of animals in distant countries.

[1] *Suprà*, p. 125. Recently another species has been described from South Africa. See Steindachner, *SB. Ac. Wien*, vol. ciii. p. 460 (1894).

APPENDIX.

LIST OF WORKS AND PAPERS MOST FREQUENTLY REFERRED TO
IN THIS VOLUME.

1. ALLEN, J. A. The Geographical Distribution of Mammals. *Bull. U. S. Geol. Survey*, vol. IV. nos. 2 and 4, pp. 313—376, (1878).

2. ——— The Geographical Distribution of North American Mammals. *Bull. Amer. Mus.*, vol. IV. pp. 199—243, (1892).

3. ANONYMOUS. Antarctica; a Supposed Former Southern Continent. *Natural Science*, vol. III., pp. 54—57, (1893).

4. ——— The Nearctic Region and its Mammals. *T. c.* pp. 288—292.

5. BEDDARD, F. E. A Text-Book of Zoogeography. (*Cambridge Natural Science Manuals*). Cambridge, 1895.

6. BLANFORD, W. T. The African Element in the Fauna of India. *Ann. Mag. Nat. Hist.* ser. 4, vol. XVIII., pp. 277—294, (1876).

7. ——— Note on the "Africa-Indien" of A. von Pelzeln, and on the Mammalian Fauna of Tibet. *Proc. Zool. Soc.* 1876, pp. 631—634.

8. ——— Anniversary Address to the Geological Society. *Proc. Geol. Soc.* 1890, pp. 43—110.

9. ——— The Age of the Himalayas. *Geol. Mag.* decade 3, vol. IX., pp. 161—168, (1892).

10. ——— On a Stag (*Cervus thoroldi*) from Tibet, and on the Mammals of the Tibetan Plateau. *Proc. Zool. Soc.* 1893, pp. 444—449.

11. BOURNS, F. S., and DEAN, C. W. Preliminary Notes on Birds and Mammals collected in the Philippine Islands. *Occasional Papers Minnesota Acad.* vol. I., pp. 1—64, (1894).

12. BULMAN, G. W. The Effect of the Glacial Period on the Fauna and Flora of the British Islands. *Natural Science*, vol. III., pp. 261—266, (1893).

13. CARPENTER, G. H. Nearctic or Sonoran? *Natural Science,* vol. V., pp. 53—57, (1894).

14. EVERETT, A. H. A Nominal List of the Mammals inhabiting the Bornean Group of Islands. *Proc. Zool. Soc.* 1893, pp. 492—496.

15. FORBES, H. O. The Chatham Islands; their Relation to a Former Southern Continent. *Supplemental Papers R. Geogr. Soc.* 1893, pp. 607—637.

16. HEDLEY, C. On the Relation of the Fauna and Flora of Australia to those of New Zealand. *Natural Science,* vol. III., pp. 187—191, (1893).

17. HEILPRIN, A. The Geographical and Geological Distribution of Animals. *International Scientific Series,* London, 1887.

18. HUXLEY, T. H. On the Classification and Distribution of Alectoromorphæ and Heteromorphæ. *Proc. Zool. Soc.* 1868, pp. 294—319.

19. MERRIAM, C. H. The Geographic Distribution of Life in North America, with special reference to the Mammalia. *Proc. Biol. Soc. Washington,* vol. VII., pp. 1—64, (1892).

20. ——— Laws of Temperature Control of the Geographic Distribution of Terrestrial Animals and Plants. *Nat. Geogr. Mag.* vol. VI., pp. 229—238, (1894).

21. NEHRING, A. Ueber Säugethiere von den Philippinen, namentlich von der Palawan-Gruppe. *Sitzber. Ges. Naturf. Berlin,* 1894, pp. 179—193.

22. OGILBY, J. D. Catalogue of Australian Mammals. Sydney, 1894.

23. OSBORN, H. F. The Rise of the Mammalia in North America. *Stud. Biol. Laborat. Columbia College, Zoology,* vol. I., art. 2, (1893).

24. ——— , and EARLE, C. Fossil Mammals of the Puerco Beds. *Bull. Amer. Mus.* vol. VII., pp. 1—70, (1895).

25. SCHARFF, R. F. On the Origin of the Irish Land and Freshwater Fauna. *Proc. Irish Acad.* ser. 3, vol. III., pp. 479—485, (1894).

26. SCLATER, P. L. On the General Geographic Distribution of the Members of the Class Aves. *Journ. Linn. Soc. Zool.* vol. II., pp. 130—145, (1858).

27. ——— The Geographical Distribution of Mammals. *Manchester Science Lectures,* ser. 5 and 6, pp. 202—219, (1874).

28. ——— W. L. The Geography of Mammals. *Geographical Journal,* 1894, 1895.

29. SHARPE, R. B. On the Zoo-Geographical Areas of the World, illustrating the Distribution of Birds. *Natural Science,* vol. III., pp. 100—108, (1893).

30. THOMAS, O. T. The Mammals of the Solomon Islands. *Proc. Zool. Soc.* 1888, pp. 470—484.

31. —————— Preliminary Diagnosis of New Mammals from Northern Luzon, collected by Mr J. Whitehead. *Ann. Mag. Nat. Hist.* ser. 6, vol. XV. pp. 160—164, (1895).

32. WALLACE, A. R. The Geographical Distribution of Animals. London, 1876, 2 vols. 8vo.

33. —————— Island Life. London, 1880, 8vo.

34. —————— What are Zoological Regions? *Nature*, vol. XLIX., pp. 610—613.

35. —————— The Palæarctic and Nearctic Regions compared as regards the Families and Genera of their Mammalia and Birds. *Natural Science*, vol. IV., pp. 433—445, (1894).

36. ZITTEL, K. A. VON. Die geologische Entwickelung, Herkunft, und Verbreitung der Säugethiere. *Sitzber. bayer. Akad.* vol. XXIII., pp. 137—198, (1893).

INDEX.

For EU product safety concerns, contact us at Calle de José Abascal, 56–1°, 28003 Madrid, Spain or eugpsr@cambridge.org.

www.ingramcontent.com/pod-product-compliance
Ingram Content Group UK Ltd.
Pitfield, Milton Keynes, MK11 3LW, UK
UKHW010850090126
466816UK00011B/137